Second Generation Biometrics:
The Ethical, Legal and Social Context

The International Library of Ethics, Law and Technology

VOLUME 11

For further volumes:
http://www.springer.com/series/7761

Emilio Mordini • Dimitrios Tzovaras
Editors

Second Generation Biometrics: The Ethical, Legal and Social Context

Editors
Emilio Mordini
Centro per la Scienza, la Societàe la
Cittadinanza (CSSC)
Piazza Capo di Ferro 23
00186 Roma, Italy

Dimitrios Tzovaras
Informatics & Telematics Institute Centre
for Research & Technology Hellas
1st km Thermi-Panorama Road
570 01 Thermi-Thessaloniki
Greece

ISSN 1875-0044 ISSN 1875-0036 (electronic book)
ISBN 978-94-007-3891-1 ISBN 978-94-007-3892-8 (eBook)
DOI 10.1007/978-94-007-3892-8
Springer Dordrecht Heidelberg New York London

Library of Congress Control Number: 2012935857

Printed on acid-free paper

Springer is part of Springer Science+Business Media (www.springer.com)

Foreword

Privacy Implications of Biometrics

The use of biometric technologies is becoming increasingly prevalent in individuals' lives. Biometrics can prove to be a useful tool for recognizing or verifying the identity of a person based on physical or behavioral characteristics. Biometrics have a practical application in both the public and private sectors, and the technology can be applied across several areas, including inter alia social security, entitlements, payments, health systems, immigration and border control. In the European Union. Some Member States use biometric features in national ID cards and according to EU law, biometrics identifiers are to be incorporated in travel documents such as passports and residence permits. Furthermore, biometrics will be processed in large-scale IT systems being developed in the area of asylum, immigration, and border control.

In recent decades there has been significant evolution in biometric technologies. Apart from the standard types of biometrics used in common applications today, e.g. face, fingerprints and iris, advances are being made in second-generation biometrics, which include among others neural wave analysis, skin luminescence, remote iris scan, advanced facial recognition, gait, speech, behavioural biometrics and so on.

Whilst recognising that such technologies can benefit society, the principal question we are facing when using biometrics is to ensure that they are designed and used in full respect of fundamental rights. Thus, when deciding to use biometric identifiers and choosing the most appropriate ones, the European Commission will take due regard for the individual rights of the persons concerned, notably the right to data protection and privacy.

In the European Union any collection and processing of personal data, including biometric data, must comply with fundamental rights laid down in the Charter of Fundamental Rights of the EU (which includes the right to the protection of personal data in its Article 8) and with the relevant EU secondary legislation, in particular Data Protection legislation such as Directive 95/46/EC. The main challenge is to put in practice and to render effective these legal safeguards. This implies, among others things, that the introduction of any new instruments requiring the collection and exchange of personal data such as biometrics pass both the necessity and proportionality tests, i.e. they must be considered necessary for the development of common policies in the public interest and be proportionate to their goals. Furthermore, careful consideration must be given to the categories of biometric data that would be stored, as well as to the specific nature of biometrics.

In addition, it is important to ensure that processed biometric data are not (re) used for purposes incompatible with those for which the data were collected and made available in a specific system. It is also important to guarantee a high level of reliability in the process of collecting and verification of the biometric data. Indeed, since biometrics establish a unique identifier of an individual, errors resulting from the collection or processing of biometrics can have severe consequences for that individual. For instance, an individual could be refused social benefits because the biometrics indicators processed do not match his/her actual biometrics elements. And because of the reliability of biometrics, it can be difficult for an individual to prove that biometrics have been erroneously processed. For that reason it is necessary that appropriate mechanisms are implemented to ensure that the individual can address those situations where biometrics data have been erroneously processed and affected his or her rights. Appropriate security and technical measures should be adopted to prevent accidental or unlawful destruction, loss, alteration, or unauthorized disclosure whenever biometric data are transmitted across any network.

The EU recognises the benefits of using biometrics as a technological tool for both increasing security and improving accuracy when verifying the identity of individuals and thus contributing to the fight against the theft of identity. At the same time it will improve and facilitate the lives of citizens and other individuals in its many uses such as automated border controls or e-government applications. The EU is also aware of the specific nature of biometrics, so it considers that whenever biometric indicators are processed, adequate procedures must be available in respect of both those individuals the collection of whose biometrics is not possible either wholly or in part, and those whose biometrics are not completely accurate as a result of faulty recording. Such procedures should be implemented and used in order to respect the dignity of persons who are not able to provide biometric data of the requisite quality, so as to avoid transferring onto them the consequences of the deficiencies of the system.

EU data protection law sets out robust principles ensuring that the use of biometrics respects fundamental rights of individuals. I think that this is essential if we want to build trust of individuals with regard to the processing of their biometric personal data and respect for their dignity.

I am grateful that this book addresses the ethical, social and privacy implications that biometrics raise. This makes it an especially attractive means of acquiring an understanding of the different aspects of biometrics.

Viviane Reding
Vice-President of the European Commission,
EU Commissioner for Justice, Fundamental Rights
and Citizenship, Brussels, Belgium

Contents

Contributors

Holly Ashton The Centre for Science, Society and Citizenship, Rome, Italy

Susan E. Barrett Lehigh University, Bethlehem, PA, USA

Iván Cester Starlab Barcelona S.L., Barcelona, Spain

Paul de Hert Vrije Universiteit Brussels and Tilburg University, Tilburg, The Netherlands

Farzin Deravi School of Engineering and Digital Arts, The University of Kent, Canterbury, Kent, UK

Anastasios Drosou Informatics and Telematics Institute, Thermi-Thessaloniki, Greece

Imperial College London, London, UK

Stephen Dunne Starlab Barcelona S.L., Barcelona, Spain

Marcello Ferro Institute of Computational Linguistics "A. Zampolli" (ILC) National Research Council (CNR) of Pisa, Pisa, Italy

Giampaolo Ghilardi Università Campus Bio-Medico di Roma, Rome, Italy

Dimosthenis Ioannidis Informatics and Telematics Institute, Thermi-Thessaloniki, Greece

Anil K. Jain Department of Computer Science and Engineering, Michigan State University, East Lansing, MI, USA

Department of Brain and Cognitive Engineering, Korea University, Seoul, South Korea

Flavio Keller Università Campus Bio-Medico di Roma, Rome, Italy

Ajay Kumar Department of Computing, The Hong Kong Polytechnic University, Hung Hom, Hong Kong

Juliet Lodge Jean Monnet European Centre of Excellence, Institute of Communication Studies, University of Leeds, Leeds, UK

Emilio Mordini The Centre for Science, Society and Citizenship, Rome, Italy

Gabriele Dalle Mura Interdepartmental Research Centre "E. Piaggio", Faculty of Engineering, University of Pisa, Pisa, Italy

Alice O'Toole University of Texas at Dallas, Richardson, TX, USA

Giovanni Pioggia Institute of Clinical Physiology of CNR, Pisa, Italy

Alejandro Riera Starlab Barcelona S.L., Barcelona, Spain

Giulio Ruffini Starlab Barcelona S.L., Barcelona, Spain

Albert Ali Salah University of Amsterdam, Amsterdam, The Netherlands

Ben A.M. Schouten Fontys University of Applied Science, Eindhoven, The Netherlands

Günter Schumacher European Commission Joint Research Centre, Institute for the Protection and Security of the Citizen, Brussels, Belgium

Annemarie Sprokkereef Vrije Universiteit Brussels and Tilburg University, Tilburg, The Netherlands

Massimo Tistarelli Computer Vision Laboratory, University of Sassari, Sassari, Italy

Alessandro Tognetti Interdepartmental Research Centre "E. Piaggio", Faculty of Engineering, University of Pisa, Pisa, Italy

Dimitrios Tzovaras Informatics and Telematics Institute, Thermi-Thessaloniki, Greece

Gaetano Valenza Interdepartmental Research Centre "E. Piaggio", Faculty of Engineering, University of Pisa, Pisa, Italy

Irma van der Ploeg Infonomics and New Media Research Centre, Zuyd University, Maastricht, The Netherlands

Rob van Kranenburg Fontys University of Applied Science, Eindhoven, The Netherlands

Chapter 1
Introduction

Emilio Mordini, Dimitrios Tzovaras, and Holly Ashton

1.1 From Identity to Identification

Identification is the action to attribute "identity" to an item. "Identity" is the state of being identical, say, to be the same, of persons or things.[1] An item can be the same in the sense that it is the same of itself, which is called *absolute identity* (A is A), or that it is the same of something else, which is called *relative identity* (A is B). When we state that A is A, we mean that there is only one A considered under two different perspectives. Conversely when we state that A is B, we actually mean that A has some attributes of B. In other words relative identity can be conceptualised as the inclusion of an individual in a set, whose members share some relevant attributes and are thus the *same* under that specific account.

We are so interested in *sameness* because we are temporal creatures. Everything which is made up by time unavoidably changes and consequently poses questions about its identity. Truly no one "ever steps in the same river twice, for it's not the same river and he's not the same man".[2] Yet in everyday life we cannot afford being

[1]The philosophical debate on the notion of identity is vast, ranging from Heraclitus, to Hume, Leibnitz, Wittgenstein, Heidegger, Quine, Derrida, to cite a few. Discussing this debate is definitely beyond the scope of this introduction.

[2]Philosophers often distinguish between perdurance and endurance. One calls "perdurance" the thesis according to which objects have temporal parts, which change over time. According to this perspective objects have different modal properties, say, they are fourth dimensional objects. This may allow explaining why they can change over time although they remain the same, being time

E. Mordini (✉) • H. Ashton
Centre for Science, Society and Citizenship, Piazza Capo di Ferro 23, Rome, 00186, Italy
e-mail: emilio.mordini@cssc.eu; holly.ashton@cssc.eu

D. Tzovaras
Informatics and Telematics Institute, 6th km Charilaou-Thermi Road,
P.O. Box 361, Thermi-Thessaloniki 57001, Greece
e-mail: dimitrios.tzovaras@iti.gr

E. Mordini and D. Tzovaras (eds.), *Second Generation Biometrics: The Ethical, Legal and Social Context*, The International Library of Ethics, Law and Technology 11,
DOI 10.1007/978-94-007-3892-8_1,

entangled in such metaphysical enquiries. We need to be able to ascertain whether two items met in different conditions can be treated as though they were the "same" item. Identity and identification are hardly the two sides of the same coin. Identity concerns *who one is* (which is a, highly controversial, metaphysical concept), while Identification concerns *how* – or *in virtue of what – one may be recognised or recognise* (which is chiefly a practical matter that regards our way to conduct everyday business).

The need for specific methods for identification is presumably connected with the birth of the first urban societies during the so-called "Neolithic Revolution".[3] From at least three million years ago, humans have lived by hunting (or fishing) and gathering edible items (such as fruit and insects). Only a few thousand years ago, the mankind developed a new economic model based on farming (cultivated crops and domesticated animals). The transition from an economy based on hunting and gathering, to an economy based on farming, implied the emergence of sedentary dwelling. Human groups gave birth to sedentary communities organised in small villages and towns. Around 9000 BC, Jericho[4] was the first known town, with a population of 2,000 or more. The city hosted an agricultural community (as it is witnessed by findings of agricultural tools and domestic cereals in rudimentary pottery), had protective walls and a tower (the first of this kind ever observed by archaeologists), and was surrounded by other tribes dependent on gathering food. Farming economy also meant the creation of food surpluses, which promoted trade of food and food related products (e.g., salt, which was probably one of first commodities because of its ability to preserve food). Growing societal complexity, alongside developments in intra- and inter-societal trade, made the identification of the stranger increasingly vital to the normal functioning of these early societies. We have indirect evidence of this, in a late poem, the Odyssey, whose plot is based on a series of recognitions (passive and active) of a hero who is travelling abroad, far from his own homeland, striving for going back home. Indeed the reading of the Odyssey allows listing the main identifiers used by early human communities. They include a description of physical appearance (e.g., body size and shape, skin and hair colour, face shape, physical deformities and particularities, wrinkles and scars, etc.) and artificial and more permanent body modifications (e.g., branding, tattooing, scarification, etc). Also tokens (e.g., a pass, a seal, a ring, etc.) and mental contents (e.g., memories, poems, music, recollection of family and tribal links, etc.) were important means to recognise and to be recognized.

---

just one the various properties of an object, like colour or shape. One calls "endurance" the opposite, more conventional, thesis according to which objects have only spatial parts, say they are three dimensional. According to this second perspective the notion of identity over time is destined to remain rather puzzling (P. T Geach, "Identity", *Review of Metaphysics* 21 (1967/1968): 3Y12).

[3] V. G. Childe "The Urban Revolution", *Town Planning Review* 21 (1950): 3–17.

[4] Charles Gates, *Ancient Cities: The Archaeology of Urban Life in the Ancient near East and Egypt, Greece and Rome* (London: Routledge, 2003).

As population densities increased, also social hierarchies developed. The birth of the first sovereign states and empires (Egyptian, Chinese and Assyrian) introduced three new important drivers for personal identification; taxation, conscription, and the administration law.[5] All these three activities were in turn related to a certain development of literacy.[6] The Roman Empire was the first cosmopolitan society in the west providing for a universal identification system through a tripartite codified name scheme, which was related to the birth of the first comprehensive legal system on property and political citizenship. The Roman name scheme remained partly operational in Europe during the Middle Ages, yet most Mediaeval "nomadic" individuals – e.g., beggars, pilgrims, merchants, professional soldiers – were identified only through passes and safe-conducts issued by religious and civil authorities, whose genuineness was witnessed by seals and handwriting, rather than their names.

The Modern Era saw increased mobility associated with new geographic discoveries, urbanization and, later on, industrialization. The need for more effective recognition schemes emerged in parallel with the development of post-Westphalia polities. The passage from feudal to modern society led to a new notion of national borders and state issued documents emerged as a new key identifier.[7] The first passports were issued in France by Luis XIV, and the first legislation in the West linking personal identities to birth registration was enacted during the French revolution. The passage from the mediaeval identification scheme based on community membership (e.g., family, guild, village, manor, parish, etc) to an identification scheme based on a document issued by the state central authority is full of meaning. The new citizen who finally emerged from this process was an unmarked individual who was reliably distinguishable only through her name, nationality, place and date of birth. Religion, ethnicity, race, cast, social condition, etc, became (at least in theory) irrelevant in order to identify individuals[8] making all human beings equal before the state. In parallel, one of the main tasks (and sources of power) of modern states became to certificate (and guarantee) citizens' identities. This was realized by establishing, and ensuring continuity to, an "identity chain", starting with civil birth

---

[5]The administration of justice is a primary way for ruling conflicts within a state's territory, which is essential for preventing state's self-disintegration. Yet justice must be administered on an individual basis, which demands identification of individuals. This is well illustrated by early myths. In ancient Greek, for instance, the sole gods who can write and read were the judges of the netherworld.

[6]J. Goody, *The Logic of Writing and the Organization of Society* (Cambridge: Cambridge University Press, 1986).

[7]Till the birth of the nation state, the concept of national borders was quite vague and chiefly related to military issues. To a certain extent, governments were scarcely interested in who inhabited within their borders. Only after European religion wars in seventeenth century, the issue of border control became relevant.

[8]Of course this is largely theoretical because they were still used in establishing ID schemes. The history of how nation states kept on using these categories largely overlaps with the main twentieth century atrocities.

registration and ending with death certificate. A citizen exists as such, only to the extent that her birth has been certified while her death has not yet.[9]

After World War One, most European countries introduced systems of identity cards, incorporating facial photography (and, in some cases, also fingerprinting), as a tool for identifying people within their state borders. Yet even in countries where identity cards did not become mandatory, a new powerful driver for personal identification emerged: the need to identify and authenticate people entitled to receive social benefits. The welfare state, which first emerged in north Europe after World War Two, is based on the provision of services via redistributionist taxation. Taxation and welfare provision both rely on robust and reliable systems of personal identification.

1.2 The Emergence of New Identification Technologies

After the agricultural, the industrial, and the welfare revolutions, we are now on the verge of a new, epochal, transition (Fig. 1.1). We call it with a nice word, "globalisation", which includes a lot of meanings, but which chiefly, points at one crucial event: human species is becoming again nomadic.

Each year, about two billion persons move across large geographic distances (not to mention people in "virtual mobility" through info-communication tech). The International Air Transport Association reported that their members carried 1.6 billion passengers in 2007, among which 699 million flew internationally.[10] The United Nations World Tourism Organization estimated 924 million international tourist arrivals in 2008.[11] International movements for permanent resettlement by immigrants, refugees, asylum seekers, or refugee claimants, and temporary movement by migrant workers and others augment the total international movements each year. The International Labour Organization stated that in 2004, an estimated 175 million persons (3% of the world's population) lived permanently outside their country of birth and that there were 81 million migrant workers (excluding refugees) globally.[12]

---

[9]In a true bureaucratic frenzy some modern governments even issue a Certificate of Life (also called a Proof of Life) to confirm that an individual is still alive. There are countless consequences of the bureaucratic approach to civil identity. The most horrible event of the twentieth century, the *Shoa*, was made possible chiefly by the existence of an effective, largely automated, bureaucratic apparatus for certifying identities, as Edwin Black has convincingly argued in *IBM and the Holocaust* (New York: Crown Publishing, 2001).

[10]International Air Transportation Association. Fact sheet: IATA – International Air Transport Association, 2010. http://www.iata.org/pressroom/facts_figures/fact_sheets/iata.htm (accessed August 18, 2011).

[11]United Nations World Tourism Organization. UNWTO world tourism barometer, 2009. http://unwto.org/facts/eng/pdf/barometer/UNWTO_Barom09_1_en_excerpt.pdf (accessed August 18, 2011).

[12]International Labour Organization. Towards a fair deal for migrant workers in the global economy. International Labour Conference, 92nd Session, 2004. Report VI. http://www.ilo.org/wcmsp5/groups/public/-dgreports/-comm./documents/meetingdocument/kd00096.pdf (accessed August 18, 2011).

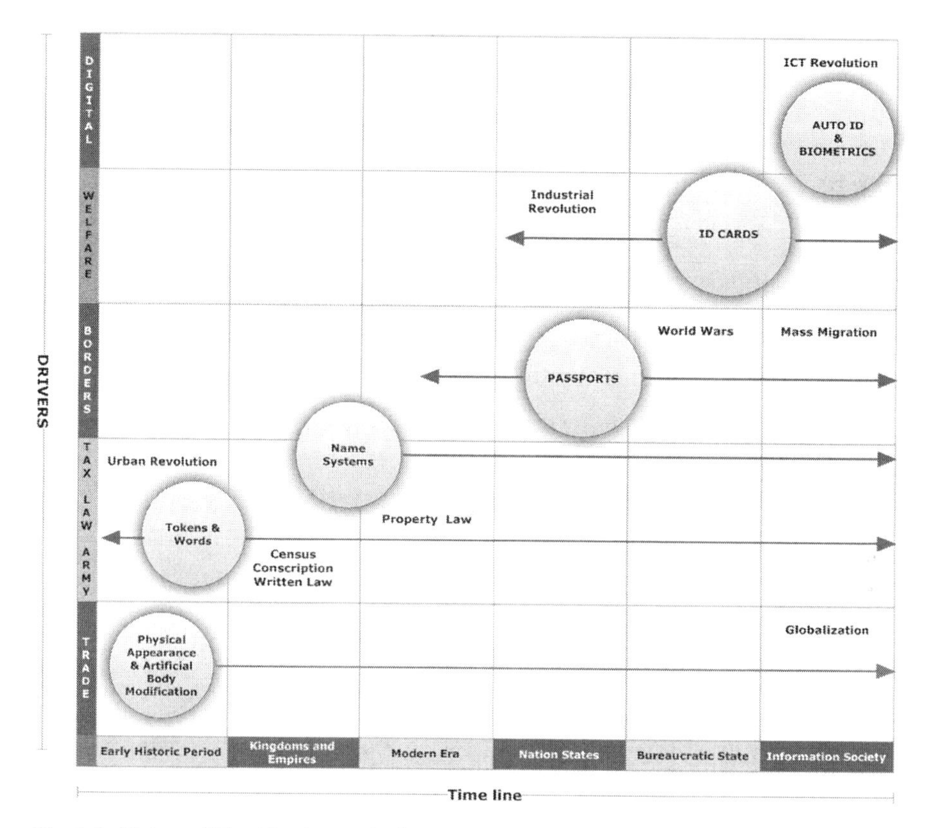

Fig. 1.1 History of identification methods

This huge amount of people on the move has often weak and unreliable identity documents and the poorer do not even have these. In 2000, UNICEF calculated that 50 million babies (41% of birth worldwide) were not registered at birth, so making totally unreliable their future passports and IDs. In this scenario, a personal identification scheme based on birth registration and state-issued passports is less and less tenable. Traditional means of proving identities are not dependable enough in most parts of the world and unfit for global digital networks.

Globalization transcends national control and regulation and this has dramatic consequences for traditional identification schemes. The tourist hoping to use her credit card in any part of the globe, the asylum seeker hoping to access social benefits in her host country, the banker hoping to move money from one stock market to another in real time—all have the same need. They must prove their identities and be certain of others'. In such a global context Automatic Identification and Data Capture technologies (AIDC, or AUTO ID) emerged as a possible solution.

Shaped on the experience of military logistics, AUTO IDs include a vast array of technologies for "tagging" and tracking vehicles, items, and individuals, such as

Bar Codes,[13] Optical Memory Cards,[14] Contact Memory Buttons,[15] Radio Frequency Identification[16] and Radio Frequency Data Capture,[17] Micro Electro Mechanical Systems,[18] Smart Cards.[19] AIDC may converge with Satellite Tracking Systems[20] (STSs) and Unmanned Aerial Vehicles (UAVs)[21] to locate the identity data within a time and place. AUTO IDs are tools which allow capturing relevant properties of an item in order to recognise it. Most AUTO IDs make use of tokens and tags, say, they don't recognise directly an item but through an object[22] which is linked with it. This object can be an electronic devise (e.g., active RFID, MEMs, Smart Cards, etc.) or a code that can be electronically read (e.g., passive RFID, Bar Codes, OMCs, etc). The main issue with tokens and tags is that can be copied, stolen, or forgotten. In other words they are not secure and often are also unpractical and demand any

[13]Bar Codes include: (a) linear bar codes, which consist of vertical black lines and white spaces that carry data. Linear bar codes are interpreted using software and special devices such as laser scanners; (b) 2 Dimensional Bar Codes, which use similar technology as linear bar codes but carry about 100 times more data.

[14]Optical Memory Cards (OMC) use optical read/write technology similar to the familiar CD-ROM production process except an OMC can have data written to it in a sequential order on many occasions until all available memory space has been filled. The standard OMC is credit card sized with a recording surface encapsulated between two protective layers that form the outside of the card. The card provides durable storage that is not affected by electromagnetic interference and the protective layer is resistant to various chemical solvents, dirt, rain, heat, cold, and severe impact. The OMC reader/writer is usually a small breadbox sized unit attached to a Personal Computer (PC).

[15]Contact Memory Buttons are similar to a floppy disc in that it has a read and write capability. In addition, their design protects them from almost all types of environmental damage. They are significantly more costly (about 600 times more than a bar code). They also require the user to touch the device when reading or writing to it.

[16]Radio Frequency Identification (RFID) is a small radio transceiver combined with a memory unit. Since each unit has a discrete address, communication is accomplished with a single device at a time.

[17]Radio Frequency Data Capture use a built-in radio, where a bar code scanner can talk directly to the host computer and pass messages back and forth, similar to real-time receipt processing.

[18]Micro Electro Mechanical Systems (MEMS) are made up by several chemical and environmental sensors on a credit card-sized radio transceiver.

[19]A smart card contains an integrated circuit chip, with a microprocessor that is able to read, write and calculate. It also may include one or more methods of storing information, such as a magnetic strip, digitized photo or bar code.

[20]Satellite tracking systems have been proposed (and used in UK and USA) to closely monitor sex offenders, persistent juvenile offenders, hooligans, illegal aliens.

[21]Unmanned Aerial Vehicles (UAVs) are remotely piloted or self-piloted aircraft that can carry cameras, sensors, communications equipment or other payloads. They have been used in a reconnaissance and intelligence-gathering role since the 1950s. Today they are also used for combat missions.

[22]The term "symbol" means "to bring together" and originally the Greek word for "symbol" meant a plank, which was broken, in order for friends to recognize each other by mail. For example, if a messenger came from a friend to ask for help, he was to bring the second part of the broken plank, and if it matched the first part, then indeed it was a meeting with a friend.

active action from the subject. The true breakthrough in the field of AUTO IDs has been the emergence of biometric technologies. Biometrics is the sole identification technologies which are based on "natural" properties, say, properties which are owned by the item itself and are not added by us. In the next chapters we will briefly introduce the notion of biometric technologies and, in doing so; we will also discuss the definition of "first" and "second", or "next", generation biometrics.

1.3 Biometric Technology

The word "biometrics" includes two concepts which are both relevant.[23] First biometrics refer to the notion of measurability (metrics). Indeed biometrics do not target "natural properties" in general, but they target only "measurable physical properties".[24] A mechanical devise (sensor) is modified by the input signal, in a way which is proportional to the magnitude of the signal. This generates an electric output. Then through the repetitive measurement of the electric output at certain interval of time, the magnitude of the voltage is turned into a proportional number, which is usually encoded as a binary number, or a gray code, or still in other ways.

The analogue-to-digital conversion is the occurrence which makes automatic biometrics radically different from any analogue, human performed biometrics. In the moment in which the representation of the qualities of an item is turned into digits, we can operate through numbers, which changes completely the "rules of the game". We are no longer in the field of recognition through physical appearance, as it could still seem at a superficial glance. There is no longer need of a human eye that evaluates the correspondence between a sample and its representation, as when we observe a picture to search correspondence with an actual face. On the contrary we are entering into a new territory, the digital. Only in such a territory, the recognition process can be fully automatic. The first passage is the creation of the template. Templates are normalized representations of the biometric sample. The need for a template is generated by the fact no physical property is actually stable enough to be captured forever in one shot. In real life there is a continuous variability, and if we want to have a model reliable enough to be used for recognizing an individual, we need to merge signals captured in different moments and under different conditions.

---

[23] Various overviews of biometric modalities and techniques are available. See, e.g., *The Handbook of Biometrics*, ed. Anil K. Jain, Patrick Flynn and Arun A. Ross (New York: Springer, 2008).

[24] Measurable physical properties are discrete elements that can be put in bi-univocal correspondence with a set of numbers. Physical properties are organized in a dimensional system built upon seven base quantities, which include Length, Mass, Time, Electric Current, Temperature, Amount of Substance, and Luminous Intensity. Other physical quantities are all derived from these base quantities by multiplication and division.

Further operations performed by a biometric system are more obvious. Templates are stored in a media which is different according to the application (it can be a central or decentralised database, or a card) and used for future reference. When the subject comes under observation, the system performs the same operations and produces a new digital representation that will be matched against the template previously stored. Systems are "tuned" to determine the tolerance level allowed and the match is considered a match within a given confidence interval. We call such a match "identification" when it occurs against a data base (1 to many) or "verification" when it occurs against only one specific template (1 to 1). In the first case we can conclude that our data base includes the person under observation; in the second case we can conclude that the person under observation is the same whose data are in the card or the other medium submitted. We can be also interested in checking that a person is not in a database (e.g., a watch list). In such a case we compare the current biometric sample against a database with the aim not to match it. In other words we are not interested in *whom you are*, but in *whom you are not* (negative identification).

1.4 Strong, Weak and Soft Biometrics

The second central concept included in the word "biometrics" is life (*bio*). Only living beings are believed to possess highly specific qualities, which are unique and stable enough to be used as identifiers.[25] As far as we use them for biometric identification, we call them "biometric features". Engineers usually distinguish between strong, weak, and soft biometric features.

"Strong biometrics" are features that can be considered unique (very unlikely to be found equal in two individuals) and permanent (enduring in time). To be sure no quality is truly unique or permanent, but some qualities can be treated as though they were. For instance, in human beings some patterns of fingerprint, iris, hand veins, and the retina, can be practically considered unique and stable.[26] This allows using these qualities as though they were almost tokens. In other words, their presence is considered almost be-univocally linked with a given subject. The early

[25] Natural properties of non living objects are often considered non-specific enough to ensure precise and conclusive identification; consequently we don't use them to identify inanimate objects. However recent research has found that most objects contain a unique physical identity code formed from microscopic imperfections in their surface. This covert 'fingerprint' can be read using a portable laser scanner, using a technique called 'Laser Surface Authentication' (P. R. Seem, J. D. R. Buchanan, and R. P. Cowburn "Impact of surface roughness on laser surface authentication signatures under linear and rotational displacements", *Optics Letters* 34 (2009): 3175–3177, ISSN:0146–9592.

[26] The same holds true for muzzle patterns of cattle, patterns of black and white feathers in penguins, footprints of fishers, zebra and tiger stripes, etc. (http://www.bromba.com/knowhow/BiometricAnimals.htm)

development of biometric technologies was chiefly based on strong biometrics. Logically speaking, identification based on strong biometrics is hardly different from identification based on artificial tokens. What changes with biometrics is that the token is no longer an external object associated to the person, but it is a bodily feature.

"Weak biometrics" are features that are "less unique" or "less stable" than those used as strong biometrics. In humans, weak biometrics include features like body shape, odours, behaviour (e.g., gestures, gait, face dynamic, etc.), voice, body sounds, electrophysiological phenomena (e.g., hearth and muscular electrical activity, brain waves). Weak biometrics can be used for identification purposes provided that they are not used in isolation. Given that one can only in part establish a be-univocal correspondence among them and a given individual, they should be used only in context (by considering also space and time coordinates) or in association. This implies that in order to use them efficaciously we should collect also other details such geo-spatial localization, moment in which the feature has been collected, and so. We can also fruitfully associate and merge two or more weak biometrics, and further associate them with soft biometrics (see below). By doing so, we can achieve a high discriminating power, which is fully satisfactory for verifying alleged identities (authentication) and also quite good for identifying people in groups of limited dimensions and for monitoring individual identities over limited period of time.

Finally, with the expression "soft biometrics" engineers refer to features which are generic and cannot be linked to a specific individual. They include categories such as gender; age; race and ethnicity; weight; height; eyes, skin and hair colour; etc. Soft biometrics can be fruitfully used to reinforce strong and weak biometrics. Basically they allow to reduce the number of odds and consequently to refine the identification process.

1.5 First and Next Generation Biometrics

The distinction between first and second, or next, generation biometrics is primarily descriptive, in other words it concerns the distinction between mature and new and emerging technologies. Advances in sensor technologies, which enable different bodily and behavioral characteristics to be captured, and emergence of potentially new biometric traits, have been the main technological driver of second generation biometrics. Generally speaking, emerging biometrics include technologies which measure "motor skills" (i.e. the ability of a human being to utilize muscles), technologies which measure electromagnetic body signals, and finally technologies which measure human-computer interaction patterns. Examples of emerging biometrics include gait recognition (analysis of walking patterns); dynamic facial features, eye blinking, lip movements, smile recognition; voice recognition (analysis of vocal behavior); signature/handwriting or other authorship based biometrics; electrocardiogram (ECG, records the electromagnetic signals produced by the heart as measured on the skin); electroencephalogram (EEG, records the electromagnetic signals generated by the brain as measured on the scalp); electrooculogram (EOG, records eye movements); electromyogram (EMG, records muscle activity); body odor recognition; keystroke or mouse dynamics; online behaviour recognition; and so. Yampolskiy[27] has proposed to categorize these biometrics under the common heading of behavior-based authentication mechanisms, which are overall characterized by "the incorporation of time dimension as a part of the behavioral signature" (p. 376). He further classifies them into: (1) Behavioral Biometrics (Authorship based, Human Computer Interaction Based, Motor Skill, and Purely Behavioral), (2) Behavioral Passwords (syntactic, semantic, one-time methods and visual memory based), (3) Biosignals (Cognitive and semi-controllable biometrics); and (4) Virtual Biometrics (representations of users in virtual worlds).

Most new biometrics require less user cooperation and can be run almost unobtrusively and in a way transparent to the subject, they can also capture signals from a distance, or on-the-move. They are consequently the ideal candidate for being integrated in ambient intelligence environments, and in any other ambient destined to be constantly automatically supervised. Finally, given their lower discriminatory capacity and lower degree of stability, new biometrics are often integrated in multi biometric systems, that can target a sole biometric feature (but using different kind of sensors), or several biometric features in parallel or sequentially.

The passage from first generation technologies to next generation technologies realizes a paradigm shift similar to the shift that characterized the passage from conventional, human eye performed, biometrics, to automated, digital, biometrics. In order to appreciate this event, it is important to understand biometrics as a semiotic activity. Semiotics is the theory of sign and meaning. A sign is something that stands for something other.[28] The American philosopher C. S. Peirce[29] distinguished between three kind of signs: (1) ICON, which is a sign similar to its object, a sign that physically resembles what it 'stands for'; (2) INDEX, which is a sign that points at something, a sign which is contiguous to its object; (3) SYMBOL, which is a sign neither similar or contiguous to its object, but only with a conventional link with it. All signs have some iconic, indexical, symbolic, aspects; it depends on the concrete context and moment in which they are used whether one of these aspects prevails on the others. We communicate indirectly through signs, and make sense of our world by interpreting signs into meaning.

The first biometric revolution was marked by the passage from iconic to indexical signs. The current revolution is leading biometrics from the indexical to the

---

[27] R.V. Yampolskiy, "Behavioral, Cognitive and Virtual Biometrics," in *Computer Analysis of Human Behavior,* ed. A.A. Salah and T. Gevers, 347–385 (London: DO Springer-Verlag London, 2011).

[28] Signs are patterns used to convey messages. "I *define a Sign as anything which is so determined by something else, called its Object, and so determines an effect upon a person*" (C. S. Peirce, 1908, A Letter to Lady Welby, SS 80–81. http://www.helsinki.fi/science/commens/dictionary.html).

[29] Although Pierce's distinction has been variously criticised and modified, it still remains the most commonly accepted taxonomy.

symbolic realm. First generation technologies chiefly rely on the indexical value of biometric signs. Body morphology was explored with the aim to find bodily signs stable and discriminating enough to be used as a "silver bullet" for people identification. They were searched chiefly for their indexical value, say, for their pointing at the subject now and here. Such a function is still searched by new biometrics technologies, but they are increasingly used also to characterize the individual. This is made possible by the dynamic, time dependant, nature of new biometrics, which is based on physiology rather on morphology. Physiology results from complex interactions of multiple control mechanisms that enable an individual to adapt to the exigencies and unpredictable changes of everyday life. As a consequence any measure of physiological functions provides information on the ever changing relation between an individual and her environment. With multiple usage of different weak and soft biometrics, the single individual appears as constituted by several contextual identities. S/he is recognized as the point of intersection between the various sets to which s/he belongs. Each set is connected to other cultural significant systems, hence achieving a proliferation of extra information and meanings. For instance, gestures used for identification unavoidably provide relevant details on cultural features; face and voice provide semantic information and important clues about the emotional state; brain waves can be related to mental activity; muscular and cardiac electric activity inform on the state of arousal; and so. Biometric signs become closer and closer to a language. Yet this promises to pose totally new ethical, legal, and social challenges.

1.6 Ethical, Social and Legal Implications

Whilst the ethical, social and legal context for first generation biometrics has been addressed in a number of ways, including several projects funded by the European Commission (e.g. RISE,[30] BITE,[31] HIDE[32]) the context for the so-called 'second generation biometrics', has yet to be adequately addressed. It is evident that in some quarters, there is a fear of biometric technologies on the whole, and this could prove problematic for technology developers and the Governments wanting to employ them. Yet, before these fears have even been adequately and conclusively addressed in relation to such things as finger-printing and iris scanning, a new wave of biometrics are emerging.

Second generation biometrics progress from asking *who* you are (the focus of first generation biometrics) to asking *how* you are; they are less interested in permanent data relating to a pure identity, and more propelled by an individuals' relationship with their environment. What are your intentions and *how* do you manifest these?

[30] www.riseproject.eu

[31] www.biteproject.org

[32] www.hideproject.eu

Of course, this has extreme social implications as it raises questions concerning surveillance and privacy in a way that first generation biometrics never quite did. However, it also promises a world which can be tailored to suit our needs. It is salient to remember that any new technology always brings risks. What therefore becomes paramount is that the people interested in investing in these technologies take heed of the atmosphere of discontent and make efforts to improve public confidence in these technologies.

One way to redress some of this confidence is surely to openly and honestly consider the ethical and legal implications of these technologies. Surprisingly, though there are of course discussions on this theme in progress, these have not been overly visible. There is a strong sense that technology is leading the way and that the ethical and legal considerations are an afterthought. Playing 'catch-up' to the technology puts these others in a vulnerable position. It also no doubt plays a part in the unease felt by privacy advocates, who feel that the general public might be expected to use the technologies before proper data protection and privacy safeguards have been put in place or even considered. Furthermore, it could turn out to be difficult to implement appropriate safeguards in systems which are already up-and-running. Therefore, it is vital that this discussion come to the forefront now in order to begin to rectify the imbalance between the progress of these technologies and the ethical and legal considerations put in to them.

This book aims to rectify this by opening up discussion in this field and we believe that it goes some way in opening up this vital discourse. It is the first book specifically dedicated to the ethical, legal and social implications of second generation biometrics. It pulls together a number of specialists from different fields including academia, law, psychology, engineering, neuroscience, and medicine and computer science to offer their expertise. The book is divided into four parts; the first provides an introductory discussion on foundational issues; the second explores some of the technology involved in a number of new biometric modalities; the third explores a number of human related aspects; and the fourth provides good round-up of key ethical and legal issues that need to be addressed concerning these technologies. Finally a brief conclusion will try to summarise the main open findings of this discussion.

In the second chapter, Giampaolo Ghilardi and Flavio Keller set the scene by providing us with the philosophical rationale for biometrics as a scientific tool. In particular, they take up the challenge of deciding on a philosophy that is appropriate for second generation biometrics. They note that for the first time in history, biometrics are allowing scientists to measure and quantify life, but, they ask, in philosophical terms, what is 'life' and can we rationally claim it to be quantifiable? Ghilardi and Keller are aware that by asking this question, they are at one of the critical junctures of western philosophy; the distinction between primary and secondary qualities. They progress from this distinction to show a possible bridge between the world of qualities and that of quantities. The chapter discusses a number of different philosophical approaches and finally notes that the emergence of second generation biometrics, which allows for the notion of 'dynamic identity', is, in their opinion, closer to a realistic epistemology which acknowledges that numerical translations of reality 'are and always will be partial and incomplete'. The authors claim that this epistemological approach to second generation biometrics leaves room for ethical considerations; it is allowing for the idea of partiality and fallibility within biometric verification, thus the gaps left by the inability to completely verify an individual in numerical terms leaves space for ethical considerations of a deeper kind. If we acknowledge that these technologies can never perfectly 'know' us, what implications could this have on their adoption as tools to be used as authenticators in security settings?

From this thought provoking start, the book progresses on to a basic overview of existing biometric technologies by Anil Jain and Ajay Kumar. This introduction to first generation biometrics is concluded with the acknowledgement that the recognition accuracy of such biometrics may not be adequate to meet the requirements of high security applications. For this purpose, the use of multimodal systems is recommended. The authors go on to discuss the expectations that society has from biometrics regarding the categories of performance, cost, user convenience, interoperability and system security and they comment on why it is that first generation biometrics fall short of these expectations – particularly where high security infrastructures are concerned. From this, they progress to considering second generation biometrics in terms of what they can offer as improvements on first generation applications but also, in particular, the challenges these face from both an engineering and social point of view. Whilst it becomes apparent that the engineering challenges are constantly being worked on in order to improve the technological capacity of these biometric systems and that large-scale projects and applications are already being created, it is equally apparent that efforts to ensure high degrees of privacy and security are not quite up to the same standard. Indeed Jain and Kumar note that many of the policies and security measures being put in place for biometric systems actually interfere with existing national data and privacy protection policies. This highlights that existing policies are not adequate for the societal development taking place and shows the vital need for serious consideration of these issues before the technology evolves to such an extent that any ethical sanctions are merely an afterthought which may have arrived too late. As one positive example, they offer the example of Smart ID cards in Hong Kong which have been successfully in place and in use for a few years now with no reported privacy concerns or instances of misuse. Such an example could provide a good initial model from which discussion on integrating privacy issues within biometric systems could be stimulated.

The fourth chapter delves more deeply into the legal implications of the biometric technologies presented by Jain and Kumar. Annemarie Sprokkereef and Paul deHert note that first generation biometrics, and the way in which they are handled, already raise fundamental discussions concerning data protection and human rights law. On top of this, we now have second generation biometrics emerging which (for reasons touched upon above) may be even more contested. Thus within this chapter, the authors assess the existing legal framework which is governing first generation biometrics (including challenges and human rights aspects) and then delve more closely into looking at the relationship between the emerging technologies and the law. They recommend that second generation biometrics necessitate a 're-assessment of the traditional data protection approach that only data relating to identified or identifiable persons have to be protected' and claim that existing European legislation is not sufficient to guarantee effective protection against profiling. This is a sobering thought and one which needs addressing properly if the growth in the use of these biometrics is to be accepted on a greater scale.

Thus from the first part of the book, the reader should be able to glean a better understanding of second generation biometrics; why they have emerged, what they are and some of the challenges they present. In the second part, we seek to enhance the readers' comprehension of some of the more technical aspects of second generation biometrics, and by doing so, to further highlight both the opportunities and risks that they present.

In the fifth chapter, Dimosthenis Ioannidis et al. discuss gait and anthropometric biometrics (a sensing seat) which are two types of second generation biometrics that could prove useful in high security settings, multimodal systems. First, the potential of these technologies to be used as biometrics is discussed. This is then followed by a slightly more in-depth look at the approaches and technological aspects behind the two types of authentication. Finally, some experimental results using both gait and the sensing seat for user authentication are presented. They demonstrate that both these biometrics do show potential to be used as person authentication tools in high security settings.

Anastasios Drosou and Dimitrios Tzovaras provide a further account of activity and event related biometrics in Chap. 6. The chapter highlights that the potential to identify and authenticate individuals from their behaviour is a key research area of recent years and is largely a response to the limits and problems relating to first generation biometrics (as noted by Jain and Kumar in Chap. 3). The chapter starts with a look at what activity related biometrics are – the concept behind them and why activities and behaviour could be good ways of authenticating an individual. The authors note that many behavioural biometrics, by themselves are not strong enough to provide reasonable competition to the traditional examples of fingerprinting or iris scanning, however they are intended to be used in multimodal systems which both strengthens their performance and means that they are more flexible than first generation biometrics. Clearly, one of the issues with considering activity related biometrics is in choosing appropriate activities to monitor, as not all would provide suitable 'signatures' for person authentication. Drosou and Tzovaras discuss a number of event related activities which display increased authentication potential and thus prove to be appropriate candidates for being used as biometrics. They go on to discuss the EU funded research project, ActiBio, which is looking at unobtrusive authentication using activity related and soft biometrics and is exploring many issues relating to finding and assessing a number of activity related biometrics suitable for use in security systems for critical infrastructures. The chapter then progresses to some slightly more technical issues such as discussing activity recognition, the tracking of anthropometric characteristics and feature extraction. Whilst presenting many opportunities in the security field, continuous authentication via the monitoring of behaviour could certainly give rise to a number of ethical concerns. Encouragingly, though not themselves experts in the ethical field, the authors acknowledge some of the ethical implications of their work in the final section of the chapter.

Following on from the second generation biometrics presented in the previous chapter, in Chap. 7, Alejandro Riera and his colleagues introduce the reader to electrophysiological biometrics. Thus, they describe various types of electrophysiological signals and discuss how they could be used as biometrics, including an overview of the advantage electrophysiological biometrics could have over classical biometric modalities. Following this comes a slightly more technological explanation of the ways biometric signals can be recorded, giving information about signals and artefacts. A more in-depth look at EEG, ECG, EOG and EMG then follows, commenting on the potential of each to individually be used as a biometric authentication tool. The chapter then progresses to looking at the technology trends and opportunities available using electrophysiological biometrics, including BCI (Brain Computer Interfacing) and Telepresence Systems. This latter is proposed as a potential solution to the problem of trust in virtual worlds where avatars are used. Of course the chapter would not be complete without an overview of the potential risks concerning the use of electrophysiological signals as biometrics. One of the key points to emerge in this section is the need for ensuring the privacy of personal health records.

Of course the potential for second generation biometrics to enhance security, should we manage to overcome the issues surrounding privacy and mistrust, are huge. In the eighth chapter, Farzin Deravi presents and suggests new approaches for the use and deployment of multibiometrics in security scenarios. He points out the problems with existing first generation biometric technologies including issues surrounding recognition accuracy and security/privacy vulnerabilities, including the potential for spoofing. He goes on to describe a number of potential technological solutions which might be able to address these challenges. These include multibiometrics, adaptive biometrics, aliveness detection, biometric sample quality and biometric encryption. Such approaches allow for the fusion of different modalities and flexibility which ensures that it is more difficult for people to bypass systems by fraudulent means. The use of encryption also ensures better protection of data by making it impossible to match original samples with data. These five technological solutions are a big feature in many of the second generation biometrics being developed and they could lead to big changes in the ways biometrics are integrated into our environments. For instance, they will allow the transition of biometrics from controlled applications to more ambient and less restrained scenarios. Though useful, this will of course bring additional challenges as more intelligent systems are needed. Deravi discusses the potential for using software agents, multi-agent systems and agent architectures in order to face the increase in complexity. He then goes on to propose that it does not take a great leap of the imagination to imagine such multimodal ambient environments also transferring to the virtual realm. One current problem with virtual worlds and online communities is that it is difficult to link virtual identities to real ones. This means that the emergence of crime is difficult to tackle in these virtual worlds. The author proposes that second generation biometrics may be a way of linking virtual and real identities and to secure online security – this position clearly links with the work of Riera et al. in the previous chapter. However, beyond this talk of increased security and more 'intelligent' technology, Deravi highlights the importance of societal and legal remedies which *"are likely to play an equally important role if we are to avoid a future where biometrics is used against the interest of the citizens or is abandoned as being unworkable or dangerous"* (Deravi, Chap. 8).

The second part of the book can therefore be seen to clarify some of the technical ideas and real implementations of second generation biometrics. From here, we move on to part three which focuses more on some of the more human aspects of these technologies.

The first chapter in this third part of the book sees Ben Schouten et al. comment that identity is based on interaction and communication rendering it fluid and active – in some ways it can be seen to follow on from the philosophy of the second chapter (by Ghilardi and Keller) with a look at the concepts of 'identity' and 'body'. The fact that second generation biometrics use behaviour (forms of interaction with others and the environment) as a way of authenticating an individual means that, in theory, social signals generated by these behaviours could be detectable, possibly providing opportunities relating to marketing and business intelligence applications. Thus the chapter goes on to discuss a number of different applications for behavioural biometrics. The authors also note that in order to adapt biometrics to ever more complex situations, multibiometrics systems are needed which can also incorporate soft biometrics. This in fact ties in with the second part of the book which noted that fusing several different biometric modalities into one biometric system could provide more opportunities for intelligent, unobtrusive ambient environments. The chapter continues with consideration of the concepts of accountability and control in relation to identity. The authors note that there exists both technological/biometric identity and also social identity. What largely differentiates these is the role of the end-user and how much control they feel they have over their identity management. Schouten et al. comment that currently with most biometric systems, control is with the authority of the system rather than with the end-user and if biometrics are to become more acceptable, we should look to ways of giving control back to the end-users so that biometrics are seen as less threatening to freedom and diversity. They argue that one way to do this would be to empower the end-user by creating more commercial and user centred applications of biometrics – indeed the authors claim that this is the only way in which biometrics will be accepted by the public on a large scale.

The importance of addressing legal and ethical issues is highlighted in Chap. 10, where Guenter Schumacher discusses behavioural biometrics and some of the risks they present. He starts by noting that currently there is a change from the long held notion of biometrics as a technology to be used purely for identification purposes; with the development of technologies relating to dynamic biometrics, it is now being suggested as a tool for potentially identifying suspicious behaviour and also for allowing ambient intelligence environments to recognise the needs of the people engaging with them. In this way, there is a clear progression from the first generation of biometrics and indeed within the chapter, Schumacher compares behavioural biometrics with more traditional ones. He then categorises them into three groups; motor skills, body signals and machine usage. He comments on the risks that each of these categories of behavioural biometrics might present and notes that in general, the key issues with behavioural biometrics relate to discrimination, stigmatisation and unwanted confrontation. Schumacher then provides discussion on the 'risk mediators' that have been proposed to deal with these problems. In particular he mentions the technological measures of Privacy Enhancing Technologies (PET) and the new concept of Transparency Enhancing Technologies (TET) which have been introduced in an attempt to overcome some of the drawbacks of PETs. The author makes it clear however, that currently technology is not up to solving the issues posed by behavioural biometrics. Training and awareness raising into both the potentials of behavioural biometrics beyond mere identification capabilities and the existing technologies to safeguard data are needed. Furthermore, in an echo of de Hert and Sprokkereef, the author points out that the fact that current legislation does not adequately cover the issue of data protection concerning behavioural information means that serious consideration must be given to this topic.

Chapter 11 sees Massimo Tistarelli et al. discussing face recognition. It starts with a general overview of face recognition in humans and how it is also closely related to emotion detection and possibly, intention detection. The chapter then goes on to comparing how far technological efforts to emulate human recognition have progressed. The authors describe various technologies which are being developed and worked on. As in the previous chapter, they note that this includes efforts to emulate emotion and intention detection. Though this could be a cause for concern, they comment that machines that could be capable of detecting emotions and intentions would not necessarily be bad – they could in fact be very useful by, for example, better fulfilling consumers' needs or offering fitting solutions for an individual at a specific moment in time. This in fact provides a link with the previous chapter by highlighting the potential of biometrics, particularly if they are used by end users and not just as a method for security and control. Human facial recognition and the interactions arising from it are a normal part of life and the authors suggest that it could be a natural and efficient aim for machines to be developed to 'behave' in a similar fashion. They note that in fact, whether or not such technology was cause for concern "depends on intentions of the men behind its application" (see Chap. 11) – an idea which links with Schouten et al.'s 'end-user' discussion in the earlier chapter. The overall message of this chapter is that the ability for machines to emulate human face recognition could be a highly beneficial addition to society, however there is a need for caution. It can be taken as another clear message concerning the importance of dealing with the legal and ethical issues concerning these technologies.

Chapter 12 is the final chapter in Part III and in it, Emilio Mordini and Holly Ashton discuss the idea that second generation biometrics may have the potential for detecting private medical information of an individual. They start with a generic look at how this could be possible and why it would be problematic. They consider the concept of medical privacy and the importance placed on the discretion of doctors as witnessed by the continuing existence of the Hippocratic Oath. They point out that there is an inconsistency between the way this issue is dealt with in the medical profession and by the current development of technologies which may be able to detect similar information and yet are not answerable to any similar oath of responsibility. They go on to specifically assess exactly what kinds of medical information different types of second generation biometrics (including behavioural biometrics, Human Computer interfacing, voice biometrics, electrophysiological biometrics and soft biometrics) might be able to detect – and it is evident that all are *capable* of disclosing a number of different medical ailments. However, what is the realistic potential of medical information being deducted from these technologies? In the conclusion, Mordini and Ashton consider this question and note that the key issue to be addressed concerns how data within these systems is handled and dealt with. Capture, storage and disposal of data are key issues which need to be fully addressed and covered by adequate legislation in order to prevent these new biometric technologies from being used for the wrong reasons – including medical observance. The chapter offers another insight into the importance of properly addressing ethical, privacy and legal issues concerning these technologies.

So the third part of the book presents some of the human concerns of using second generation biometric technologies by, amongst other things, discussing possible alternative uses of the technology (both positive and negative), as well as ways in which it might have the potential to disclose sensitive, personal information about the individuals using it. It adds to the consensus from the previous parts that more work is needed on addressing the ethical, legal and privacy context of these technologies. Thus the final part of the book focuses on the real issues at stake and provides a clearer overview of policy issues and needs relating to this topic.

Irma Van der Ploeg starts Chap. 13 with a look at how recent technological developments (which have led to the emergence of second generation biometrics) are giving rise to a new set of ethical and socio-political issues. In particular she considers two specific developments which have featured in previous chapters; the emergence of new, multimodal biometrics including soft biometrics and a move towards embedded, ambient biometric systems. She notes that these introduce new critical issues and policy challenges which need to be addressed. For instance, the incorporation of soft biometrics within biometrics systems encourages the categorisation of individuals into groups which in turn raises the potential danger of exclusion or automatic classification of certain individuals which could certainly generate a number of problems. Similarly, embedded intelligent systems, though likely to improve convenience and efficiency on one hand, also have the potential to become ultimate tools for tracking and traceability and raise questions concerning the issue of privacy. Van der Ploeg raises some important issues concerning such matters as anticipatory conformity, intention interpretation, discrimination and the very concept of a 'free society'. She then explores the 'assumption of availability' inherent in biometric systems. That is, the presupposition that the body is available to be used for particular purposes. She highlights why this is another matter for concern which again threatens certain ethical and legal principles. The chapter concludes with a highlighting of the importance of interdisciplinary study in this field, particularly concerning the 'deeply political and ethical nature of the incremental informatization of the body we witness today' (Chap. 13).

In keeping with this theme of policy issues, in the final chapter of this book, Juliet Lodge focuses on the challenges facing those with the task of demonstrating public accountability concerning the way identity verification techniques are used for public policy purposes. She starts by pointing out that new technologies are often promoted in a positive light as a means of boosting convenience and efficiency however the ethical implications are rarely debated. She discusses the role of biometrics within the concepts of security and liberty and goes on to explore defining and using them in line with these terms. Although there is a tendency to assume that due to the uniqueness of biometric traits, they must be infallible in terms of identifying an individual, and therefore can only serve to increase security, there are still problems relating to the technology used and even environmental factors which may influence the authentication process. Further to this, second generation biometrics which by nature are often continuous and unobtrusive may generate an atmosphere of constant anxiety and alertness. Both of these factors contribute to raising questions about the actual potential impact of these technologies on society. In a part on biometrics and governance, Prof. Lodge points out that "legislation at all levels is still playing catch-up to scientific and technological innovation" (see Chap. 14). To highlight this point, she goes on to discuss various issues including decision-making, accountability and the impact of group think and explores the impact on society of these factors in relation to biometrics. The chapter highlights that current safeguards such as encryption are not adequate for providing sufficient security, though Lodge is careful to note that, "the problem lies not with the biometric per se but rather with their contextualisation and their use as policy tools". The chapter concludes with what could be seen as a challenge for action, with Lodge noting that accountability mechanisms as they currently stand are not yet sufficient to the task of retaining public trust in either biometric technologies, or the people deploying them in the name of security. Ethical considerations must be taken more seriously.

As we can see from this overview of the chapters included in this book, although no-one denies that there could be great advantages to be gleaned from adopting second generation biometrics in terms of convenience, efficiency and security, there are many ethical and social implications to them which have not (until now) been sufficiently addressed. By presenting the thoughts of a number of specialists in this field, this book aims to open up sensible dialogue and encourage honest consideration of the potential 'dark side' to these technologies. Sticking our heads in the sand won't make uncomfortable issues go away and indeed, could prove to have a devastating impact on the society in which we live. Careful contemplation and confrontation of these issues is needed now so that as we move in to the future, we can embrace the potential of second generation biometrics in order to use them in the positive ways envisaged, rather than shying away from them out of fear and mistrust. We hope that you find this book useful and interesting.

Part I
Foundations and Issues

Chapter 2
Epistemological Foundation of Biometrics

Giampaolo Ghilardi and Flavio Keller

2.1 Introduction

Biometrics cannot be considered just as a set of technologies any longer, it rather has to be considered as an established discipline with its own specific methodology. In the present chapter we are going to outline the epistemological issues biometrics needs to deal with, in order to answer the question: what kind of discipline is this? We will try to argue that, even if the common approach to this topic shares a positivistic orientation, biometricians are overcoming the limitations of that perspective, gaining a more appropriate attitude and a better epistemological awareness. Second generation biometrics, e.g., by performing a continuous authentication check in, endorses a different concept of identity, seen as a dynamic process rather than something static.

The chapter has been divided into six sections: the first one sketches the role played by biometrics in the history of science; the second one points out which unit of measurement can be used to deal with biometrics and what the epistemological meaning of collecting data through sensors is; the third one develops the epistemology required by a realistic approach to biometrics, from action to being; the fourth one outlines the link between intentionality and emotions analysed both from the philosophical and the physiological perspectives; the fifth one addresses

G. Ghilardi (✉)
FAST Istituto di Filosofia dell'Agire Scientifico e Tecnologico, Università Campus Bio-medico - Roma, Via Álvaro del Portillo 21, 00128 Rome, Italy
e-mail: g.ghilardi@unicampus.it

F. Keller
Istituto di Neuroscienze, Università Campus Bio-medico - Roma, Via Álvaro del Portillo 21, 00128 Rome, Italy
e-mail: f.keller@unicampus.it

E. Mordini and D. Tzovaras (eds.), *Second Generation Biometrics: The Ethical, Legal and Social Context*, The International Library of Ethics, Law and Technology 11,
DOI 10.1007/978-94-007-3892-8_2,

epistemological issues about intentionality detectability; the sixth one analyses the meaning of identity digitalization process from both a philosophical and an epistemological perspective.

2.2 Biometrics in the History of Science

The term biometrics derives from the Greek words "*bios*" (life) and "*metrics*" (to measure). Automated biometric systems have only become available over the last few decades, due to the significant advances in the field of computer processing. Many of these new automated techniques, however, are based on ideas that were originally conceived hundreds, even thousands of years ago.[1]

Sir Francis Galton (1822–1911) can be considered a mile stone under many aspects in the field of modern biometrics. First of all, together with Karl Pearson, he was one of the founders of the journal *Biometrika* in 1901. It was the first journal completely devoted to a discipline not yet recognized as such. He was also the first one who gave a modern and scientific ground to this new discipline, defining it (quoted from Galton's preface to the first volume) "the application to biology of the modern method of statistics".[2] The role played by statistics in Galton's works is fundamental, as it was the method that allowed him to look into biology in a scientific way.

There is at least another important reason why Galton has to be considered a father of modern biometrics and a mile stone of the modern science. To track down the beginning of the modern science, we have to go back to Galileo, who was the one who grounded this science on the terrain of quantification, and we can say that Galton's biometrics fulfilled that task by measuring and quantifying a field like biology, which was mainly a qualitative discipline before his approach. The way Galton used statistics in the biological field was revolutionary in many senses. He found a new way to quantify a domain of reality like biology, by studying differences between individuals, according to Darwin's evolutionary theory, that is by investigating individual variations among the members of a given race or species.[3]

It would be beyond the scope of the present study to show how statistics has changed the way science was carried out in the previous century, it is enough to say that what Galton did in biometrics also happened in other fields of knowledge, such as sociology at the end of the nineteenth century, where Durkheim,[4] e.g., found a similar way to investigate an issue apparently impossible to statistically analyse such as suicide. Even in this case, a topic dealt just by a qualitative psychological approach, became available to sociological and consequently mathematical inquiry through statistical methods. There is no need to underline how the authors of these changes claimed to be the only ones to deal scientifically with these topics as compared

---

[1]Cfr. http://www.Biometricscatalog.org/Introduction/default.aspx

[2]Stigler (2000).

[3]Ruiz et al. (2002).

[4]Durkheim (2002).

to other approaches. While this is a well known chapter of the philosophy of science that goes under the definition of positivism, the role played by statistics in this development is less well known.

In his classical quotation that reveals the *Geistzeit* of this period about the role of measurement in science, but even more so in knowledge, Lord Kelvin states: "*when you can measure what you are speaking about, and express it in numbers, you know something about it; but when you cannot measure it, when you cannot express it in numbers, your knowledge is of a meagre and unsatisfactory kind; it may be the beginning of knowledge, but you have scarcely in your thoughts advanced to the state of Science, whatever the matter may be*".[5] Galton's work can be read in the light of this assumption as a huge scientific improvement in the direction shown by Galileo and Kelvin.

We can see that biometrics is at the centre of a secular movement that started in the seventeenth century and it has not yet fulfilled its path. Biometrics, as a complex concept made up of life and measure, attempts to synthesize two irreducible elements which have each one its own long history. The risk of biometrics is to flatten the delicate balance between the two poles of this relation on the ground of measure, to lose the awareness about the reduction that measuring implies to life. This is more than a hypothetical risk, whereby knowledge is defined just in mathematical terms, when quantity is seen as the mark of scientific perspective, and consequently science has been promoted to the unique consistent knowledge, as we have seen in Kelvin's words.

In this sense the father of modern Science, Galileo, was epistemologically correct when he distinguished primary qualities (measurable) and secondary qualities (non-measurable). He never said that only primary qualities were the real ones, he just claimed not being able to investigate other qualities of reality; he did not say that they did not exist, or that a non-quantitative knowledge is a lesser form of knowledge.

The path taken by Galileo's epigones has been in the direction of maximizing the range of the quantitative domain, attempting to quantify even what at a first glance would not seem to be measurable at all, such as Life. Statistics was the instrument that allowed such a task in biometrics and in other fields.

The term "biology" made its first appearance as a concept endowed with scientific meaning just at the beginning of the nineteenth century, in 1801, when J. B. Lamarck coined it, to react to a certain baroque progress of botanic and biology that were turning into just classificatory disciplines.[6] Lamarck attempted to find a common link between all the living beings, and he thought he had identified it in the concept of organization, which underlies the microscopic structure of the cell. The path from the bio-*logy* of the earlier nineteenth century to the bio-*metrics* of the end of the nineteenth century was just a further step in the same direction. In the same way as scientific biology looked for an organization that could unify all the living beings, biometrics provided a useful tool to deal with Life in mathematical terms, fulfilling the ancient task of reducing quality to quantity, so as to be able to deal with it in a scientific way.

---

[5]Thomson Kelvin (1883).

[6]Ilich (1992).

The logic of the biology of the earlier nineteenth century was to find a link between all living beings, previously catalogued and divided according to the Aristotelian classification, that was found not to be satisfactory anymore. On the other hand, the metrics of biometrics was supposed to measure individual variations among the members of a given race/species. These measurements were another attempt to find a unity, or at least the degree of unity on a temporal scale, among living beings.

2.3 Which Unit of Measurement for Life?

From a philosophical point of view, the problem of measuring life, which is the specific nature of biometrics, is not a trivial issue. This question would bring us beyond the scope of the present analysis though, since it would require a closer investigation about what life is and what measure is, and what it means to measure life. These are deep philosophical problems which would open a too large discussion field. Therefore we will assume that life is somehow measurable and we will focus on a more specific problem of biometrics: the unit of measurement of life.

The question we would like to address now will turn into: what do you measure when you measure life? At a first glance we can recognize that the action of measuring life does not really measure life itself, it rather measures some aspects of life, some quantitative aspects of it. We need to further investigate these aspects to better understand what kind of information we can gather through a quantitative approach to life.

From a historical point of view, the categorization and classification of living beings, which are the first forms of reasoned measure applied to life, have been a philosophical matter dealt in different ways according to the different philosophical frameworks adopted. The first one to deal with this issue in a systematic way was Aristotle,[7] who gave birth to the discipline of the classification of living beings. He started to think about the qualities that are specific to living beings allowing to assess how they could be categorized.

What we need to point out here from a philosophic point of view is how some life qualities started to be considered more meaningful than others and why this happened. The need to classify raised from the need to deal with living beings in a more rational way than previously. Aristotle's goal was to find a coherent unit of measurement that allowed him to divide the reign of the living beings in a logical form. It is worth noting that Linnaeus, the recognized father of the systematic approach to botany, used Aristotelian categories to found his taxonomic binomial nomenclature, that is still in use as classification criterion among botanists.

Greek philosophers started to approach the phenomenon of life looking at it as something hidden under its sensory qualities shown at a first glance. They had the problem of the so called sensory fallacy. A classical example done to clarify this issue was to

---

[7]Aristotle did it mainly in his works *De Generatione et Corruptione Animalium* and in *De Anima*.

consider an oar in water: it looks as if it were flexed, while everybody knows it is not. This simple observation was enough to make Greek philosophers uncomfortable with sensory qualities, that they considered somehow deceptive; the essence of the oar is to be straight and it is not achievable by looking at it while dipped in water. Hence the beginning of a sort of diffidence to the sensory world of perception.

These remarks allowed the main distinction between primary and secondary qualities, a distinction that will be susceptible of some changes, and a real inversion in a given time, along the history of philosophy and science. Primary qualities were the ones that fall under the reign of sensory observation: what is perceivable by the senses. These qualities are named primary because they are the first access to the living being. Under the definition of secondary qualities, on the other hand, fell the geometrical properties underlying living beings. Democritus can be considered the first thinker who pointed out this kind of qualities. A large part of the ancient philosophy was devoted to answer the question raised from the deception that senses can generate, yet they do not doubt the fact that the senses are a legitimate approach to life.

Starting with Plato, who was in this respect the heir of Pythagoras' tradition, this conception began to change. Plato looked at the senses as just a source of deception, he was the one who clearly stated an essential difference between the data we gather via the senses and the ideal form underlying every living being. He had no doubt that primary qualities were to be considered these latter ones, that are only achievable intellectually, regardless of the information the senses provide us.

In spite of the platonic thought, senses and primary sensory qualities were still considered the main source of information to study the living beings, and even Aristotle strongly relied on them in his foundation of the classification of living beings, and so did the main philosophical and scientific school until sixteenth century.

The whole history of philosophy, and under a certain perspective the history of science too, can be read as a debate around the primacy to accord to these two kinds of qualities. This is also clearly visible in Galileo's vocabulary that defines *primary* and *secondary* qualities. These two different kinds of qualities are considered primary or secondary depending on the pre-eminence accorded to Plato's or Aristotle's point of view.

We can see an echo of this debate along the history of art, in the tractates of the art of painting written by Italian renaissance authors like Leon Battista Alberti,[8] Ghiberti,[9] Piero della Francesca,[10] and Leonard.[11] All these authors revived the platonic approach, developing a new sensitivity to prospective and geometric patterns of reality. Galileo's work can be seen as a translation in the scientific field of this platonic revival that started in the previous two centuries in humanistic fields.[12]

[8] Alberti (1980), in this work he establishes the principles of prospective and remarks the main role played by geometry in painting.

[9] Ghiberti's *Commentarii* (1998), the third book is devoted to the theory of painting.

[10] Piero della Francesca (not surely datable between 1460 and 1482) De Prospectiva Pingendi.

[11] Leonard's Treatise on painting [Codex urbinas latinus 1270].

[12] On this theme see Rodolfo Papa (2005).

What did it mean to categorize a living being through a sensitive quality? It meant that ancient thinkers tried to grasp the "essence" of a given living being through its sensory qualities. In Galileo's terms e.g. the essence of a plume lies in its *virtus solleticativa*, that means its capacity to tickle someone. And this quality was said to be in the tickled subject not in the plume itself, that's why Galileo demoted sensory qualities at secondary rank: in his words they are not objective.

Galileo refused the previous scholastic qualitative approach that relied on sensory perception. Aristotelian thinkers, as medieval men mainly were, thought that the essence of the being lies in the function that the being is able to perform. In the plume example its essence is to tickle, and this is a qualitative aspect un-reducible in mathematical terms, but it is not even a simple sensitive quality, since to tickle someone is not a colour, nor a sound, nor a smell. Although this qualitative approach is not quantitative, functions cannot be directly measured, while length, weight, and mass can.

The qualitative approach strongly depends on observation and primary qualities can be defined as the qualities that fall under the range of perceptual senses.

Current science storiography locates in Galileo the change of paradigm in this respect, Galileo is seen as the father of modern science precisely because of the inversion of this observation ratio. He claimed that the meaningful aspects of reality were the measurable ones, and not the qualitative un-measurable ones. So, in Galileo's terms, the primary qualities were length, weight, and mass, that are objectively measurable, while secondary qualities like colours, sounds, and smells are subjective and not measurable, the first ones are scientific attributes while the second ones are not.

If it is indisputable that Galileo shifted the attention to objective qualities from the so called subjective ones, it is less true that this was a totally new approach in the history of science. In fact, as we have seen, the platonic thought had already considered this perspective assuming that the real essence of the being is its formal archetypical structure, therefore its mathematical form. So Galileo, in its own way, recalled this platonic perspective, and changed, turning upside down, the previous distinction between primary and secondary qualities. Paradoxically. Galileo, who is considered the father of the empirical approach, was the one who revived the platonic thought in science, while Plato has always been celebrated as one of the bigger metaphysical philosopher, and surely not an empiricist.

This paradox is less bewildering if we look at the historical period of seventeenth century, where platonism, pythagorism and hermetism in every field of knowledge were getting a new life.

Going back to our biometric issue: "what do we measure when we measure life?", we can see that starting from Galileo's approach we measure quantitative aspects of life, such as length, weight, volume and so on, and these aspects were considered the meaningful ones, in opposition to the qualitative "functional" Aristotelian perspective.

It is interesting to note that biometrics technologies use both qualitative and quantitative analysis: soft biometrics concerning skin, hair or eyes' colour aim to recognize sensory qualities – the secondary ones in Galileo vocabulary –, but at the same time they are recorded in mathematical terms. Colours can be gathered as a wavelength but still they need to be interpreted somehow to be recognized as a given colour. We find here the historical and philosophical issue concerning how numerical series can be taken as reliable quality indicators. Eyes do not measure a wavelength when they "see" a colour, and same goes for ears, that do not count the sonic wave when they "hear" a sound. Sounds and colours need to be somehow re-interpreted in order to be recognized as such in their sensory form, and this task cannot be accomplished by a mere computational approach.

Galileo claimed that secondary qualities can be susceptible to the scientific approach just if they can be turned into mathematical terms, and this is what, under certain aspects, biometrics does. But then again, are these mathematical aspects real aspects of the beings we are going to investigate? Is the mathematical aspect of life, its quantitative side, the only scientific aspect of it, as stated by Galileo?

To translate living beings in mathematical terms, analysing them just from a quantitative perspective, could be a legitimate approach. It is of course scientific, but it is un-demonstrable that this is the only legitimate scientific approach, unless we want to state that mathematics is the only legitimate science vocabulary, as it has been claimed in modern time by Galileo's epigones.

Once stated that mathematics is a legitimate investigating approach to life, because quantitative aspect really pertains to life, it has to be pointed out that this thesis is not so neutral as it could look at a first glance. When we observe a living being we don't see a quantitative object, we see something different that could be seen as a source of quantitative data, gatherable in mathematical terms, but it has to be clearly noted that it is our methodological approach that drives us to translate it into a numerical series. In other words: it is our method that allows us to talk in mathematical terms about this object, it is not the object itself that shows that aspect. So we have to say that the quantitative aspects connected with living beings are potentially (in an Aristotle potency meaning) inherent to these beings, but they are in act just in our minds.

So all the emphasis accorded to the empirical value of mathematical observation has to be nuanced since mathematical relationships are surely inter-subjective and verifiable measurement, but they have this status because of their subjective filter. Living beings do not present themselves as quantitative objects, we need to reduce them in that form in order to deal with them. Thus we achieve their mathematical structure because we have previously applied a methodological filter that allowed us to see that structure. The role played by the subject in this reduction is therefore clear and deep, and for this reason the distinction between primary qualities – mathematical and objective – and secondary ones – sensory and subjective – is not so legitimate: both are somehow subjective, yet this does not mean arbitrary.

The paradigm shift from sensory qualities to objective ones, defined as empirical, needs to be understood too. The empirical approach fostered by Galileo and Bacon relied on sensory data filtered by their quantitative aspect, to allow their objective determination and inter-subjective verification. This perspective on one hand allows us to deal with data in a mathematical and operational way, but on the other does not allow us to know the qualities we are dealing with. It is not empirically possible to talk properly about colours, sounds, or smells.

The question about the unit of measurement is crucial because it reveals the role played by the subject in the empirical methodology. The unit of measurement is unavoidably a subjective choice, it can be more or less appropriate to the being we are going to investigate, but it still is a subjective reduction operated with a given goal, which will prejudice the observation.

It is a commonplace that we have facts on one side and theories on the other side, and these last ones can be verified through empirical observations. Empirical observations, as demonstrated by Popper, are theoretically filtered, so they cannot play the role of verification.

Once this point has been recognized, we need to accept that the empirical approach is not essentially different from the sensory one, since both of them fall under the reign of subjectivity. This does not mean we are condemned to relativism or to any other form of arbitrarism. Subjectivity just means that we cannot avoid to deal with the subject every time we want to talk about something. Hence we have to focus on new questions: "How does the subject collect data?", "What is the reliability degree of our senses?", "What is the role played by sensors in the data collection?."

2.3.1 Biometrics Sensors

In order to develop any kind of biometrics, we need a sensor to record data related to the object we want to measure. The sensors that will be used tell us what we are going to measure in the same way as we can say what we are going to perceive when we focus our attention on a particular sense rather than on another one. We can recognize someone, for example, by hearing her/his voice, by looking at her/his appearance and so on, and we need to use our senses differently to perform each kind of identification. The same will work for biometric sensors.

The *sensorium* theory tells us how, even when we focus on an individual sense to recognize something, all the other senses are involved in that task: all the perceptions we receive from them are hierarchically organized according to the dominant sense we are focusing on.

According to Aquinas's statement: *nihil est in intellectu quod prius non fuerit in sensu*,[13] the senses are the first step to rationalize the object of our knowledge. In the same way sensors are the first step to measure the object that we need to identify.

First generation biometrics uses simple sensors, able to capture and store some physical features of the object to recognize. They can be optical scanners in order to read fingerprints, for example, or voice analysers to record and evaluate the vocal tracks. This kind of sensors do not work like a normal human perception, since we have seen that a typical perception involves every sense, even if the focus is accorded just to one of them, while in this case sensors do not work together.

[13]It means: nothing is in the understanding that was not earlier in the senses. Cf. Saint Thomas from Aquino (2005), De veritate, Q2, art.3 ad XIX.

In second generation biometrics there is an improvement of the perceptive field, which is becoming more similar to the human one that always arises from different kinds of senses. As we have seen, this new kind of biometrics relies on a dynamical approach: it is increasingly based on multimodality, multiple biometrics, soft biometrics and behavioural biometrics.

From an epistemological point of view, the sensorial perception is not a one way relation, where we can gather some data through neutral devices. The way we collect data is always a bi-directional relation involving an active role of the sensor that is being used. E.g. when we are looking at something, we don't just gather some data about the observed object, we are also able to know the perspective we're looking from, the kind of light that allows us such a recognition and other contextual information. Biometric sensors work in a similar way. When a sensing device captures data, it unavoidably generates data about time and location, so the action of capturing data is always something more than a passive record; moreover, it can generate a problem about the redundancy of the data collection. This situation is even more evident in second generation biometrics, where data are gathered from multimodal sensors, and each one of them generates different data in the capture task.[14]

The advantage of the *sensorium* theory is to make clear that there is no recognition way which works as if the knowledge were something that flows simply from somewhere without any relation to the actor of this action. Any knowledge relationship implies a mutual change in the object and in the subject of such a relation.

In terms of the theory of relativistic knowledge, it could be said that knowledge is a perceptive invariance, and this invariance is what we could say biometrics is looking for.[15] To look at biometrics in terms of invariance detection is not only consistent with its historical source, according to Galton's research about hereditary variations, but it also shows the intimate and deep meaning of identity we are looking for. The concept of identity, indeed, has to deal with something which keeps changing during lifetime, but at the same time shows a certain degree of unity during this change. This concept is well expressed from a metrical point of view by the notion of invariance.

Since we have introduced the term *sensorium,* we need to quote the father of the *sensorium* concept: Marshall McLuhan, who developed a theory[16] to explain the role of the media in our senses, and how they can manipulate the ratio of the senses. E.g. the alphabet stresses the sense of sight, which in turn leads us to think in linear, objective terms: the medium of the alphabet thus has the effect of reshaping the way in we, collectively and individually, perceive and understand our environment.[17]

---

[14] Moral issues connected with this topic are developed in Mordini (2008).

[15] Bohm (1996).

[16] Carpenter and McLuhan (1960).

[17] McLuhan (1978).

2.4 From Action to Being

The main approach of biometrics is to provide a reliable way of recognition of an individual from his/her own characteristics or behaviour. Most techniques involved in second generation biometrics perform such a recognition through a dynamical approach, while techniques of the first generation did it through collection of static information, even though this distinction is not so harsh and clear cut.

First of all it has to be pointed out, from an epistemological point of view, that such an approach needs to be grounded in its logical presuppositions. It has to be made clear that the identification of someone from his/her bodily features is possible and reliable. In order to develop such an issue from an epistemological perspective, two arguments have to be shown: on one hand how a single part could be reliably representative of the whole; on the other, how a behaviour could be reliably reported to one and just one person. Both of these statements could be made, of course, but they need to be demonstrated, and it has to be shown what else derives from them.

Starting from the second point it could be noticed that it is a medieval sentence that allows us to say that a certain behaviour is an expression of a given identity: *agere sequitur esse* (to act follows being), so that even the reverse path, i.e. from action to the being, is in principle possible. Even if this quotation from Saint Thomas Aquinas was originally stated to mean that behaviour, or better, action follows being, in order to ground his ethics, it can be further pointed out that such an axiom also has a relevant epistemological meaning; that is, it concerns the intelligibility of the being. The argument could be set in this way: if actions are expression of a given being, this being then could be investigated starting from its actions. So this is the general philosophical framework that grounded the link between the being and the actions.

This philosophical truth is also consistent with empirical findings. Recent experimental observations point to the possibility of identifying individual characteristics of the human person through kinematic analysis of the person's actions. For example, through appropriate vectorial analysis of wrist movements performed by subjects who are required to point at different targets with a virtual laser pointer, it is possible to classify the mathematical surfaces (so-called Donders surfaces) representing the vectors describing the pointing movements into discrete categories. Such categories could be interpreted as "*motor styles*".[18] Even though it remains to be seen whether or not these motor styles are constant in time in the same individual, they open up interesting perspectives for second generation biometrics. Furthermore, many researchers have found that, when subjects perform actions consisting of a chain of individual motor acts (think for example of a simple action of grasping an object to put it in a container, or a more complex action like drawing a match from a box to light a candle), when the final goal of the action changes, it is not just the last motor act that changes, but the whole chain of motor acts is adapted to the new goal; this effect is observed in adults as

[18] Campolo et al. (2009).

well as in infants. Interestingly, children and adults with autism spectrum disorder (ASD) are less prone to adapt their actions when the goal has changed, which indicates a defect in action planning in ASD.[19]

Taken together, these observations indicate that actions are object-oriented, or to say it in philosophical terms, normally have intentional character.

The key concept of these tasks is "*motor style*". Everyone has their own motor style that can be used as biometric sign of identification; it should also be noted that this style can be affected by feelings, emotions, health etc. and the challenge of biometrics is to encode this style into a numerical parameter so to reliably recognize one person from another. The link between being and actions, in philosophical terms, is a strong and particular kind of relation. Actions can also modify being under determinate circumstances. As we have seen, the path between being and action can be taken in both directions: from being to action, and this is the ontological level, and from the action to being, which is the gnoseological level. Now we need to add that the path from actions to being has not just a gnoseological value, but also an ontological one: the way we behave or we perform an action qualifies our being. Behaviours and actions do not just show who we are, they rather constitute a sort of second nature: this is the meaning according to which we can say actions and behaviours mark our being. This is a well known principle of moral philosophy that is usually employed to explain the birth of habits. In the present argumentation motor style plays the role of the habits and makes it clear how a behaviour/action is not just a reliable identity marker but also an identity maker.

Behaviours/actions can be reliably used to perform identification tasks and this implies a deep philosophical meaning we need to further investigate. It could seem that the identification task is something happening for the sake of someone else, while, according to the ontological meaning we have highlighted before, it should be noted that the first one involved in the identification task is the one who is behaving/acting, since it is his/her identity that is going to be reshaped by his/her behaviours. So the *identification*, according to its etymological root (*identitas facere*, that means to make identity), is first of all something concerning the person himself/herself, rather than an operation performed by someone else. This aspect of the relationship being-actions explains why these motor styles need to be taken very carefully: a motor style involved in identification is somehow a source of reshaping of self-identity as well as a useful pattern of identity recognition. This means that identity is not a static thing, but rather a dynamic process: a person can change his/her motor style along life, or better that motor style can cooperate in changing the self-identity, and hence even the patterns of recognition can change throughout life.

It is not possible here to exhaustively examine a complex concept such as identity, it is enough to say that it is a naïve error to consider it a static thing, it rather is a multidimensional reality that keeps changing, preserving its unity. In this respect second-generation, activity-related biometric technologies, such as the ones developed in the ACTIBIO project, share an approach that takes into account the

---

[19] Fabbri-Destro et al. (2009).

changing reality of living beings. In these technologies identity is not considered as something that can be guaranteed once and forever, on the contrary the identification process is something that keeps going. This way to deal with identity, that is a sort of continuous check-in, is respectful of identity itself, which is something emerging from an ever-changing set of biological parameters. In this way we reach a coherence between the method and its "object".

In summary, the identity biometrics is looking for, shares some characteristics with the concepts of being and action: identity is also a relational concept that can be mutually influenced by the way we're going to determine it, in a similar way as the quantum physical principle claims that measurement influences what is measured.

2.5 Intentionality, Intentions and Emotions

What role, if any, could detection of emotions play in second generation biometrics? These technologies are not meant to measure emotions, but to measure bodily actions, which may give away informations about the internal emotional state of the subject. However, since emotions are inextricably connected with bodily actions, such a relationship cannot be discounted *a priori*. An extensive discussion of emotions and how they can be measured is beyond the scope of this chapter, and the reader can find a good introduction in Bradley (2007).[20] However, we should consider that the concept of "emotion" is deeply related to the concept of action (emotion stems from the latin verb *ex-movere*, moving from). Emotions are bodily activations in response to particular stimuli. In emotionally charged situations the organism is put into action: it approaches a stimulus with a positive value (something good, a gain, a reward), or it avoids a stimulus with a negative value (something bad, a loss, a punishment). The overall strength of the bodily activation has been often quantified with the term of *arousal*. The motor component of emotions serves two aims: the first one is to put the organism in the condition of acting (by mobilizing energy reserves, for example); the second aim is to signal the internal emotional state to other organisms, for example to an attacker (consider e.g. the classical fight-or-flight response of a cat attacked by a dog, consisting in acceleration of the heart rate, widening of the pupils, increased adrenalin secretion, spitting, raising the fur, curving the back etc.). Importantly, in comparison to simple reflex responses, emotions represent more complex responses, involving higher-order brain circuits. They are more plastic than reflexes. The organism can choose different strategies to obtain what it considers good, or avoid what it considers bad. Emotions bear therefore a closer relationship to personal identity than simple, stereotypical reflexes, thus raising the interesting possibility that they could be used in biometric authentication. The relationship between the bodily component of emotions and the corresponding subjective experience has been the object of a long and

[20] For a review on this topic see e.g. Bradley et al. (2007).

still ongoing controversy, known as the James-Cannon debate. According to the James-Lange theory, emotions *are* peripheral alterations of the organ/systems (heart, vessels, muscles, gut etc.) in response to specific stimuli, which we feel as emotion, much in the same way as we feel warm when heat stimulates skin thermoreceptors. In contrast, the Cannon theory affirms that the central nervous system plays an essential role in the subjective emotional feelings.

While technologies of second generation biometrics are not meant to measure emotions, emotional states could affect person identification. Think for example of an authentication method based on the recording of the heart beat: it would be important to make sure that such an authentication method is not too sensitive to changes in heart rate, as they can occur not only during the performance of many tasks, but also in emotional states such as intense fear or joy. There is however, also a risk of false-positive results: e.g. a racing heartbeat could imply intense fear, *or* that an espresso coffee has just been consumed! It should be added that emotional states manifest themselves simultaneously not just in one organ, such as heart rate, but in many organs at once, (as the fight-or-flight response demonstrates), implying that the use of multiple sensors could help to distinguish the effects of pharmacological agents (e.g. caffeine in the above example) from true emotions of fear or joy.

The concept of intentionality has a rich and glorious history. It has been a central idea of the medieval knowledge theory that allowed the scholastic philosophers to overcome both the nominalism and a certain conceptualistic perspective akin to idealism. More recently it has been rediscovered by the philosopher F. Brentano, Husserl's master, in order to overcome the so called "bridge" problem inherited from the modern dualistic philosophy. In philosophy intentionality is distinguished from intention: the first one is the reference driven by thought to reality, while the second one has a moral connotation and in this respect it is an aspect of the voluntary action. Moderns never succeeded in solving the problem about the relation between world and subject. Such impasse was overcome by the recovery of the intentionality that allows to bind the subject and her/his object to a realistic ground, neither subjectivistic nor objectivistic.

We have briefly sketched here the history of the concept of intentionality since it plays a similar role also in the field of biometrics, that is, to link two different domains of being, namely, the subject and his/her goals. The detection of people's intentions, or at least the detection of when their intentions deviate from what they are supposed to be in a specific context, is one of the most important and challenging purposes of biometric technologies, for example in security tasks.

The problem of detecting intentions is an ancient challenge. From a theological point of view, e.g., it is said that it is not possible to judge intentions since only God can see them, assuming that every human attempt to read into man's heart will fail. From a certain perspective this is correct. There is, indeed, no way to be sure to obtain a correct inference between a behaviour, an action, or any given physiological parameter, and the underlying intention. Nevertheless, even if there is not a safe path from a *phenomenon* to a given intention, we can reach a certain degree of reliability arguing that an intention (conscious or not) is always underpinning a human *phenomenon*.[21] So we are stuck in a situation where we know that there is a link between intentions and their *phenomena*, but we lack a reliable path from *phenomena* to their underlying intentions. On the other side, the notion of intention, in spite of the problems concerning its detectability, is still a central and unavoidable notion of everyday life. Even in the legal field we cannot avoid to use it in a relevant way. Just think about the role played by intentionality in determining the difference between an intentional and premeditated murder and an unintentional manslaughter, for example. Since, for an external observer, the last action of pulling the trigger is the same in both situations, we need to go back to the events and actions that have led to pulling the trigger. As we have outlined in Sect. 2.4 ("From Action to Being"), changing the last action of a chain of motor commands leads to a reprogramming of all motor commands, not just the last one. It would therefore be interesting to investigate the kinematics of those apparently identical actions, to see if different intentions would change somehow even the execution of the shooting, and this would be a real challenge for biometrics. These remarks imply that we cannot avoid dealing with the concept of intention and trying to measure it somehow.

In order to find a way to accomplish this task, we need to give up the hope of achieving an exact correspondence between intention and the related action and try to find a probabilistic way to interpret this correlation, bearing in mind that it is not possible to achieve more than one probability. Moreover this is a general issue concerning biometrics due to its particular epistemological status, as we have seen.

As we have briefly sketched in the second paragraph by drawing a metaphysical framework to ground the possibility of biometrics, we resort to the scholastic sentence according to which *agere sequitur esse*. Besides providing a foundation to the issue of the recognition in itself, this sentence has, historically, made it possible to talk about intentionality in a reasonable way.

We just need to highlight how intention can be assumed as an "action". It is indeed something different from an action, even if it is strictly connected to it.[22]

Once we recognize that actions are in some way connected with intentions, and this kind of link should be further investigated, we finally get something visible to start a recognition task. So we are able to reveal or at least to indicate an invisible concept such as intentions.

Going back to the medieval sentence, the main issue in this task is probably related to that "*sequitur*" or, in other words, to the relationship between being and its manifestations.

---

[21] We are using the etymological sense of the term "*phenomenon*" that is: what shows itself.

[22] Of course it has been remarked that not all actions, not even behaviours, are intentional, if we assume intentionality as something essentially conscious, even if this assumption is not so unquestionable as it could seem. For our argument it is enough to recognize that all intentions are supported by actions or they could be seen in actions, even if the opposite were not true. Either from an historical point of view or from a logical one in our opinion it is not possible to reduce the intentionality to the domain of the consciousness. As a matter of fact also an unconscious intentionality has been studied. It is not possible to discuss here in detail this thesis. It has been analysed in Freud (1900), and in Trincia (2008). Cf. Madioni http://www.rivistacomprendre.org/rivista//uploads/9739ca1c-b78b-8b87.pdf.

The deep philosophical problem connected with this issue is how something like intention works, or plays a role in the body's shape. What's their relationship? This is, from another point of view, a new version of the body-mind problem taken from the point of view of the visibility of their link. Intention can be seen as a link between body and mind since it shares both mental and bodily aspects.

2.6 Epistemological Issues About Detectability of Intentionality

Epistemology is the study of knowledge and justified beliefs. It is usually considered a branch of the philosophy of knowledge, precisely the discipline in charge to show the degree of reliability of a given knowledge. *Episteme* in ancient greek is a well grounded kind of knowledge, its contrary is *doxa*: opinion; therefore epistemology is a reflection about the basis of knowledge: How do we know what we know? How is it acquired? What's the bond between what we know and the truth? What are the limitations of our knowledge about a given discipline? What reliability do the different sciences have? These are some of the questions epistemology is in charge to deal with.

From an epistemological point of view and according to the idea behind biometrics, we must start to deal with the issue of intentions detectability from the body perspective, which is the visible pole of our relation. There have been several ways to outline the mind/brain or soul/body problem and their relationship. All these attempts could be summarized in three main schools that include monistic, dualistic and hylomorphical concepts.[23]

The first school reduces one of the elements of the relationship to the other, so we can have a mentalistic approach, when the bodily pole is reduced to the mental one,[24] or we can have a material approach, when on the contrary the mental pole is reduced to the bodily one. In both these situations we have a refusal of the relationship between the two poles of our argument because one of these poles is considered just an illusion. This is the monistic approach. And it is useless for our purpose to detect the intentionality because in this context the intention *is* the body, so we don't need any kind of recognition; or, again, the body is an *epiphenomenon* of intention,[25] but again we don't have any ground to start the detection of intention from, since there is only the intention and there is no body in this perspective. The monistic school with all its different variants is actually the most commonly shared point of view in the modern debate

[23]For a critical analysis of all these schools Cf. Sanguineti (2007). Another antological and useful study, even if not up to date, is: Moravia (1988).

[24]Eleatic school e.g. while affirming the true characteristic of being on the one hand, on the other hand denied to the material world any kind of existence, and so the body was considered nothing but an illusion.

[25]This is actually just an academic hypothesis, without supporters as long as we know. Maybe Berkley could be seen as an archetypical exponent of this approach according to his spiritual monism flowing from his famous sentence: *esse est percipi* (to be is to be perceived).

about the body-mind problem. It goes under the name of panpsychism (a kind of spiritual monism),[26] or behaviourism,[27] or the identity theory that it is better to call neurologism,[28] or neural biologism,[29] or emergentism,[30] or supervenience,[31] or the functionalisms that can be divided in representational functionalism,[32] causal functionalism[33] and in the computabilistic ones.[34] These are just some of the modern monistic positions. It is not the goal of the current work to give a complete review of this long and complex debate: it is enough for our purpose to outline it.

The dualistic point of view on the other hand claims in general terms the real division between soul and body, or mind and brain in contemporary words. This real division makes any kind of relationship between the poles of our pair impossible, because of their total heterogeneity. The world of mind lies under its own laws, and so does the world of matter, in our case the brain, with no points of contact between each other. This model gave birth to the "bridge" problem, the one we have sketched before: on the one hand how to draw a connection between these two domains of being, and on the other hand how to explain the harmony between these two worlds, which are doubtlessly somehow connected. These issues are the main topics around which the debate goes on all along.

The classical root of dualism is usually traced back to Plato's philosophy,[35] who defined the relationship between soul and body as if the body was the soul's cage, claiming an irreconcilable difference between them. The soul shares the immortal status of the ideas, while the body is corruptible like every material object. Plato inherited this view, which is not as simplistic as it could seem at first sight,[36] from

---

[26]Cf. Chalmers (1996).

[27]The father of this stream, according to which there are no interior psychological phenomena but just observable behaviours, is Ryle (1949).

[28]There are many authors who share this conception, some of the more radical are Churchland (1986, 1996, 1998). Among others, this thesis has been supported by H. Feigl, U. Place, J.J. Smart, and in a more sophisticated way near to functionalism by D.K. Lewis and D. Armstrong.

[29]Changeux (1983), Gazzaniga (1998), Damasio (2000), Idem (2003), Edelman (1992), Idem (1989), Tononi and Edelman (2000).

[30]Bunge (1980), Margolis (1978), Searle (1984), Idem (1992), Idem (2002).

[31]Cf. Kim (1996). A critical discussion about this thesis can be found in Murphy (1999), Seifert (1989).

[32]Cf. Fodor (1983), Idem (1994).

[33]Cf. Lycan (1987).

[34]Cf. Putnam (1975), Idem (1993), Minsky (1988), Dennett (1987), Boden (1977), Idem (1989/1990), Turing (1936, 1950). This is the classic where Touring purposes his famous and eponymous experiment named Turing test. A critical and important article written against this conception, that had a huge impact between the philosophers of mind was, Nagel (1974).

[35]The famous passages related to this topic are *Phaedo* 62D and *Ph.* 66C – 67B.

[36]Cf. Szlesak (1993), Reale (1994); both these philosophers share a new interpretation of the Plato's thought, according to the which the so called platonic dualism has to be attenuated, because of the fact that even the matter share something of the first principles, which are the One and the Diad. Therefore the body is not just a mere jail of the soul, even if it is true that the goal of the soul needs to get rid of the body in order to be achieved.

the orphic-pitagoric tradition. The one who is (maybe somewhat unjustly) considered the emblem of the dualistic school is Descartes, who quite clearly stated the difference between soul, which is a *res cogitans* in cartesian vocabulary, and body, *res extensa*. The fact that Descartes talks about *res* (thing), makes it clear that he established a substantial difference between the two reigns of being.

Since Descartes' times, many solutions have been proposed to the body-mind problem. The first one who tried to find a solution was Malebranche, who hypothesized his occasionalism theory,[37] claiming a sort of ontological parallelism between the mental world and the physical one, in order to explain the concordance between those two "things".

This "exit strategy" from the bridge problem has been further used but in an epistemological sense rather than in the ontological one. A common parable about the understanding of the issue shows that, once the correlation between mind and brain starts to be considered like two different ways to investigate the same object, the ontological monism underlying this conception will prevail, and from the dualistic methodological starting point we fall back to the monism field that is the most common and shared one.

If, according to the monistic perspective, the relationship between mind and brain fails because of the lack of one of the elements of the relation, from the dualistic point of view the relationship fails too because of the heterogeneity of the two elements of the relation, since they are unrelated to each other. In both these approaches even the epistemological conditions are not needed to consider the topic of intention detectability, because in neither of them there is anything to be discovered.

The third way to approach this topic is hylomorphism,[38] which, as we will argue, is the only one that allows to deal with the relation between a visible body and an invisible intention to be detected. Hylomorphism is a metaphysical framework thought to explain the nature either in its material aspect or in its intelligibility. Key to the understanding of this model is the word *synolon*, a Greek term to indicate the object as a composition of "*form*" and "*matter*". From this perspective, the body is essentially composed of matter and form, and this metaphysical composition makes it possible to read formal meanings in the bodily matter. Form and matter have a long and complex philosophical history, which we cannot discuss here. For our purpose it is enough to point out how the relationship between body and soul, or brain and mind, could be read in these terms. In other words, we ask ourselves if mind is the form of body.[39] This statement in Aristotelian terms means that the body is able to receive the mental form from its ontological status, and on the other side that form (mind) in its essence cannot avoid to shape body in some way.

---

[37] Cf. Malebranche (1946).

[38] The school that adopted this perspective has historically been the thomistic one, some of the authors who developed such theme are: Sanguineti (2005), Maldamé (1998), Aa. Vv (1989), Basti (1991), Possenti (2000), Idem (2000), Borghi (1992), Cross (2002), Haldane (1994), Kenny (1989), Idem (1993).

[39] Cf. Aristotle, *On the Soul* 412b5-7, 413a1-3, 414a15-18.

These remarks about the intimate nature of form and matter in the hylomorphic model are important because they point out the necessary link between form and matter, and consequently between body and mind. So in our analogy the intention, which is in the domain of the mind, cannot avoid to express itself in a bodily feature, and this makes finally viable our goal to pursue a path from body to intention. Moreover this path has an ontological foundation, able to give reasons about the "bond of being" between intentions and their expressions, and these reasons don't just lie on a theoretical plan, but they can be seen in the reality of the object we wish to talk about.

Once we have grounded the problem of intentionality on the proper metaphysical coordinates, we should also be able to see the epistemological limitations of the knowledge about our object. Indeed the hylomorphical approach relies on an analogical framework that shows us the terms and the nature of the reachable knowledge. To talk about analogy related to our topic means to underline the intimate structure of being underlying these short Aristotelian metaphysical descriptions, which is analogical as well. This means that we are not in the computational domain, where the relations are mostly univocal, but rather in the reality domain where the relations are analogical, since they have got many irreducible meaning levels to which we can relate. This is also the proper meaning of the Greek term *ana-logos*, that could be translated as a "superior and previous *ratio*" (*anà* means up in a spatial context and logos means reason, so when the Greek man talks about analogy of being he means there are several and inter-related levels of reality, which are in analogical relation with one another).

So the analogical identification that can be achieved cannot aim to be univocal and indefectible. However, it can be reached with some degree of reliability, and it is always better than the sceptical outcome which at first sight it seemed designed for. This conclusion is coherent with the object of biometrics: identity, which is mainly an analogical concept.

2.7 Identity Digitalization

The remarks made above should alert us that the digitalization, which is the condition of possibility of all modern biometric technologies, is not a neutral process in the data collection, because it changes the sensorium in a substantial way.

From a certain point of view the digitalization in data collection is nothing but a symbolic reduction of something into a more feasible way to work with. Every scientific attempt to study an object is a kind of reduction. From another point of view, the fact that we can fragment the identity features of someone into digital records, and then encode these records to store her/his digital identity in a database, should be further analysed to understand which kind of change in the ratio-sense[40] is going to occur by performing this operation.

---

[40] The concept of ratio-sense comes from McLuhan vocabulary and means the way a sense plays a sort of dominance both in perception tasks and in rational recognition.

It has been correctly pointed out that: "digital representations always imply a certain degree of simplification, which modifies the nature of the represented object. By digitalizing representations of body parts and behaviours, biometric technologies tend to remove from them all dimensions but those which are relevant to recognition. Ideally, biometrics aims to turn persons into mere living objects, which can be measured and matched with similar living objects".[41] The effect of digitalization is not just to reduce an object to a quantitative dimension, but also to deal with this new dimension in a more rational way.

This "more rational way" is what we have to understand more deeply. It is true, of course, that once we can deal with a number, which represents someone's encoded identity, it is possible to perform a large amount of operations about that identity. But it is not true that reason lies just in the computational sense. Computability is just one of the meanings of reason, and it is not the main one. The risk of dealing with an encoded identity is to take that code as really representative of the original identity. This is not just a theoretical risk, because, as stated by Irma van der Ploeg: "the dominant view of what the body is, what it is made of and how it functions, is determined and defined by the practices, technologies and knowledge production methods applied to it".[42]

This is, from another perspective, what McLuhan expressed in his *sensorium* theory. There is a strong link between an object and its representation and it is possible to further investigate such connection. The representation could be seen as a way to access the object, so the identity of a given object is primarily depending on the representation we are going to adopt to analyse it. This conclusion leads us to recognize how digitalization is not a trivial matter. It would be a naïve error to consider the identity of the object we're dealing with as something unrelated to the way we're going "to read" it. In epistemological terms it could be said that the model according to which we are studying a given reality is in itself an important part of the examined reality, and the researcher needs to be aware of that, otherwise it is quite easy to mistake the model for the reality and the working hypothesis for the truth. We can even consider that a list of numbers with no indication about their mutual relationship is totally meaningless, and it would not allow to identify anyone.

Sensors are a legitimate instrument of reduction only if those who are going to use them are aware of the reduction connected with their use. What does this awareness mean in practical terms? It does not just mean that the identity of the reduced object does not lie in its code, which is quite obvious, but also that the recognition pattern developed through that reductive procedure is in itself quite partial and always in need of integration. This issue is not avoidable at all, because every kind of technology has to rely on a particular way of reduction. Reductionism is essential to every kind of metric: it is just something scientists have to learn to deal with. This means that we have to accept a sort of fallibilism, to use Popper's vocabulary.

---

[41] van der Ploeg (2005), quoted in Mordini and Massari (2008).

[42] Cfr. *Ibidem* (2008).

From phenomenology, on one side, and quantum physics on the other side, we learn how much the observer is involved in the relationship with the object observed, and we learn even more how the gathering of data is all but a neutral action, that's why the sensors are so important in the biometrics technology and in the epistemological inquiry about it.

Even the multimodal biometric technology, relying on different kinds of sensors able to capture different features of people's identity, encodes these features in a computational way, in order to deal with the ever growing amount of data. We have seen what it means to translate the multidimensional ontology of an identity into a mono-dimensional, numerical one. To stick to McLuhan's vocabulary, it means we're according to the linear ratio a sort of supremacy on other possible sensorial perceptions, and of course this supremacy will reshape the whole context of the perception. Therefore, the presence of this bias accorded to the linear computational approach has to be pointed out, and then how suitable it is for the task in itself.

It could be said that, in order to accomplish a recognition task, it is enough to achieve a conventionally previously defined threshold connected to the matching system. This is indeed the normal way things go in this field. What needs to be added is that this conventional threshold can just provide a conventional degree of the identity that is going to be recognized. At the same time what comes out from this statement is that the recognition we are going to achieve through this process can be just a conventional recognition, in need of integration, if we aim to a reliable recognition. The adoption of such kind of models to perform a recognition task can only lead us to obtain a hypothesis about the object of our identification, a hypothesis that needs to be further investigated if our goal is the reality.

A useful concept to clarify the context of perception is introduced by C. Fabro in his masterpiece "Phenomenology of Perception". While analysing the illusion phenomena, he points out that: "*just in given situations we are able to find ourselves in objective conformity with what we are investigating and to know the reasons of this persuasion*".[43] This means that pure perceptions do not exist, as well as pure data cannot exist. Every perception is somehow influenced by the context in which it is performed, just as every datum is connoted both by the sense/sensors that are involved in capturing it, and by the context. So a certain degree of conventionality in this field is unavoidable.

Another interesting conclusion drawn by C. Fabro in his work about perception is that: "*we have to admit that our immediate knowledge has several degrees of correspondence with its own object [...] The existence of illusions or inadequate perceptions means that there are other kinds of knowledge which are not illusions, and thus are considered adequate to their objects*".[44] This observation strengthens our remark about the provisional nature of the results obtained from every kind of sensors used in biometrics.

As we have seen in paragraph 5.2, biometrics deals with quantitative aspects of life, which have been considered the primary quality of living beings. We have also

[43] Fabro (2006). [Translation of mine].

[44] Idem (2006).

recognized how this statement according a primacy just to quantitative data has to be reconsidered. So we can observe now from an epistemological perspective that quantitative traits measured by biometricians can provide just a partial identification that needs to be reinterpreted in order be a real identification.

There is a sort of parallelism between the history of the philosophy of science and biometrics epistemology. There has been a period in philosophy of science where it was considered that data, quantitative data, had to be considered the only reliable facts, and these facts would have provided the verification or falsification of theories. This assumption, that still holds a certain degree of credit among scientists, has been criticized by Popper and many others, who observed how facts do not exist, every single fact, indeed, comes from an observation, and every observation comes from a method, a particular technique, a theory. So the clear distinction between facts on one side and theories to be verified by facts on the other side is totally wrong. When someone is trying to verify/falsify a theory using facts they are simply struggling with different theories taken under different aspects. Same goes with biometrics verification process. There are no facts/quantitative data that can provide a real verification, quantitative data obtained by biometrics measurements can provide a useful tool to proceed with further interpretation. Interpretation is then the real verification key. Just as a colour is not describable from its numerical frequency, but needs to be interpreted to grasp its specific qualitative difference, identity needs to be interpreted and grasped from its quantitative reduction as well.

2.8 Conclusions

In the current work we have sketched a brief history of biometrics, paying attention to its epistemological status; we have pointed out the role played by the statistic approach in the birth of biometry, and we have been able to appreciate the role played by biometrics in the history of science. From this perspective we have noted the continuity between a particular conception of science started with Galileo, and the growing field of biometrics technologies, concerning the problem of measuring life.

Starting from this historical remark we have then further analysed the problem of the unit of measurement in biometrics, addressing questions like: What do we measure when measuring life? What kind of qualities are achievable through biometrical measure? What is the meaning of these qualities? To answer these questions we have studied the history of the approaches to sensory world and we have been able to locate the modern science approach in a larger philosophical context, which allowed us to better understand the underlying epistemology connected with it. This clarification allowed us to also understand the unavoidable role played by subjectivity in collecting data, hence we further looked into biometric sensors, in order to analyse more properly the role played by sensors in data gathering. From this analysis we gained a better understanding of this process, that must not be thought as a passive data recording but as an active data reconstruction.

Once we have recognized the major role played by sensors in data gathering, we have moved to analyse the link between action and being, since identification tasks aim to establish a reliable connection between gathered data and the being behind them. Our analysis led us to the conclusion that new generation biometrics, due to its continuous authentication system, is more coherent with a dynamical conception of identity, and for this reason it is epistemologically better grounded and aware of its limitations than the previous one.

Once a reliable path between action and being was established, we have focused on emotions and intentionality, which are the specifications of identity features in action. We have considered what they are both from a physiological and a philosophical point of view, and what their role is in biometrics tasks. In this chapter we have run the same path as the previous one but from the opposite side: while in Sect. 4.3 we have considered the link between action and being, aiming to reach the being moving from the action, in Sect. 4.4 we have analysed the link between being and action moving from emotions and intentions, which can be considered the being's attribute. In philosophical terms the first movement is named ontological, because its goal is to reach the *being*, while the second one is named gnoseological, because it aims *to know* the being's qualities.

Once the path from identity to intention was grounded, we have worked on the epistemological issues related to intentions detectability; we have noticed which logical and ontological implications flow when we assume that intentions are a reliable path to identity. We went over the different philosophical schools that dealt with this problem looking at the hylomorphical paradigm as the one that allows us to overcome the limitations of reductionism.

From this perspective we have been able to devote the last part of our work to the issues connected with identity digitalization, which is somehow the essence of biometrics. We have discussed the limits of a biometrical approach to identity, we have pointed out what kind of identity can be detected by a quantitative approach and moreover what it means for someone to be identified. Identification has been considered as an active process wherein the self identity is the first actor involved in reshaping his/her own identity, hence this process cannot be considered a neutral or a passive technological system anymore.

These philosophical and epistemological remarks should alert us about the limits connected with the nature of biometrics, that is a legitimate instrument of knowledge but with precise limitations. First of all the quantitative approach is a partial perspective in need of a qualitative interpretation to be really effective, so we have to be careful when taking for granted a claim of objectivity for metrical technologies. Identity is not a neutral or passive object we can deal with, since we have seen how it is constitutively involved in measuring and being measured. Thus, on the one hand to measure is not a neutral action, and on the other hand identity suffers from some changes when being measured; these two statements should warn us not to abuse of identification processes since these processes will reshape the identity: to measure is somehow an objectification of a subject, and this is not respectful of the dignity inherent to every subjective identity.

It is worth noting that these remarks have been gained just on an epistemological and philosophical terrain, not on the moral one. We have remarked the limited range of a quantitative knowledge, and now a prudential reflection can start its analysis from a moral perspective. Our notes just have the role to open the field to moral and legal studies, and this field has been opened from its inside. We wouldn't like to think of moral philosophy as something that comes after science, but as a reasoning that goes with it; a prudential attitude, in fact, is not something extraneous to scientific works, science in itself needs it, since scientific knowledge, as we have tried to argue, has its own limitations. Hence a good science has to be cautious in its own conclusions and needs to leave room to considerations of different kind, since the quantitative approach cannot claim to be more rational than other ones.

References

(No authors listed). 1965. Sir Francis Galton statistician of eugenics. *JAMA* 194(6): 666–667

Aa, Vv. 1989. *Homo loquens, Uomo e linguaggio. Pensieri, cervelli e macchine*. Bologna: Edizioni Studio Domenicano.

Alberti, L.B. 1980. *De Pictura*, ed. C. Grayson. Bari: Laterza. [1435] Orig. Ed.

Audi, R. 1999. *The Cambridge dictionary of philosophy*. Cambridge/New York: Cambridge University Press.

Basti, G. 1991. *Il rapporto mente corpo nella filosofia e nella scienza*. Bologna: Edizioni Studio Domenicano.

Boden, M. 1977. *Artificial intelligence and natural mind*. Brighton (Sussex): Harvester Press Lim.

Boden, M. 1989/1990. *The philosophy of artificial intelligence*. Oxford/New York: Oxford University Press.

Bohm, D. 1996. *The special theory of relativity*. London: Routledge.

Borghi, L. 1992. Antropologia tomista e il body mind problem (alla ricerca di un contributo mancante). *Acta Philosophica* 1: 279–92.

Bourke, Vernon J. 1962. Rationalism. In *Dictionary of Philosophy*, ed. D. Runes, 263. Totowa, New Jersey: Littlefield, Adams, and Company.

Bradley, M.M. 2007. Emotion and motivation. In *Handbook of psychophysiology*, 3rd ed, ed. J.T. Cacioppo, L.G. Tassinary, and G.G. Berntson. Cambridge/New York: Cambridge University Press.

Bunge, M. 1980. *The mind-body problem. A psychobiological approach*. Oxford: Pergamon Press.

Campolo, D. et al. 2009. Kinematic analysis of the human wrist during pointing tasks. *Experimental Brain Research*, E-pub 2010 Mar; 201(3): 561–73.

Caplan, R.M. 1990. How fingerprints came in use for personal identification. *Journal of the American Academy of Dermatology* 23: 109–14.

Carpenter, E., and M. McLuhan (eds.). 1960. *Explorations in communication*. Boston: Beacon.

Chalmers, D.J. 1996. *The conscious mind*. Oxford: Oxford University Press.

Changeux, P. 1983. *L'Homme neuronal*. Paris: Fayard.

Churchland, P.S. 1986. *Neurophilosophy. Toward a unified science of the mind/brain*. Cambridge: MIT Press.

Churchland, P.M. 1996. *The engine of reason, the seat of the soul*. Cambridge: MIT Press.

Churchland, P.M. 1998. *Matter and consciousness*. Cambridge: MIT Press.

Cross, R. 2002. Aquinas and the body mind problem. In *Mind, metaphysics, and value in the thomistic and analytical traditions*, ed. J. Haldane, 36–53. Notre Dame: University of Notre Dame Press.

Crow, J.F. 1993. Francis Galton: Count and measure, measure and count. *Genetics* 135: 1–4.
Damasio, A. 2000. *The feeling of what happens: Body and emotion in the making of consciousness*. New York: Harcourt Brace & Company.
Damasio, A. 2003. *Looking for Spinoza: Joy, sorrow, and the feeling brain*. Orlando: Harcourt Inc.
Dennet, D. 1987. *The intentional stance*. Cambridge: MIT Press.
Durkheim, É. 2002. *Il Suicidio*, ed. R. Scramaglia. Milano: Biblioteca Universale Rizzoli.
Edelman, G. 1989. *The remembered present*. New York: Basic Books.
Edelman, G. 1992. *Bright air, brilliant fire, on the matter of the mind*. New York: Basic Books.
Fabbri-Destro, M., et al. 2009. Planning actions in autism. *Experimental Brain Research* 192: 521–5.
Fabro, C. 2006. *Fenomenologia della percezione*, 232. Rome: EDIVI.
Fodor, J.A. 1983. *The modularity of mind*. Cambridge: MIT Press.
Fodor, J.A. 1994. *The elm and the expert: Mentalese and its semantics*. Cambridge: MIT Press.
Freud, S. 1900. *Die Traumdeutung*. Leipzig-Wien: Franz Deuticke.
Galton, F. 1892. *Fingerprints*. London: Macmillan & Co.
Galton, F. 1907. *Probability, the foundation of eugenics*. Oxford: Clarendon.
Gazzaniga, M.S. 1998. *The mind' past*. Berkeley: The Regents of the University of California.
Ghiberti, L. 1998. *I Commentarii*, ed. L. Bartoli. Firenze: Giunti. [1477] Orig. Ed.
Haldane, J. ed. 1994. Breakdown of philosophy of mind. In *Mind, metaphysics, and value in the thomistic and analytical traditions*, 54–75. Notre Dame University Press, Notre Dame, Indiana
Haldane, J. 2004. Analythical philosophy on the nature of mind: Time for another rebirth. In *The mind body problem: A guide to the current debate*, ed. R. Warner and T. Szubka, 195–203. Oxford: Blackwell.
Ilich, I. 1992. *In the mirror of the past: Lectures and addresses 1978–1990*. New York: Marion Boyars.
Kenny, A. 1989. *The metaphysic of mind*. Oxford: Clarendon.
Kim, J. 1996. *The philosophy of mind*. Boulder: Westview Press.
Lycan, W. 1987. *Consciousness*. Cambridge: MIT Press.
Madioni, F. 2008. Una teoria del soggetto in Freud tra Brentano e Husserl. http://www.rivistacomprendre.org/rivista//uploads/9739ca1c-b78b-8b87.pdf.
Maldamé, J.M. 1998. Sciences cognitives, neurosciences et âme humaine. *Revue Thomiste* 106: 282–322.
Malebranche, N. 1946. In *La Recherche de la vérité*, vol. VI, ed. G. Lewis. Paris: Vrin.
Margolis, J. 1978. *Persons and minds*. Dordrecht: Reidel.
McLuhan, M. 1978. The brain and the media: The Western Hemisphere. *Journal of Communication* 28(4): 54–60. Autumn.
Minsky, M. 1988. *The society of mind*. New York: Simon & Schuster.
Moravia, S. 1988. *L'enigma della mente*. Bari: Laterza.
Mordini, E., and S. Massari. 2008. Body, Biometrics and Identity. *Bioethics* 22(9): 488–98.
Murphy, N. 1999. Supervenience and the downward efficacy of the mental, a non reductive physicalist account of human action. In *Neuroscience and the person, scientific perspectives on divine action*, ed. R.J. Russel et al. Vatican City State: Vatican Observatory Publication.
Nagel, T. 1974. What is it like to be a bat? *The Philosophical Review* 83: 435–50.
Papa, R. 2005. *La scienza della pittura. Analisi del "Libro di pittura" di Leonardo da Vinci*. Milan: Medusa.
Ploeg, I. 2005. *The machine readable-body. Essays on biometrics and the informatizations of the body*. Herzogenrath: Shaker.
Possenti, V. 2000. Il problema mente corpo. In *Annuario di filosofia. Corpo ed anima*, ed. V. Possenti, 265–318. Milan: Mondadori.
Putnam, H. 1975. *Mind, language and reality. Philosophical papers*, vol. II. Cambridge: Cambridge University Press.
Putnam, H. 1988. *Representation and reality*. Cambridge: MIT Press.
Reale, G. 1994. *Per una nuova interpretazione di Platone*. Milan: Vita e Pensiero.

Ruiz, R., et al. 2002. Eugenesia, Herencia, Selecciòn y Biometrìa en la Obra de Francis Galton. *ILUIL* 25: 85–107.
Ryle, G. 1949. *The concept of mind*. London: Hutchinson of London.
Saint Thomas Aquinas. 2005. *Sulla verità*, ed. F. Fiorentino. Milan: Bompiani.
Sanguineti, J.J. 2005. Operazioni cognitive: un approccio ontologico al problema mente cervello. *Acta Philosophica* 14: 233–58.
Sanguineti, J.J. 2007. *Filosofia della mente, una prospettiva ontologica e antropologica*. Rome: Edusc.
Searle, J. 1984. *Minds, brains, and science. The 1984 Reith lectures*. Cambridge: Harvard University Press.
Searle, J. 1992. *The rediscovery of the mind*. Cambridge: MIT Press.
Searle, J. 2002. *Consciousness and language*. Cambridge: Cambridge University Press.
Seifert, J. 1989. *Das Leib Seele Problem und die gegenwärtige philosophische Diskussion*. Darmstadt: Wissentschaftliche Buchgesellschaft.
Stigler, S.M. 2000. The problematic unity of biometrics. *Biometrics* 56: 653–8.
Szlesak, Th. 1993. *Platon lessen*. Stuttgart: Frommann-Holzboog.
Thomson Kelvin, W.T. 1883. Lecture on "Electrical Units of Measurement" (3 May 1883), published in *Popular lectures*, vol. I, 73; quoted in *Encyclopaedia of occupational health and safety* (1998) by Jeanne Mager Stellman, p 1973.
Tononi, A., and G. Edelman. 2000. *A universe of consciousness: How matter becomes imagination*. New York: Basic Books.
Trincia, F.S. 2008. *Husserl, Freud e il problema dell'inconscio*. Brescia: Morcelliana.
Turing, A. 1936. On computable numbers, with an application to the Entscheidungsproblem. The *Proceedings of the London Mathematical Society* (ISSN 0024-6115) is published monthly by Oxford University Press, Oxford, UK on behalf of the London Mathematical Society. Ser. 2, vol. 42, 230–265, 1937.
Turing, A. 1950. Computing machinery and intelligence. *Mind* 59: 433-460.

Chapter 3
Biometric Recognition: An Overview

Anil K. Jain and Ajay Kumar

3.1 Introduction

Human identification leads to mutual trust that is essential for the proper functioning of society. We have been identifying fellow humans based on their voice, appearance, or gait for thousands of years. However, a systematic and scientific basis for human identification started in the nineteenth century when Alphonse Bertillon (Rhodes and Henry 1956) introduced the use of a number of anthropomorphic measurements to identify habitual criminals. The Bertillon system was short-lived: soon after its introduction, the distinctiveness of human fingerprints was established. Since the early 1900s, fingerprints have been an accepted method in forensic investigations to identify suspects and repeat criminals. Now, virtually all law enforcement agencies worldwide use Automatic Fingerprint Identification Systems (AFIS). With growing concerns about terrorist activities, security breaches, and financial fraud, other physiological and behavioral human characteristics have been used for person identification. These distinctive characteristics, or biometric traits, include features such as face, iris, palmprint, and voice. Biometrics (Jain et al. 2006, 2007) is now a mature technology that is widely used in a variety of applications ranging from border crossings (e.g., the US-VISIT program) to visiting Walt Disney Parks.

A.K. Jain (✉)
Department of Computer Science and Engineering, Michigan State University, East Lansing, MI 48824-1226, USA

Department of Brain and Cognitive Engineering, Korea University, Seoul, South Korea
e-mail: jain@cse.msu.edu

A. Kumar
Department of Computing, The Hong Kong Polytechnic University, Hung Hom, Hong Kong
e-mail: ajaykr@ieee.org; csajaykr@comp.polyu.edu.hk

E. Mordini and D. Tzovaras (eds.), *Second Generation Biometrics: The Ethical, Legal and Social Context*, The International Library of Ethics, Law and Technology 11,
DOI 10.1007/978-94-007-3892-8_3,

Biometric recognition is based on two fundamental premises about body traits: *distinctiveness* and *permanence* (Jain et al. 2006, 2007). The applicability and identification accuracy of a specific biometric trait essentially depends to what extent these two premises hold true for the population at hand. Fingerprints, face, and iris are amongst the most popular physiological characteristics used in commercial biometric systems, with fingerprint alone capturing over 50% of the civilian market share (Biometrics Market Intelligence 2010). Distinctiveness as well as the permanence of many of the behavioural characteristics proposed in the literature (such as signature, gait, and keystroke dynamics) is weak. As such, very few operational systems based on these traits have been deployed so far. The choice of a specific biometric modality typically depends on the nature and requirements of the intended identification application. As an example, voice biometric is appropriate in authentication applications involving mobile phones since a sensor for capturing voice (microphone) is already embedded in the phone. Fingerprint is the most popular biometric for accessing laptops, mobile phones and PDAs since low cost, small footprint fingerprint sweep sensors can be easily embedded in these devices. Some of the traits, for example, hand geometry, are more appropriate for verification applications (1:1 matching) whereas others like fingerprint, iris, and face have sufficient discriminating power to be applicable in large-scale identification applications (1:N matching). One of the unique applications of biometrics is in the negative identification, *i.e.*, the person is not the one who has already been registered/enrolled in the system. The negative identification is required to prevent multiple enrolments of the same person which is critical for large scale biometric applications, *e.g.* claiming social benefits from the government sponsored programs. Therefore, even in verification applications, identification capabilities for the negative identification are necessary. We now briefly introduce some of the popular biometric modalities.

(a) Face: Humans have a remarkable ability to recognize fellow beings based on facial appearance. So, face is a natural human trait for automated biometric recognition. Face recognition systems typically utilize the spatial relationship among the locations of facial features such as eyes, nose, lips, chin, and the global appearance of a face. The forensic and civilian applications of face recognition technologies pose a number of technical challenges both for static mugshot photograph matching (e.g., for ensuring that the same person is not requesting multiple passports) to unconstrained video streams acquired in visible or near-infrared illumination (e.g., in surveillance). An excellent survey of existing face recognition technologies and challenges is available in Abate et al. (2007). The problems associated with illumination, gesture, facial makeup, occlusion, and pose variations adversely affect the face recognition performance. While face recognition is non-intrusive, has high user acceptance, and provides acceptable levels of recognition performance in controlled environments, robust face recognition in non-ideal situations continues to pose challenges.

(b) Fingerprint: Fingerprint-based recognition has been the longest serving, most successful and popular method for person identification. There are numerous historical accounts which suggest that fingerprints have been used in business transactions as early as 500 B.C. in Babylon (NSTC 2010) and later by Chinese officials to seal the official documents in the third century B.C. (Fingerprinting 2010). Fingerprints consist of a regular texture pattern composed of ridges and valleys. These ridges are characterized by several landmark points, known as minutiae, which are mostly in the form of ridge endings and ridge bifurcations. The spatial distribution of these minutiae points is claimed to be unique to each finger; it is the collection of minutiae points in a fingerprint that is primarily employed for matching two fingerprints. In addition to minutiae points, there are sweat pores and other details (referred to as extended or level three features) which can be acquired in high resolution (1,000 ppi) fingerprint images. These extended features are receiving increased attention since forensics experts seem to utilize them particularly for latent and poor quality fingerprint images. Nearly all forensics and law enforcement agencies worldwide utilize Automatic Fingerprint Identification Systems (AFIS). Emergence of low cost and compact fingerprint readers has made fingerprint modality a preferred choice in many civil and commercial applications.

(c) Iris: The iris is the colored annular ring that surrounds the pupil. Iris images acquired under infrared illumination consist of complex texture pattern with numerous individual attributes, *e.g.* stripes, pits, and furrows, which allow for highly reliable personal identification. The iris is a protected internal organ whose texture is stable and distinctive, even among identical twins (similar to fingerprints), and extremely difficult to surgically spoof. An excellent survey on the current iris recognition technologies and future research challenges is available in (Bowyer et al. 2008). First invented by Daugman (University of Cambridge 2010), both the accuracy and matching speed of currently available iris recognition systems is very high. Iris recognition has been integrated in several large-scale personal identification systems (e.g., border crossing system in the United Arab Emirates (Daugman and Malhas 2004)). Several efforts are also being made to capture iris at a distance (Matey et al. 2008; Proença et al. 2009). However, relatively high sensor cost, along with relatively large failure to enrol (FTE) rate reported in some studies, and lack of legacy iris databases may limit its usage in some large-scale government applications.

(d) Palmprint: The image of a human palm consists of palmar friction ridges and flexion creases (Adler 2004). Latent palmprint identification is of growing importance (Jain and Feng 2009) in forensic applications since around 30% of the latent prints lifted from crime scenes (from knifes, guns, steering wheels) are of palms rather than of fingers (Dewan 2003). Similar to fingerprints, latent palmprint systems utilize minutiae and creases for matching. While law enforcement and forensics agencies have always collected fingerprints, it is only in recent years that large palmprint databases are becoming available. Based on the success of fingerprints in civilian applications, some attempts have been made to utilize low resolution palmprint images (about 75 dpi) for access control applications (Sun et al. 2005; Zhang et al. 2003; Kumar 2008).These systems utilize texture features which are quite similar to those employed for iris recognition. To our knowledge, palmprint recognition systems have not yet been deployed for civilian applications (e.g., access control), mainly due to their large physical size and the fact that fingerprint identification based on compact and embedded sensors works quite well for such applications.

(e) Hand Geometry: It is claimed that individuals can be discriminated based on the shape of their hands. Person identification using hand geometry utilizes low resolution (~20 ppi) hand images to extract a number of geometrical features such as finger length, width, thickness, perimeter, and finger area. The discriminatory power of these features is quite limited, and therefore hand geometry systems are employed only for verification applications (1:1 matching) in low security access control and time-and-attendance applications. The hand geometry systems have large physical size, so they cannot be easily embedded in existing security systems.

(f) Voice: Speech or voice-based recognition systems identify a person based on their spoken words. The generation of human voice involves a combination of behavioral and physiological features. The physiological component of voice generation depends on the shape and size of vocal tracts, lips, nasal cavities, and mouth. The movement of lips, jaws, tongue, velum, and larynx constitute the behavioral component of voice which can vary over time due to person's age and medical condition (e.g., common cold). The spectral content of the voice is analyzed to extract its intensity, duration, quality, and pitch information, which is used to build a model (typically the Hidden Markov Model) for speaker recognition. Speaker recognition is highly suitable for applications like tele-banking but it is quite sensitive to background noise and playback spoofing. Again, voice biometric is primarily used in verification mode.

(g) Signature: Signature is a behavioral biometric modality that is used in daily business transactions (e.g., credit card purchase). However, attempts to develop highly accurate signature recognition systems have not been successful. This is primarily due to the large *intra-class variations* in a person's signature over time. Attempts have been made to improve the signature recognition performance by capturing dynamic or online signatures that require pressure-sensitive pen-pad. Dynamic signatures help in acquiring the shape, speed, acceleration, pen pressure, order and speed of strokes, during the actual act of signing. This additional information seems to improve the verification performance (over static signatures) as well as circumvent signature forgeries. Still, very few automatic signature verification systems have been deployed.

(h) DNA: The DNA is an acronym for deoxyribonucleic acid which is present in nucleus of every cell in human body and therefore a highly stable biometric identifier that represents physiological characteristic (DNA Fingerprint Identification 2010). The DNA structure of every human is unique, except from identical twins, and is composed of genes that determine physical characteristics (like eye or hair color). Human DNA samples can be acquired from a wide variety of sources; from hair, finger nails, saliva and blood samples. Identification based on DNA requires first isolating from source/samples, amplifying it to create multiple copies of *target sequence,* followed by sequencing that generates a unique DNA profile. The DNA matching is quite popular for forensic and

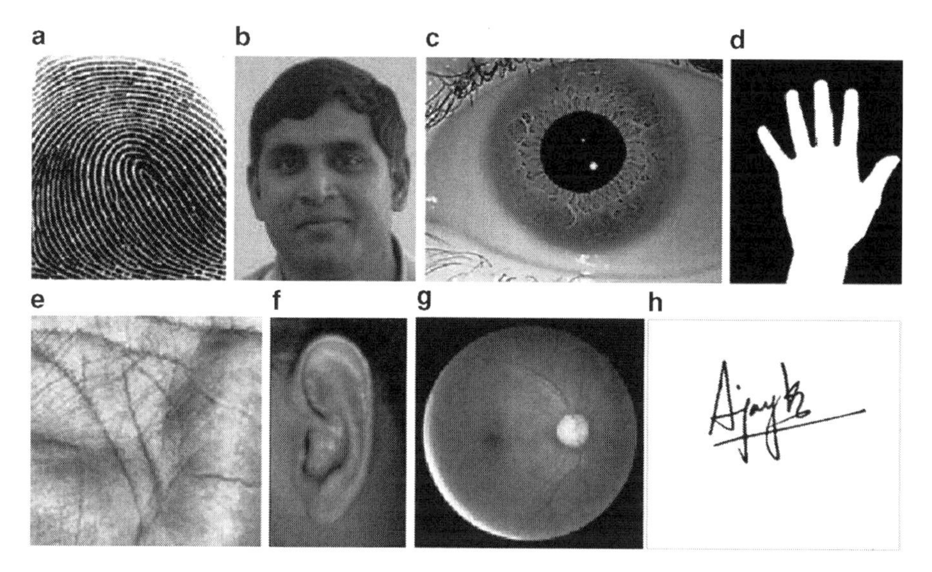

Fig. 3.1 Commonly used biometric traits: (**a**) fingerprint, (**b**) face, (**c**) iris, (**d**) hand shape, (**e**) palmprint, (**f**) ear, (**g**) retina, and (**h**) signature

law enforcement applications. However, it requires tangible samples and cannot yet be done in real time. Currently, not all the steps in DNA matching are automated and therefore results can be skewed if the process is not conducted properly or the DNA samples themselves get contaminated. In summary, the DNA matching process is expensive, time consuming and therefore not yet suitable for large scale biometrics applications for civilian usage.

(i) Hand Veins: The pattern of blood vessels hidden underneath the skin is quite distinct in individuals, even among identical twins and stable over long period of time. The primary function of veins is to carry blood from one part of the body to another and therefore vascular pattern is spread throughout the body. The veins that are present in hands, *i.e.* palm, finger and palm dorsal surface, are easy to acquire (using near infrared illumination) and have been employed for the biometric identification (Kumar and Prathyusha 2009). The vein patterns are generally stable for adults (age of 20–50 years) but begin to shrink later due to decline in strength of bones and muscles. There are several diseases, like diabetes, atherosclerosis, or tumors, which can influence the vein patterns and make them thick or thin. Biometric authentication devices using finger and palm vein imaging are now available for some commercial applications (PalmSecure 2010) to the best of our knowledge, there is no known large scale vascular biometric system. This could be primarily due to concerns about the system cost and lack of large scale studies on vein individuality and stability. On the plus side, these vascular systems are touchless which often appeals to the user (Fig. 3.1).

The recognition accuracy of individual biometric traits outlined above may not be adequate to meet the requirements of some high security applications. The low individuality or uniqueness and lack of adequate quality of individual biometric traits for some users in the target population can also pose problems in large scale applications. The biometric modality employed for large-scale deployments demands high universality among the user population. It was reported (NIST Report to the United States Congress 2002) that about 2% of the population does not have usable fingerprints for enrolment in fingerprint identification systems (Note that this figure can vary significantly from one target population to the other). Therefore the combination of different biometric modalities needs to be employed to ensure desired level of security and flexibility in some applications. Another advantage of multimodal systems is that they can potentially offer protection against spoof attacks as it is extremely difficult to spoof multiple modalities simultaneously.

3.2 Expectations from Biometrics Technologies

Increasing requirements for security in many sectors of our society have generated a tremendous interest in biometrics. This has also raised expectations from biometric technologies. These expectations can be summarized into five categories: performance, cost, user convenience, interoperability, and system security.

(i) **Performance**: The recognition performance achievable from a biometric system is of utmost interest in the deployment of biometric systems. A biometric system is prone to numerous errors; failure to enrol (FTE), false accept rate (FAR), and false reject rate (FRR). The system performance is further characterized in terms of transaction time or throughput. The accuracy of a biometric system is not static, but it is data dependent and influenced by several factors: (a) biometric quality, which is related to the quality of sensed signal/image, (b) composition of target user population (e.g., gender, race, age, and profession), (c) size of database (i.e., number of subjects enrolled in the system), (d) time interval between enrolment and verification data, (e) variations in the operating environment (e.g., temperature, humidity, and illumination), (f) distinctiveness of biometric modality, and (g) robustness of employed algorithms (namely, segmentation, feature extraction, and matching algorithms). A biometrics authentication system can make two types of errors: a *false match*, in which the matcher declares a match between images from two different fingers, and a *false non-match*, in which it does not identify images from the same finger as a match. A system's false match rate (FMR) and false non-match rate (FNMR) depend on the operating threshold; a large threshold score leads to a small FMR at the expense of a high FNMR. For a given biometrics system, it is not possible to reduce both these errors simultaneously.

(ii) **Cost**: The cost of deploying a biometric system is often estimated from its direct and indirect components. The direct component includes hardware components (sensor, processor, memory) and the software modules (GUI and matcher). The sensor should be low cost and it should be easy to embed it in the existing security infrastructure. There are multifaceted components that constitute the indirect cost in the usage of biometric system. These include system installation, training/maintenance requirements, and most importantly, user acceptance. In the end, return on investment or the cost-benefit analysis is critical for making a case for biometric systems in most applications.

(iii) **Interoperability**: As biometrics systems are being increasingly deployed in a wide range of applications, it is necessary that the system be interoperable among different biometrics technologies (sensors/algorithms/vendors). A biometric system can no longer operate under the assumption that the same sensor, same algorithms, or same operating conditions will always be available during its lifetime. The biometric system should be highly interoperable to authenticate individuals using sensors from different vendors and on varying hardware/software platforms. The system should employ usage/development/deployment of common data exchange facilities and the formats to exchange the biometric data/features between different vendors, from different geographical locations. This would significantly reduce the need for additional software development and bring all the associated advantages (cost savings and efficiency).

(iv) **User Convenience**: A biometrics system should be user friendly. Any perceived health or hygienic concerns with the continuous usage of biometric sensors can influence user acceptance. Hygiene as well as security has been one of the motivations for developing touchless fingerprint sensors. Some biometric modalities are easier to acquire than others and require less user cooperation during data acquisition. Human factors and ergonomic issues will continue to play a major role in widespread deployment of biometric systems in non-government applications (such as physical and logical access control).

(v) **Security**: Biometric systems are vulnerable to potential security breaches from spoof and malicious attacks. These systems should therefore offer a high degree of protection to various vulnerabilities resulting from intrinsic failures and adversary attacks (Jain et al. 2008b). One of the major system security concerns deals with biometric template security. The access protocols and the storage of biometric and other user specific data should be provided the highest level of security.

Based on the above considerations, the second generation biometric systems should be easy to use, have low cost, be easy to embed and integrate in the target security application and be robust, secure, and highly accurate in their matching performance.

3.3 First Generation Biometrics

Early applications of biometrics, mainly fingerprints, were primarily in forensics and law enforcement agencies. Even though fingerprints were first used over 100 years ago to convict a criminal (South Wales Police 2010) the first generation of automatic fingerprint identification systems (AFIS) for law enforcement agencies

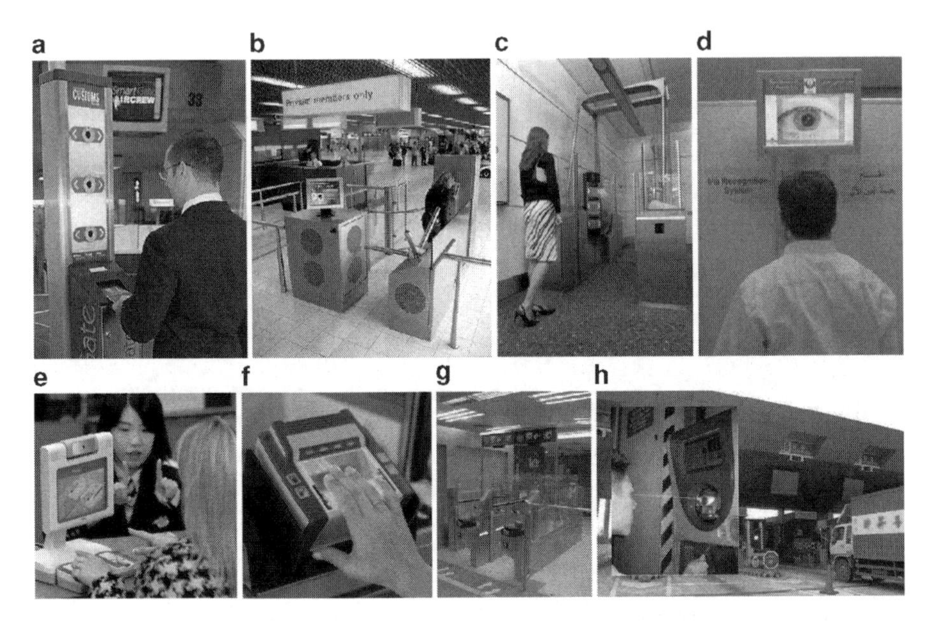

Fig. 3.2 Deployment of biometrics systems at border crossings for immigration control; (**a**) face recognition system (SmartGate) at Sydney airport (Australian Government), (**b**) iris recognition system at Amsterdam Schiphol airport (Schiphol), (**c**) at Manchester airport (UK) (Ranger 2006) and (**d**) at UAE airport (Daugman and Malhas 2004); (**e**) fingerprint recognition using index fingers at airports in Japan (Higaki 2007); (**f**) ten fingerprint acquisition at airports in the United States (US Department of Homeland Security); (**g**) fingerprint based immigration clearance for passengers at Hong Kong airport; (**h**) vehicular clearance using fingerprint and face in Hong Kong (E-Channel)

did not become available until the 1970s. We refer to these systems as the *zeroth generation biometric systems* because of their limited performance and lack of interconnectivity (stand alone). We also place the hand geometry systems which were used in several access control applications, including The Immigration and Naturalization Service Accelerated Service System (INSPASS US Department of Homeland Security 2010) in this early generation. The INSPASS system, installed at some of the major airports in the U.S. in mid 1990s, was later abandoned due to its limited user enrollment and weak performance (Fig. 3.2).

We use the term *first generation biometric systems* to describe biometric readers and systems developed and deployed over the last decade. These systems include a variety of fingerprint, iris, and face recognition systems that have found their applications in a wide range of civilian and commercial systems. Some examples include: the US-VISIT system based on fingerprints (US Department of Homeland Security 2010), the Privium system at Amsterdam's Schiphol airport based on iris (Schiphol 2009), and the SmartGate system at the Sydney airport (Australian Government 2010) based on face. These are examples of major first generation systems used at international border crossings. Other examples include the fingerprint-based system at Walt Disney Parks (Walt Disney World Resort 2010) and face-recognition-based cigarette vending machines installed at some locations in Japan (PROnetworks 2010). Continuing advances in the sensing technologies, computational speed, operating environment, and storage capabilities over the past decade have spearheaded the development and deployment of first generation biometric systems. This has helped permeate biometric authentication in our daily lives, as evident from laptops that come embedded with fingerprint sensors. These advances have also tremendously improved the speed and accuracy of fingerprint matching in forensics and law enforcement. As an example, the FBI's IAFIS system has a database of ten print fingerprint images for about 80 million subjects and handles close to 80,000 searches per day (CJIS 2010). However, even with this impressive throughput of IAFIS system, it may not be adequate to meet the increasing workload and real-time requirements for several online applications (*e.g.* for border crossings and law enforcement systems). This problem is even worse for matching latent fingerprints where a substantial amount of human expertise is required both for feature marking and evaluating the matches returned by the system. The ongoing efforts by NIST under MINEX (The Minutiae Interoperability Exchange Test) program are focused at improving the template based interoperability. The recent interoperability test (Grother et al. 2009) from NIST has suggested several limitations of state-of-the-art fingerprint matching algorithms in coping with minutiae interoperability. In addition to poor interoperability, the first generation biometrics systems are vulnerability to spoofing (e.g. Serkan 2009) and face increasing challenges in ensuring template security and privacy from sophisticated attacks. In summary, the major limitations of the first generation biometrics system can be summarized as: achievable performance, security, and privacy. This demonstrates the need to develop second generation biometrics system which is described below.

3.4 Second Generation Biometrics

The deployment of first generation biometrics solutions has highlighted several challenges in the management of human identity. The *second generation biometrics systems* must confront these challenges and develop novel techniques for sensing, signal/image representation, and matching. The challenges posed to the second generation biometric technologies can be put in two categories: (i) challenges from *engineering perspective*, which are focused on problems related to security, accuracy, speed, ergonomics, and size of the application; (ii) challenges from the *social perspective*, which include the privacy protection policies, ethical and health related concerns, and cultural biases.

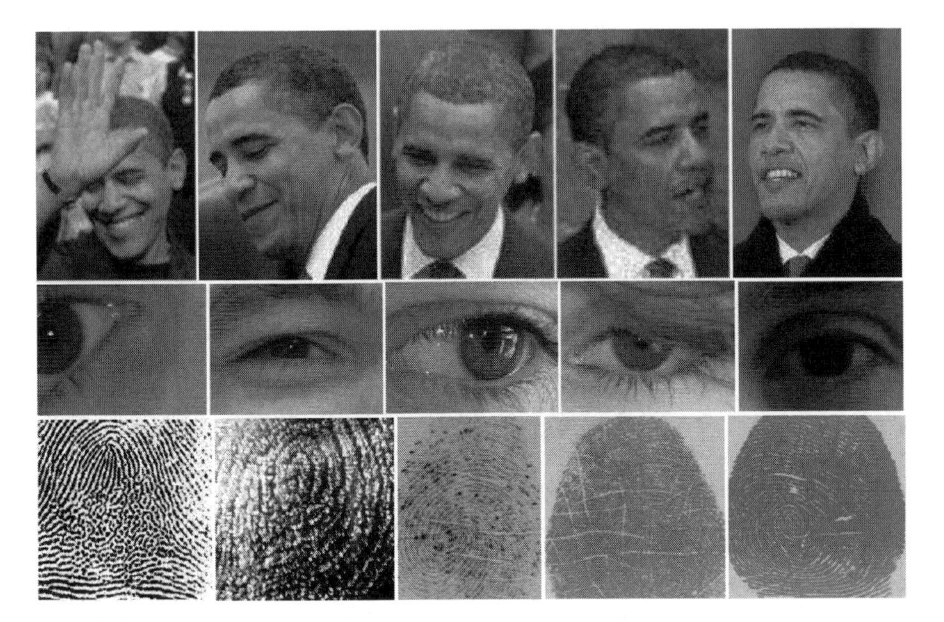

Fig. 3.3 Typical examples of face (FANPIX.net), iris (Proença et al. 2009) and fingerprint images (Hong 1998) acquired in less than ideal conditions

3.4.1 Engineering Perspective

3.4.1.1 Data Acquisition Environment

The performance of matching algorithm critically depends on the quality of biometric data. Sensor design and deployment faces two contradictory requirements: high quality data for improved accuracy vs. flexible data acquisition protocol with the least amount of user cooperation for high user acceptability. We now outline two such challenges facing the second generation biometrics system to improve the data acquisition environment using new sensing technologies.

Improving User Convenience

Most biometric technologies are able to provide satisfactory matching performance in controlled situations where the user is cooperative and data acquisition conditions and environment can be controlled (e.g., face image captured for passport photos). However, in many applications, biometric data is acquired in less than ideal conditions (Fig. 3.3), such as matching latent prints lifted from crime scenes or recognizing faces in surveillance videos. The unsatisfactory performance of biometrics technologies in these relatively uncontrolled situations has limited their deployment; as a result, these applications still heavily rely on human intervention. A significant improvement in recognition performance in less controlled situations is one of the main challenges facing biometric technologies.

There are an increasing number of research and development efforts to expand the scope of personal identification *at-a-distance* and *on-the-move*. Médioni et al. (2009) have demonstrated the feasibility of non-cooperative personal identification using face images at a distance of 6–9 m. There have been some efforts to achieve iris identification on-the-move, as well as at-a-distance to enable capture of iris images of sufficient quality while the subjects are moving at a normal walking speed (Matey et al. 2008, 2006; Proença et al. 2009). However, biometric identification at-a-distance and on-the-move is still in the research domain and not yet sufficiently mature (Biometric Technology Today 2009) for deployment. Human face images are highly pose, view and illumination dependent (Fig. 3.3) and the performance evaluation conducted by NIST suggested that the recognition accuracy falls to 47% for the best system in less constrained outdoor conditions (NIST 2010). The increased user convenience therefore requires the development of robust matching algorithms and noise elimination techniques that can handle wide range of pose and illumination variations in biometric imaging.

Improving Data Acquisition Quality

The next generation biometrics sensors that can acquire high quality of biometric data will be required to facilitate the significantly higher level of identification accuracy required in wide range of large scale applications. High resolution fingerprint sensors that can facilitate use of extended features for more accurate identification are being adopted as a standard in law enforcement. Similarly, biometrics sensors that can simultaneously acquire 2D/3D face data can evolve as an essential component of face recognition applications. The current biometrics systems are predominantly focused on 2D imaging and the use of 3D image acquisition has not delivered its promise due to technological limitations posed by speed, cost, resolution, and size of 3D imagers/scanners as well as the representation and matching issues. Therefore, continued design and development of multimodal biometric sensors that can simultaneously acquire 2D and 3D images is another key challenge in the development of second generation biometric technologies.

3.4.1.2 Handling Poor Quality Data

Consider the case of latent fingerprints that are imaged through a number of techniques ranging from simply photographing the impression to more complex dusting or chemical processing (Lee and Gaensslen 2001). As latent prints provide important clues for investigating crime scenes and acts of terrorism, matching latents against reference prints (rolled prints) of known persons is routinely performed in law enforcement applications. Compared to plain/rolled fingerprint matching, latent matching is a much more challenging problem because latent prints have complex background, small image area, unclear ridge structure, and large distortion. The accuracy of automatic latent matching is significantly lower than plain/rolled fingerprint matching (Jain et al. 2008a). As a result, manual intervention is essential in the latent matching process, which leads to low throughput and introduces an element of subjectivity. There has been substantial effort in government, industry, and academia to achieve significant improvement in both the accuracy and degree of automation (*lights out* capability) (Jain et al. 2008a; NIST 2007; Indovina et al. 2009). To improve the matching accuracy, extended fingerprint feature set (EFS) has been utilized in addition to minutiae (Jain et al. 2008a). However, manually marking EFS is very tedious and therefore robust automatic extraction algorithms need to be developed. The increased capabilities to handle poor quality data for biometric identification is not only required for improving latent matching accuracy but is also essential for range of biometric systems employed for commercial applications. The failure to enroll rate (FTE) and the achievable throughput from the deployed biometrics system can also be further improved by imparting new capabilities that can handle poor quality biometric data.

There are a number of applications and scenarios where multiple levels of security and/or throughput are expected in the deployed biometric systems. There have been some efforts (Veeramachaneni et al. 2005; Kumar et al. 2010; Tronci et al. 2007; Poh et al. 2009) in developing multimodal biometric systems to achieve such dynamic security requirements. However, it is not easy to design adaptive multimodal biometrics systems that are flexible enough to consider user preference for biometric modalities, user constraints, and/or varying biometric image quality. In this context, Nandakumar et al. (2009) suggested that the likelihood ratio-based fusion can effectively handle the problem of missing biometric modality/data, which could also be perceived as user preference in adaptive multimodal biometric systems.

New user enrolments in a large-scale biometric system will typically require periodic re-training or updating of the matcher. Therefore, another aspect of an adaptive biometric system is *online learning*, which can periodically update the matcher (Singh et al. 2009). A semi-supervised learning approach for adaptive biometric systems is proposed in (Poh et al. 2009).

3.4.1.3 Biometric System Security

The security ensured by the deployed biometric systems can itself be compromised. A number of studies have analyzed the likelihood of such security breaches and potential approaches (Reddy et al. 2008) to counter these vulnerabilities. The general analysis of a biometric system for vulnerability assessment determines the extent to which an impostor can compromise the security offered by the biometric system. Ratha et al. (2001) identified the potential points of an adversary attack on the biometric system as shown in Fig. 3.4. While many of these attacks are applicable to any information system, the attacks using *fake biometric* and template modification are unique to biometrics systems. We briefly discuss the characteristics of such attacks, which need to be effectively thwarted in second generation biometrics systems.

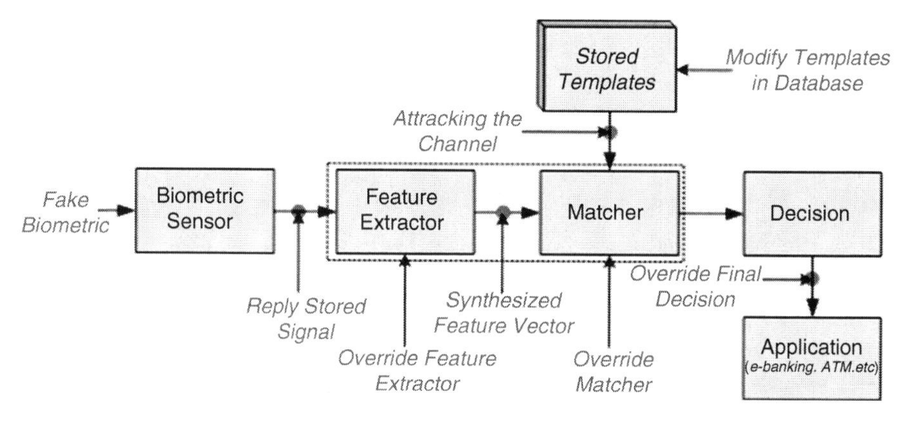

Fig. 3.4 Typical attack points in a biometric system (Adapted from Ratha et al. 2001)

(i) Sensor level attack: A *fake* biometric sample can be presented at the sensor to gain access. A fake biometric can be generated by covertly acquiring the biometric characteristics of a genuine user, e.g. lifting fingerprint impressions from objects touched by persons.

(ii) Replay attack: It is possible for an adversary to intercept or acquire a digital copy of the stored biometric sample and replay this signal bypassing the biometric sensor (Pink Tentacle 2010).

(iii) Trojan Horse[1] attack: The feature extractor can be replaced by a program which generates the desired feature set.

(iv) Spoofing the features: The feature vectors generated from the biometric samples are replaced by the set of synthetically generated (fake) features.

(v) Attack on matcher: The matcher can also be subjected to a Trojan Horse attack that always produces high (or low) match scores irrespective of which user presents the biometric at the sensor.

(vi) Attack on template: The template generated during the user enrolment/registration can be either locally stored or at some central location. This type of attack can either modify the stored template or replaces it with a new template.

(vii) Attack on communication channel: The data being transferred through a communication channel can be intercepted for malicious reasons and then modified and inserted back into the system.

(viii) Attack on decision module: The final decision generated by the biometric system can be overridden by a Trojan Horse program.

[1] Virus program(s) that hide within a set of seemingly useful software programs to facilitate unauthorized access to a hacker

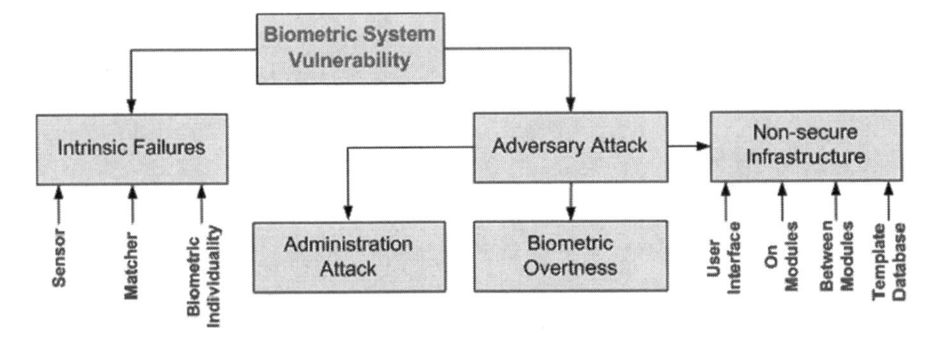

Fig. 3.5 Biometric system vulnerabilities

A biometric matcher is typically only a part of a larger information and security management system. Therefore, the non-biometric modules in the overall system can also introduce some security flaws. An example is the iris-based access control system in a New Jersey School (Cohn 2006). Sometimes, a user will prop the door open, so anyone can enter the school, bypassing the security offered by the biometric module. Another scenario could involve disabling the power supply or damaging the sensor that can make the whole biometric security system ineffective (Abdullayeva et al. 2010). Jain et al. (2008b) provide extended discussion on typical biometrics system vulnerabilities, which are also summarized in Fig. 3.5.

Biometrics Alteration and Spoof Detection

The border control officials are seeing an increased use of altered fingerprints (see Fig. 3.6), used by individuals who do not want to be identified because they have prior criminal records (Singh 2008). Several biometrics technologies deployed today are susceptible to attacks in which static facial images (Pink Tentacle 2010), fake fingerprints, and static iris images can be used successfully as biometric samples. These fraudulent samples are processed by the biometric sensors as original biometric sample from the registered users and then attempted to be verified as the enrolled users. The use of spoof detection technologies is increasingly becoming an essential component of biometrics systems. The liveness detection and exploitation of biological properties of the biometrics sample is the heart of these approaches; for example, Daugman (University of Cambridge 2010) has identified several approaches for the detection of spoof iris samples, including changes in the wavelengths offered by live tissues to the incident infrared illumination; the fingerprint spoof attacks has been successfully detected from a range of approaches (Reddy et al. 2008; Antonelli et al. 2006), including the measurement of percentage of oxygen in the blood and skin distortion analysis. The second generation biometrics technologies face increasing challenges to develop significantly enhanced capabilities in identifying and rejecting altered and/or spoof biometrics samples using increasingly sophisticated techniques, *i.e.* surgery, fabrication, and simulation.

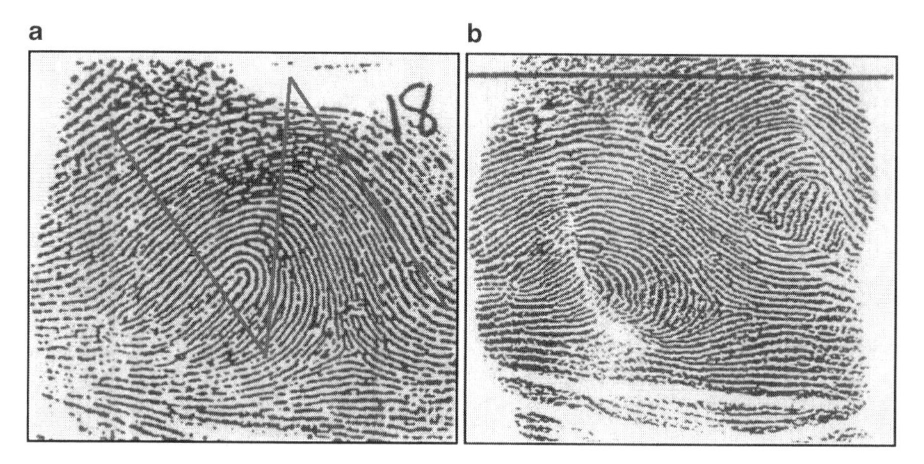

Fig. 3.6 Fingerprint alteration. (**a**) Original fingerprint and (**b**) an instance of an altered fingerprint. The criminal made a "Z" shaped incision (illustrated in the *left figure*) into each of his fingers, switched two triangles, and stitched them back into the finger

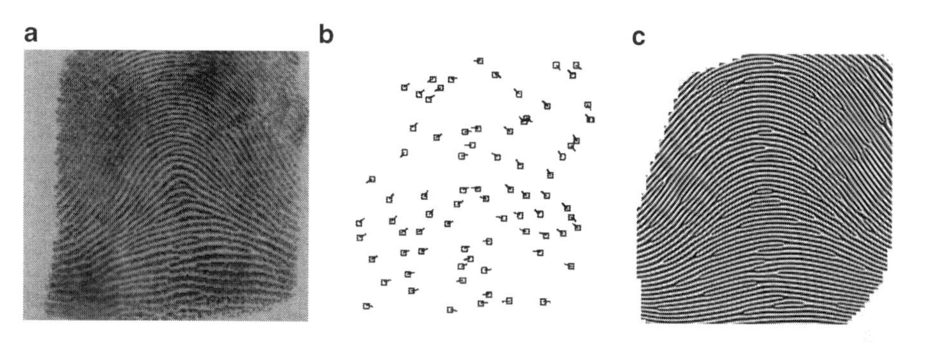

Fig. 3.7 Reconstructing a fingerprint image from minutiae template (Feng and Jain 2009). (**a**) Original image, (**b**) minutiae template, (**c**) reconstructed fingerprint. Images in (**a**) and (**c**) can be matched with high accuracy

Template Protection

A template is essentially a compact representation (a set of invariant features) of the biometric sample that is stored in system database. If the security of stored templates is compromised, the attacker can fabricate physical spoof samples to gain unauthorized access. Such efforts have been detailed in Adler (2004), Ross et al. (2007), and Feng and Jain (2009)). The stolen templates can also be abused for other unintended purposes, *e.g.* performing unauthorized credit-card transactions or accessing health related records. Figure 3.7c shows an example of a reconstructed fingerprint image from its minutiae representation (b), which is typically employed for fingerprint templates.

An ideal template protection scheme for a biometric system should have the following four properties (Jain et al. 2008b): (a) Diversity: The cross-matching of

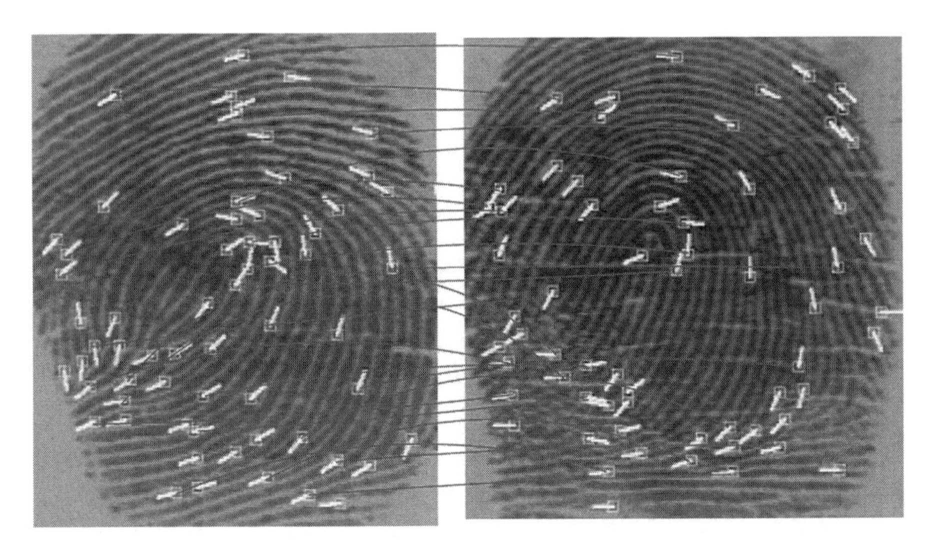

Fig. 3.8 Two fingerprint images from the same finger showing the variability in minutiae localization (Pankanti et al. 2002)

secured templates should be ensured in such a manner that the privacy of the true owner of the template is ensured; (b) Revocability: When the biometric template is compromised, it should be possible to revoke the compromised template and reissue a new template based on the same biometric trait; (c) Security: It should be extremely difficult to generate the original biometric feature set from the protected biometric templates, (d) Performance: The template protection scheme should not degrade the matching performance.

One of the key challenges for second generation biometric systems relates to the development of a template protection scheme that can simultaneously meet all the four requirements. The *intra-class variability* (Fig. 3.8) in the feature vectors from successive biometric samples of the same user limits the usage of standard encryption techniques (RSA, AES, *etc.*). The available template protection techniques (LaCous 2008) can be broadly classified into two categories (see Jain et al. 2008b): *feature transformation approach* and *biometric cryptosystem*. The characteristics of the transformation function can be used to further categorize the techniques into *salting* or *non-invertible transforms*. The salting schemes are capable of recovering original templates but only when a secret key is made available. The *non-invertible transforms* are one-way functions which make it extremely difficult to recover the original template from the transformed template even if the secret key is made available. In biometric cryptosystems, some public information about the templates, also referred to as *helper data*, is made available. While the helper data does not reveal any significant information about the original template, it is useful during the matching process to generate the cryptographic keys. The biometric cryptosystems can be further classified into *key binding* or *key generation* depending on derivation/extraction of helper data. In the *key binding* cryptosystem, the helper data is obtained by

binding the biometric template with the key. The template p
fuzzy vault (Jules and Sudan 2002), shielding functions (
distributed source coding (Draper et al. 2007) are typical ex
cryptosystem. The fuzzy vault schemes for template prot
mented for a number of biometric modalities; fingerprint (
face (Feng and Yuen 2006), iris (Lee et al. 2007), palmpr
2009), and signature (Freire-Santos et al. 2006). In a *key g*
the helper data is only derived from the original biometric template while cryp-
graphic keys are generated from the helper data and query biometric template. While direct key generation from the biometric feature is an attractive scheme, it suffers from low discriminability, *i.e.* it is difficult to simultaneously achieve high key entropy and high key stability (Jain et al. 2008b).

The available template protection schemes cannot yet simultaneously meet all the four requirements of revocability, diversity, security, and high performance. They are also not yet mature enough for large-scale deployment and require extensive analysis of their *cryptographic strength* (Nagar and Jain 2009). It is unlikely that any single template protection scheme can meet all the application requirements. Therefore, hybrid scheme, which can avail the advantages of different template protection schemes, should be pursued.

3.4.1.4 Large-Scale Applications

Biometric systems that can effectively and efficiently operate in ultra large-scale applications, *i.e.* those capable of supporting hundreds of millions of registered users, have a number of potential opportunities. Such systems will be able to support National ID programs or improve homeland security, e-commerce, and more effective implementation of social welfare programs in countries with large population (e.g., India, China and United States). The expectations from biometrics systems for such large-scale applications can be summarized as follows: (i) high accuracy and throughput under varying operating conditions and user composition, (ii) sensor interoperability, (iii) rapid collection of biometric data in harsh operating environments with virtually no failure to enrol rate, (iv) high levels of privacy and template protection, and (v) secure supporting information/operating systems.

The selection of biometric modality for large-scale applications is a judicious compromise between performance, convenience/ease in acquisition, cost, compatibility with legacy databases, and application constraints. In order to consider the requirements of large-scale identification, consider the following example involving fingerprint-based identification. Fingerprint identification system performance is measured in terms of its *false positive identification* rate (FPIR) and *false negative identification* rate (FNIR). A false positive identification occurs when the system finds a hit for a query fingerprint that is not enrolled in the system. A false negative identification occurs when it finds no hit or a wrong hit for a query fingerprint enrolled in the system. The FPIR is related to the FMR of the fingerprint matcher as $FPIR = 1-(1-FMR)^N$, where N is the number of users enrolled in the system. Hence, umber of enrolled users grows, the FMR of the fingerprint matcher needs to xtremely low for the identification system to be effective. For example, if a PIR of 1% is required in a fingerprint identification system with 100 million enrolled users, the FMR of the corresponding fingerprint matcher must be of the order of 1 in 10 billion. To meet such a stringent FMR requirement, it is necessary to use a multimodal biometric system (e.g., all the ten fingerprints of a person or a combination of fingerprint and palmprint) or to use demographic data to filter the database. This illustrates the need to continuously decrease the error rates of fingerprint matchers for successful deployment of large-scale identification systems.

The four most popular biometric modalities deployed today are face, fingerprint, iris and hand geometry. The first generation biometric systems typically employed a single modality, primarily fingerprint (ten prints), for large-scale applications. While the automated border crossing in several countries (*e.g.* USA and Japan) requires authentication using fingerprints, the system in the United Kingdom is based on iris (UK Border Agency 2010) and the one in Australia is based on face (Australian Government 2010). The FBI has (Nakashima 2007) embarked on its New Generation Identification (NGI) project for law enforcement applications that will fuse fingerprints, face, and palmprints along with some *soft biometrics* such as scars, marks, and tattoos (SMT) (Lee et al. 2008).

The government of India has recently announced a new project, called Unique ID (UIDAI 2010) to deliver multipurpose unique identification number to its over one billion citizens. This ongoing project is expected to create the largest biometric database on the planet and, on its successful completion, it can become a model of very large-scale usage of biometrics in *e*-governance. It is generally believed that the retrieval of biometrics templates from the database of India's billion plus population will require highly efficient indexing techniques (Mhatre et al. 2005) for biometric data. Therefore, the design and development of efficient and effective large-scale indexing techniques for the (multi-) biometric data is another challenge in the efficient usage/deployment of large-scale biometric systems. The *individuality* or the achievable recognition performance from the chosen biometric modality is another important criterion when millions of identities need to be discriminated. The presence of identical twins is also a problem that needs consideration in large-scale applications (Sun et al. 2010). In this context, *multimodal biometric systems* that can simultaneously employ multiple biometric modalities (*e.g.*, multiple fingers, two irises, finger and iris, *etc.*) are expected to offer higher accuracy and can also address the problem of non-universality. There is range of biometric fusion methodologies (Ross et al. 2006) proposed, but, it appears that the simple *sum rule*, with proper normalization of matching scores, does an adequate job in most cases. Biometrics systems that can simultaneously acquire multiple modalities are expected to become popular in large-scale deployments with data collection in the field. A prototype (Fig. 3.9b, c) of an acquisition device that can *simultaneously* acquire five fingerprints, palmprint, and shape of a hand is described in Rowe et al. (2007) while Printrak Division 2010 details a device to acquire fingerprints, palmprints, face images, iris images, signature details and soft biometrics data such as scars and tattoos (Fig. 3.9).

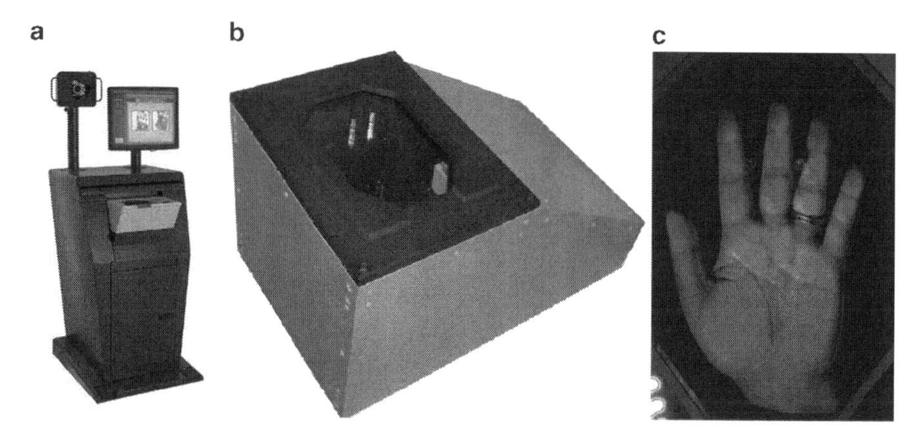

Fig. 3.9 Multibiometric data acquisition. (**a**) A multibiometrics acquisition system from Printrak (Printrak Division) to acquire fingerprints, palmprints, face images, iris images, signature details and soft biometrics data such as scars and tattoos; (**b**) full hand multispectral scanner from Lumidigm (Rowe et al. 2007) and (**c**) a sensed hand image from (**b**)

3.4.1.5 Soft Biometrics

S*oft biometrics* are those human characteristics that provide some information about the individual, but lack the distinctiveness and permanence to sufficiently differentiate any two individuals (Jain et al. 2004a, b). Examples of soft biometrics used in law enforcement include scars, marks, tattoos, color of eye, and hair color. In several biometric system deployments, such *ancillary information* is acquired and stored along with the primary biometric during the enrollment phase. These characteristics can be potentially exploited in three ways: (i) to supplement as the features in an existing biometrics system, (ii) to enable fast retrieval from a large database, and (iii) to enable matching or retrieval from a partial or profile face image with soft biometric attributes, *i.e.*, facial marks.

There are several studies in the literature which have demonstrated the effective usage of soft biometric characteristics for performance improvement. Wayman (1997) proposed the use of soft biometric traits like gender and age for filtering a large biometric database. Jain et al. (2004b) demonstrated that the performance of a fingerprint matching system can be effectively improved (~5%) by incorporating additional user information like gender, ethnicity, and height. Jain and Park (2009) have utilized micro-level facial marks, *e.g.* freckles, moles, scars, *etc.*, to achieve face recognition performance improvement on an operational database. Scars, marks and tattoos (SMT) are the imprints that are typically employed by law enforcement agencies for identification of suspects. The SMT provide more discriminative information, as compared to other personal indicators such as age, height, gender, and race, and can be effectively used in assisting suspect identification. Lee et al. (2008) have conducted a study to employ these imprints for content-based tattoo image retrieval.

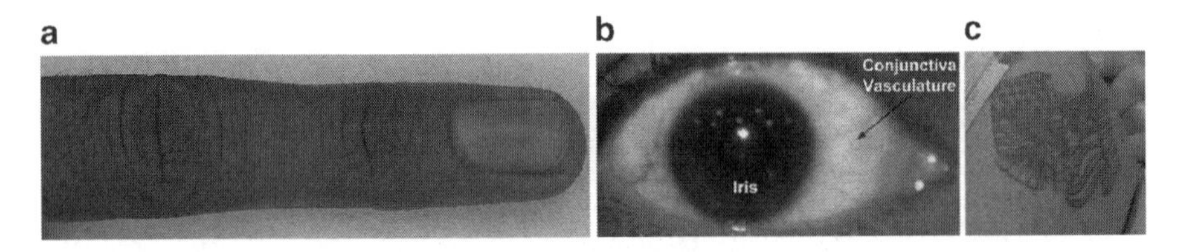

Fig. 3.10 Emerging soft biometrics. (**a**) Image sample for knuckle biometrics, (**b**) conjunctival vasculature, and (**c**) tattoo image

There have been some efforts to extract novel anatomical, physiological, and behavioural characteristics (gait (Kellokumpu et al. 2009; Tao et al. 2007), ear (Yan and Bowyer 2007; Bhanu and Chen 2008), footprints (Nakajima et al. 2000; Uhl and Wild 2008), periocular (Park et al. 2009), finger knuckle (Woodard and Flynn 2005; Kumar and Ravikanth 2009; Kumar and Zhou 2009), keystroke dynamics (Bender and Postley 2007), and nose shape (Drira et al. 2009; Song et al. 2009)) and investigate their potential to support human identification (see Fig. 3.10). Several other physiological characteristics that have been extracted for the purpose of biometric identification include arterial pulse (Joshi et al. 2008; Irvine et al. 2008), fingernails (Topping et al. 1998), odour (Ramus and Eichenbaum 2000), bioelectric potential (Hirobayashi et al. 2007), knee x-rays (Shamir et al. 2009), frontal sinus (Falguera et al. 2008; Tabor et al. 2009), and otoacoustic emissions (Swabey et al. 2004). These efforts explore and identify additional sources of *soft biometrics* to either improve the performance of traditional (hard) biometric modalities or to help provide identification in the absence of primary biometric attributes. Some of these characteristics can be simultaneously acquired along with the more popular biometric modalities; for example conjunctival scans (Crihalmeanu et al. 2009) can be acquired along with iris while periocular biometrics (Park et al. 2009) is more suitable for simultaneous acquisition with face images. However, the *persistence* and *permanence* of these body attributes and behavioural characteristics is not yet known.

3.4.2 Application Perspective

Biometric systems are often felt invasive, since the sensors directly interact with the human body to capture person-specific data that is considered *privileged*. The stigma of forensic and criminal investigations has been known to influence the user acceptance of the first generation biometric systems. Biometric traits are part of human body and behaviour and therefore, releasing this information to a biometric system during enrolment or verification can threaten the personal privacy of some users. Biometric traits can be used to track a person over time; this can be technologically possible when biometric recognition at-a-distance becomes mature and when we learn to *mine* and *link* vast amounts of sensor and demographic data. In addition, by linking the biometric database with other databases

Table 3.1 Privacy concerns with biometric modalities

Biometric modality	Possible indicators of user's health
Retina	Eye related disease (e.g. diabetic retinopathy)
DNA	Genetic diseases or susceptibility to specific disease, gender
Palmprint	Prediction of congenital heart disease and laryngoscopy in diabetics
Face	Facial thermograms for fever and related medical conditions/diseases
Gait	Physical disability

(*e.g.* user's credit card transactions), we know where the person has been and at what time. In addition to the personal privacy, there are also concerns that biometric data can be exploited to reveal a user's medical conditions. Such information is privileged that could be potentially used to discriminate some users for employment or benefits purposes (*e.g.*, health insurance). Table 3.1 lists some of the known medical indicators that are believed to have some association with the corresponding biometric modalities.

The deployment of biometric systems by several countries to safeguard border security has necessitated the adoption of new policies and security measures. These measures, including the use of biometric technology often interfere with the existing national data and privacy protection policies (Clarke 2000). The development of technical standards is generally perceived as a sign of maturity in the protection and exchange of electronic data. In this context, the biometrics standards support interchange ability and interoperability, can ensure high degree of privacy and security, while reducing the development and maintenance costs for biometrics technologies. The current efforts in developing such biometric standards (Biometrics.gov 2010; NIST; Department of Defense; BioAPI 2010) are focussed on specifications for collection, storage, exchange and transmission of biometric data, file formats, technical interface, performance evaluation and reporting standards for biometrics related solutions. The International Civil Aviation Organization (ICAO) mandates that the biometric data (fingerprint/face/iris) in *e*-passports should conform to SC37 (Standing Committee 2010) biometric data interchange format. The BioAPI (BioAPI Consort 2010) consortium, on the other hand, is the collective efforts of more than 120 companies to develop specifications for a standardized application programming interface (API) which is compatible with wide variety of biometrics products. It is hoped that the *second generation biometrics technologies* will further promote such efforts in the formulation of new international standards for uniform practices pertaining to biometric data collection (in less than ideal conditions), usage, storage, exchange for cross-border and inter-organization applications.

There are also some concerns that biometrics systems may exclude some potential user groups in the society. The definition and the context of personal privacy vary in different societies and are somewhat related to the cultural practices. Some social practices, especially in rural population (which should be the target group for providing social and economic privileges by providing them legal identity (UIDAI 2010)), may not relate biometrics with personal privacy concerns. Some cultural groups or religious exercises/convictions/practices may not permit blood samples

Table 3.2 Rating technology risks for personal privacy

Privacy impact	Key areas of biometrics system deployment				
High	Identification	Covert	Physiological	Biometric images	Centralized/large database
Low	Verification	Overt	Behavioural	Encrypted templates	Localized/small database

Table 3.3 Incidents of privacy and social challenges

Technology deployment	Reference	Benefits	Concerns
Street View from Google	Bangeman (2009)	Remote view, security/ traffic	Personal privacy
Privacy controls in Facebook	Denham (2009)	Sharing images, contacts	Child safety, personal privacy
DNA tests for asylum seekers	Doward (2009)	Immigration security	Profiling, discrimination
Mandatory DNA tests for orphans	(CBS News)	Finding natural parents	Privacy rights, discrimination
Fingerprint for attendance	(MIS ASIA)	Efficient/accurate monitoring	Personal privacy, social concerns
Age verification in facial images	(Pink Tentacle)	Supervision, tobacco control	Personal privacy, spoofing

for DNA extraction (Volokh 2007) while some religious practices may forbid 'complete' (multiple) biometrics enrolment. Wickins (2007) has explored the vulnerabilities of a typical user population falling in to six groups: people with (i) physical and/or learning disability and (ii) mental illness, people of certain (iv) race and (v) religion, and those that are (iii) elderly and (vi) homeless. The second generation biometric technologies need to ensure that such user groups do not suffer disproportionately as a result of deployment of biometric systems.

Policy formulations in the selection and deployment of biometrics technologies can also have different impact on the privacy concerns. Some technologies are more likely to be associated with privacy-invasive requirements, *e.g.* covert biometric identification, and therefore more prone to personal privacy risks than others. Privacy groups such as (BioPrivacy Initiative 2010) provide privacy risk assessment for various biometrics technologies in four key areas. The privacy risks from the selection and deployment of biometrics technologies can however be extended into five areas and is summarized in Table 3.2. This table outlines the widely perceived concerns related to template security and on the use of large centralized databases, which often includes function creep.

In view of potential legal and ethical challenges, several privacy commissioners and data protection offices have now formulated new guidelines for the deployment and usage of biometrics technologies in government organizations and private sectors, *e.g.* Office of the Privacy Commissioner for Personal Data Protection 2010. Table 3.3 lists some usage of identification technologies, including biometrics, that have raised privacy concerns. Table 3.3 highlights concerns about the use of

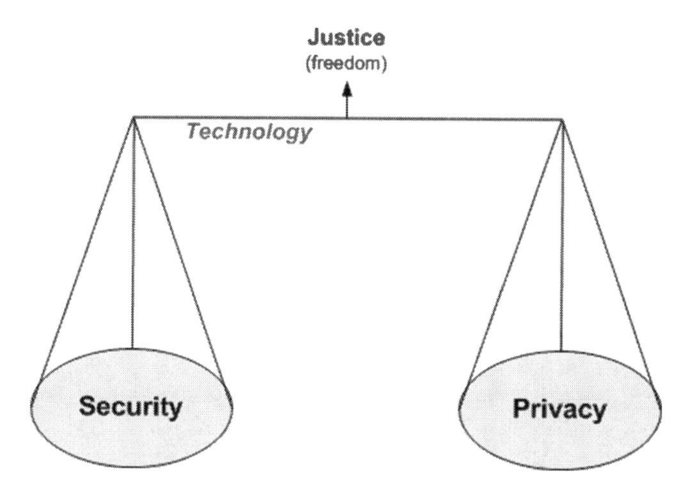

Fig. 3.11 Second generation biometric technologies need to ensure a balance between privacy and security

advanced technologies, including biometrics and resulting controversy on the standard policies to protect fundamental ethical values. These values are primarily concerned with privacy, trust, liberty, autonomy, equality, informed consent that is widely perceived to be available to all the citizens in a democracy. While the primary obligation of a state is to ensure the safety and security of its citizens, it is also necessary to protect and respect fundamental rights and values. Some of these rights are also legally enforceable and include right to respect for personal/private life and the right for equal treatment. The ongoing ethical and legal debate in the deployment of biometrics technologies has suggested (Nuffield Council on Bioethics 2007) that any interference with these rights must be proportionate.

It is expected that over time, the concerns and demands for protecting security and privacy will actually increase. Biometric technology has the potential to offer and ensure individual freedom and therefore, one of the key challenges for the second generation biometric technologies is to provide an appropriate balance between privacy and security (Fig. 3.11).

3.4.2.1 The Hong Kong Smart ID Card Experience

The Hong Kong international airport is one of the busiest airports in the world and annually handles about 31 million landing or departing passengers (Hong Kong Immigration Department 2007). The Hong Kong immigration department's automated fingerprint based passenger clearance system, *i.e. e*-Channel, provides one of the successful examples of high-speed immigration control at border crossings. The biometric border crossing between Hong Kong and Shenzhen has an enrolment of more than 1,600,000 users and handles about 400,000 border crossings every day. This is one of the busiest border crossings in the world and has resulted in detention or arrest of more than 50,000 persons since 2005. The deployment of 361 passenger *e*-Channels at all border crossings in Hong Kong ensures a maximum of 15 min of waiting time for 95% of Hong Kong residents and for 92% of visitors (E-Channel 2010). The usage of biometrics based high-speed border clearance at borders is not limited only for the passenger traffic. The automated vehicle clearance system employed at all border control points in Hong Kong has provided effective solution to mounting traffic needs using fingerprint based smart identity cards for 80 *e*-Channels used only for vehicular traffic (Fig. 3.2h). The fingerprint based mandatory Hong Kong identity cards issued to all Hong Kong residents are not limited for their usage in border controls but find wide range of applications in e-governance from authentication and access to government online services, hospitals, banks, social welfare schemes to property transactions. In addition, these smart cards also promote secured *e*-transactions (Hong Kong Post e-Cert 2010), *i.e.* threats from hackers and online theft, as the residents can store their own encrypted public keys to ensure secured online digital identity.

A small fraction of population can sometime find difficulties to successfully clear the e-Channels due to dry, wet or poor quality fingers. There are, however, alternative manual or regular channels in the vicinity to avoid delay in passenger clearance. It is worth noting that, to the best of our knowledge, privacy concerns and misuse of Hong Kong identity cards have not been reported. The benefits offered by the Hong Kong identity cards apparently outweigh the potential privacy concerns and the Hong Kong residents do not seem overly concerned with the privacy issues (MacManus 2009) which are more effectively regulated by the office of privacy commissioner (Privacy Commissioner Hong Kong 2010). In summary, the successful usage of *smart* Hong Kong identity cards since 2005 provides a model for the effective deployment of biometrics technologies for the benefits of citizens in e-governance, e-commerce and in high-speed border crossings.

3.5 Concluding Remarks

There are four technological developments that will lead to evolution of second generation biometrics systems; (i) emergence of potentially new biometric traits, (ii) added value offered by soft biometrics, (iii) effective use of multiple biometric traits for large-scale human identification, and (iv) technologies to ensure a high degree of privacy, security and flexibility in the usage of biometrics systems. The expectations and the challenges for the second generation biometrics technologies are huge. The development of second generation biometrics technologies is going to be cumulative and continuous effort, rather than resulting from a single novel invention. The low cost of biometrics sensors and acceptable matching performance have been the dominating factors in the popularity of fingerprint modality for commercial usage. Continued improvements in the matching performance and gradual reduction in cost of biometrics sensors can be cumulative enough to alter the selection of biometrics modalities in future. The development of smart sensing technologies will allow the researchers to effectively exploit extended biometric features and develop high performance matchers using efficient noise elimination techniques. Such multifaceted efforts can achieve the much needed gains from the second generation biometrics technologies at faster pace.

We believe that social and privacy concerns associated with biometrics technologies can be effectively handled with a two-fold approach. Firstly, the personal privacy should be regarded as an essential component of biometrics technologies. Policy makers, system developers and system integrators must ensure that these technologies are used properly. Secondly, the policy issues (ethical and legal framework) relating to the deployment of biometrics technologies should be clearly formulated to demarcate the conflict of interests among the stakeholders. The development of widely acceptable biometrics standards, practices and policies should address not only the problems relating to *identity thefts* but also ensure that the advantages of biometrics technologies reaches, particularly to the underprivileged segments of society (UIDAI 2010) who have been largely suffering from *identity hacking*. In our opinion, based on the current biometric deployments, the security, and benefits they offer far outweigh the apparent social concerns relating to personal privacy. Hong Kong identity cards should be a promising model to judge the benefits and concerns in future deployments of biometrics technologies.

It is widely expected that sensing, storage, and computational capabilities of biometric systems will continue to improve. While this will significantly improve the throughput and usability, there are still fundamental issues related to (i) biometric representation, (ii) robust matching, and (iii) adaptive multimodal systems. These efforts along with the capability to automatically extract behavioural traits may be necessary for deployment for surveillance and many large scale identification applications.

Acknowledgement The authors thankfully acknowledge Dr. Jianjiang Feng, Tsinghua University, Dr. Salil Prabhakar, DigitalPersona, Dr. Karthik Nandakumar, Institute for Infocomm Research (I2R), and Abhishek Nagar, Michigan State University for their constructive comments and suggestions. Part of Anil Jain's research was supported by WCU(World Class University) program through the National Research Foundation of Korea funded by the Ministry of Education, Science and Technology (R31-10008).

References

Abate, A.F., M. Nappi, D. Riccio, and G. Sabatino. 2007. 2D and 3D face recognition: A survey. *Pattern Recognition Letters* 28(14): 1885–1906.

Abdullayeva, F., Imamverdiyev, V. Musayev, and J. Wayman. 2008. Analysis of security vulnerabilities in biometrics systems. *Proceedings of the Second International Conference on "Problems of Cybernetics and Informatics"*, September 10–12, Baku, Azerbaijan. http://danishbiometrics.files.wordpress.com/2009/08/1-13.pdf. Accessed 30 Jan 2010.

Adler, A. 2004. Images can be reconstructed from quantized biometric match score data. *Proceedings of the Canadian Conference on Electrical and Computer Engineering*, Niagara Falls, 469–472.

Antonelli, A., R. Cappelli, D. Maio, et al. 2006. Fake finger detection by skin distortion analysis. *IEEE Transactions on Information Forensics and Security* 1: 360–373.

Australian Government. Introducing SmartGate. http://www.customs.gov.au/webdata/resources/files/BR_introSmrtGt0409.pdf. Accessed 30 Jan 2010.

Bangeman, E. 2009. Swiss privacy commissioner says "nein" to Google Street View. Ars technical. http://arstechnica.com/tech-policy/news/2009/08/swiss-privacy-commissioner-says-nein-to-google-street-view-swiss-privacy-commissioner-says-nein-to-google-street-view.ars. Accessed 30 Jan 2010.

Bender, S., and H. Postley. 2007. Key sequence rhythm recognition system and method. US Patent No. 7206938.

Bhanu, B., and H. Chen. 2008. *Human ear recognition by computer*. New York: Springer.

BioAPI Consort. http://www.bioapi.org. Accessed 30 Jan 2010.

Biometrics Market Intelligence. http://www.biometricsmi.com. Accessed 30 Jan 2010.

Biometrics.gov. Registry of USG recommended biometric standards. http://www.biometrics.gov/Standards/StandardsRegistry.pdf. Accessed 30 Jan 2010.

BioPrivacy Initiative. Best practices for privacy-sympathetic biometric deployment. http://www.bioprivacy.org. Accessed 30 Jan 2010.

Bowyer, K.W., K. Hollingsworth, and P.J. Flynn. 2008. Image understanding for iris biometrics: A survey. *Computer Vision and Image Understanding* 110(2): 281–307.

CBS News. Argentina forces dirty war orphans to provide DNA. http://www.cbsnews.com/stories/2009/11/21/ap/world/main5727307.shtml. Accessed 30 Jan 2010.

CJIS. Integrated automated fingerprint identification system or IAFIS. http://www.fbi.gov/hq/cjisd/iafis.htm. Accessed 30 Jan 2010.

Clarke, R. 2000. Beyond the OECD guidelines: Privacy protection for the 21st century. Roger Clarke's Web-Site. http://www.rogerclarke.com/DV/PP21C.html. Accessed 30 Jan 2010.

Cohn, J.P. 2006. Keeping an eye on school security: The iris recognition project in New Jersey Schools. National Institute of Justice J, No. 254. http://www.ojp.usdoj.gov/nij/journals/254/iris_recognition.html. Accessed 30 Jan 2010.

Crihalmeanu, S., A. Ross, and R. Derakhshani. 2009. Enhancement and registration schemes for matching conjunctival vasculature. *Proceedings of the 3rd IAPR/IEEE International Conference on Biometrics*, Alghero, Italy.

Daugman, J., and I. Malhas. 2004. Iris recognition border-crossing system in the UAE. Biometrics. http://www.cl.cam.ac.uk/~jgd1000/UAEdeployment.pdf. Accessed 30 Jan 2010.

Denham, E. 2009. Findings under the personal information protection and electronic documents act (PIPEDA). Office of the Privacy Commissioner of Canada. http://www.priv.gc.ca/cf-dc/2009/2009_008_0716_e.cfm. Accessed 30 Jan 2010.

Department of Defense. Electronic biometric transmission specifications. http://www.biometrics.gov/Standards/DoD_ABIS_EBTS_v1.2.pdf. Accessed 30 Jan 2010.

Dewan, S.K. 2003. Elementary, Watson: Scan a palm, find a clue. The New York Times. http://www.nytimes.com. Accessed 30 Jan 2010.

DNA Fingerprint Identification. http://www.fingerprinting.com/dna-fingerprint-identification.php. Accessed 20 Feb 2010.

Doward, J. 2009. DNA tests for asylum seekers 'deeply flawed'. guardian.co.uk. http://www.guardian.co.uk/world/2009/sep/20/asylum-seeker-dna-tests. Accessed 30 Jan 2010.

Draper, S.C., A. Khisti, E. Martinian, et al. 2007. Using distributed source coding to secure fingerprint biometrics. ICASSP, Hawaii, 129–132.

Drira H., B.B. Amor, M. Daoudi, and A. Srivastava. 2009. Nasal region contribution in 3D face biometrics using shape analysis framework. *Proceedings of the 3rd IAPR/IEEE International Conference on Biometrics*, Alghero, Italy, 357–366.

E-Channel. Immigration Department, Hong Kong. http://www.immd.gov.hk/ehtml/20041216.htm. Accessed 30 Jan 2010.

Editor. 2009. Iris at a distance not yet mature enough, says UAE. *Biometric Technology Today* 17(2): 1.

Falguera, J.R., F.P.S. Falguera, and A.N. Marana. 2008. Frontal sinus recognition for human identification. *Proceedings of SPIE*, vol. 6944, Orlando, FL, 69440S.
FANPIX.net. Barack Obama pictures. http://www.fanpix.net/gallery/barack-obama-pictures.htm. Accessed 30 Jan 2010.
Feng, Y.C., and P.C Yuen. 2006. Protecting face biometric data on smartcard and Reed-Soloman Code. *Proceedings of CVPR Workshop on Biometrics*, New York, 29.
Feng. J., and A.K. Jain. 2009. FM model based fingerprint reconstruction from minutiae template. *Proceedings of ICB 2009*, Alghero, Italy, 544–553.
Fingerprinting. History of fingerprinting. http://www.fingerprinting.com/history-of-fingerprinting.php. Accessed 30 Jan 2010.
Freire-Santos, M., J. Fierrez-Aguilar, and J. Ortega-Gracia. 2006. Cryptographic key generation using handwritten signature. *Proceedings of SPIE Conference on Biometric Technologies for Human Identification*, vol. 6202, Orlando, 225–231.
Grother, P., W. Salamon, C. Watson et al. 2009. MINEX II – Performance of match on card algorithms Phase II/III report. NIST Interagency Report 7477. http://fingerprint.nist.gov/minexII/minex_report.pdf. Accessed 30 Jan 2010.
Higaki, T. 2007 Quotes of the day. TIME. http://www.time.com/time/quotes/0,26174,1685967,00.html. Accessed 30 Jan 2010.
Hirobayashi, S., Y. Tamura, T. Yamabuchi, and T. Yoshizawa. 2007. Verification of individual identification method using bioelectric potential of plant during human walking. *Japanese Journal of Applied Physics* 46(4A): 1768–1773.
Hong, L. 1998. Automatic personal identification using fingerprints. PhD thesis. Michigan State University.
Hong Kong Immigration Department. Hong Kong immigration department annual report 2007–2008, 2009. http://www.immd.gov.hk/a_report_07-08/eng/chapter02/index.htm. Accessed 30 Jan 2010.
Hong Kong Post e-Cert. http://www.hongkongpost.gov.hk/index.html. Accessed 30 Jan 2010.
Indovina M., et al. 2009. ELFT Phase II-An Evaluation of Automated Latent Fingerprint Identification Technologies. NISTIR 7577.
Irvine, J.M., S.A. Israel, W.T. Scruggs, and W.J. Worek. 2008. Eigenpulse: Robust human identification from cardiovascular function. *Pattern Recognition* 41(11): 3427–3435.
Jain, A.K., and J. Feng. 2009. Latent palmprint matching. *IEEE Transactions on Pattern Analysis and Machine Intelligence* 31(6): 1032–1047.
Jain, A.K., and U. Park. 2009. Facial marks: Soft biometric for face recognition. *Proceedings of the International Conference on Image Process*, Cairo, Egypt.
Jain, A.K., S.C. Dass, and K. Nandakumar. 2004a. Can soft biometric traits assist user recognition? *Proceedings of SPIE*, vol. 5404, Biometric Technology for Human Identification, Orlando, 561–572.
Jain, A.K., S.C. Dass, and K. Nandakumar. 2004b. Soft biometric traits for personal recognition systems. *Proceedings of International Conference on Biometric Authentication*, Hong Kong, 731–738.
Jain, A.K., A. Ross, and S. Pankanti. 2006. Biometrics: A tool for information security. *IEEE Transactions on Information Forensics and Security* 1(2): 125–143.
Jain, A.K., P.J. Flynn, and A. Ross, eds. 2007. *Handbook of biometrics*. New York: Springer.
Jain, A.K., J. Feng, A. Nagar, K. Nandakumar. 2008a. On matching latent fingerprints. *Proceedings of CVPR Workshop on Biometrics*, Alaska, 1–8.
Jain, A.K., K. Nandakumar, and A. Nagar. 2008b. Biometric template security. EURASIP J Advances in Signal Processing, Special issue on Biometrics.
Joshi, A.J., S. Chandran, V.K. Jayaraman, and B.D. Kulkarni. 2008. Arterial pulse rate variability analysis for diagnoses. *Proceedings of ICPR*, Tampa, FL, 1–4.
Jules, A., and M. Sudan. 2002. A fuzzy vault scheme. *Proceedings of the IEEE International Symposium on Information Theory*, Lausanne, Switzerland, 408.
Kellokumpu, V., G. Zhao, S.Z. Li, et al. 2009. Dynamic texture based gait recognition. *Proceedings of the ICB 2009*, Alghero, Italy, 1000–1009.

Kumar, A. 2008. Incorporating cohort information for reliable palmprint authentication. *Proceeding of ICVGIP 2008*, Bhubaneswar, 583–590.
Kumar, A., and A. Kumar. 2009. Development of a new cryptographic construct using palmprint based fuzzy vault. *EURASIP Journal on Advances in Signal Processing* 2009: 1–12. ASP/967046.
Kumar, A., and K.V. Prathyusha. 2009. Personal authentication using hand vein triangulation and knuckle shape. *IEEE Transactions on Image Processing* 38(9): 2127–2136.
Kumar, A., and Ch Ravikanth. 2009. Personal authentication using finger knuckle surface. *IEEE Transactions on Information Forensics and Security* 4(1): 98–110.
Kumar, A., and Y. Zhou. 2009. Human identification using knucklecodes. *Proceeding of BTAS 2009*, Washington, DC.
Kumar, A., V. Kanhangad, and D. Zhang. 2010. A new framework for adaptive multimodal biometrics management. IEEE Transactions on Information Security Forensics, available online (to appear).
LaCous, M.K. 2008. Match template protection within biometric security systems. US Patent No. 7454624.
Lee, H.C., and R.E. Gaensslen. 2001. *Advances in fingerprint technology*. Boca Raton: CRC Press.
Lee, Y.J., K. Bae, S.J. Lee, et al. 2007. Biometric key binding: Fuzzy vault based on iris images. *Proceedings of the International Conference on Biometrics*, Seoul, 800–808.
Lee, J.-E., A.K. Jain, and R. Jin. 2008. Scars, marks and tattoos (SMT): Soft biometric for suspect and victim identification. *Proceedings of the Biometric Symposium, Biometric Consort Conference*, Tampa, FL.
MacManus, R. 2009. Hong Kong's octopus card: Utility outweighs privacy concerns. Read WriteWeb. http://www.readwriteweb.com/archives/hong_kongs_octopus_card.php. Accessed 30 Jan 2010.
Matey, J.R., O. Naroditsky, K. Hanna, et al. 2006. Iris on the move: Acquisition of images for iris recognition in less constrained environments. *Proceedings of the IEEE*, vol. 94, 1936–1947.
Matey, J.R., D. Ackerman, J. Bergen, and M. Tinker. 2008. Iris recognition in less constrained environments. In *Advances in biometrics sensors, algorithms and systems*, ed. N.K. Ratha and V. Govindaraju, London: Springer, 107–131.
Médioni, G., J. Choi, C.-H. Kuo, and D. Fidaleo. 2009. Identifying noncooperative subjects at a distance using face images and infrared three-dimensional face models. *IEEE Transactions on Systems, Man, Cybernetics. Part A: Systems and Humans* 39(1): 12–24.
Mhatre, A., S. Palla, S. Chikkerur, et al. 2005. Efficient search and retrieval in biometric databases. *Proceedings of SPIE Defense and Security Symposium*, vol. 5779, 265–273.
MIS ASIA. HK privacy commissioner: Fingerprint collection excessive. http://www.mis-asia.com/news/articles/hk-privacy-commissioner-fingerprint-collection-excessive. Accessed 30 Jan 2010.
Nagar, A., and A.K. Jain. 2009. On the security of non-invertible fingerprint template transforms. *Proceedings of the First IEEE International Workshop on Information Forensics & Security (WIFS 2009)*, London.
Nakajima, K., Y. Mizukami, K. Tanka, and T. Tamura. 2000. Footprint-based personal recognition. *IEEE Transactions on Biomedical Engineering* 47(11): 1534–1537.
Nakashima, E. 2007. FBI prepares vast database of biometrics. The Washington Post. http://www.washingtonpost.com/wp-dyn/content/article/2007/12/21/AR2007122102544.html. Accessed 30 Jan 2010.
Nandakumar, K., A.K. Jain, and A. Ross. 2009. Fusion in multibiometric identification systems: what about the missing data? *Proceedings of the ICB*, Alghero, Italy, 743–752.
NIST Report to the United States Congress. 2002. Summary of NIST standards for biometric accuracy, tamper resistance and interoperability.
NIST. Evaluation of Latent Fingerprint Technologies 2007. http://fingerprint.nist.gov/latent/elft07/. Accessed 30 Jan 2010.
NIST. National and international biometric standards. http://www.itl.nist.gov/div893/biometrics/standards.html. Accessed 30 Jan 2010.

NIST. Summary of NIST standards for biometric accuracy, tamper resistance, and interoperability. http://www.itl.nist.gov/iad/894.03/NISTAPP_Nov02.pdf. Accessed 30 Jan 2010.
NSTC. Biometrics history. http://www.biometrics.gov/Documents/BioHistory.pdf. Accessed 30 Jan 2010.
Nuffield Council on Bioethics. 2007. *The forensics use of bioinformatics: Ethical issues*. Cambridge/London: Cambridge Publishes.
Office of the Privacy Commissioner for Personal Data Protection. Privacy Commissioner Responds to Public Enquiries about the Issue of "Employer Collecting Employees' Fingerprint Data for Attendance Purpose". http://www.pcpd.org.hk/english/infocentre/press_20090716.html. Accessed 30 Jan 2010.
PalmSecure. http://www.fujitsu.com/us/services/biometrics/palm-vein. Accessed 20 Feb 2010.
Pankanti, S., S. Prabhakar, and A.K. Jain. 2002. On the individuality of fingerprints. *IEEE Transactions on Pattern Analysis and Machine Intelligence* 24(8): 1010–1025.
Park, U., A. Ross, and A.K. Jain. 2009. Periocular biometrics in the visible spectrum: A feasibility study. *Proceedings of the BTAS*, Washington, DC.
Pink Tentacle. Magazine photos fool age-verification cameras. http://pinktentacle.com/2008/06/magazine-photos-fool-age-verification-cameras. Accessed 30 Jan 2010.
Poh, N., R. Wong, J. Kittler, et al. 2009. Challenges and research directions for adaptive biometric recognition systems. *Proceedings of the ICB*, Alghero, Italy.
Printrak Division. Printrak LiveScan 4000 Ruggedized. http://www.morpho.com/MorphoTrak/PrinTrak/prnt_prod/pp_Ls-4000.html. Accessed 30 Jan 2010.
Privacy Commissioner of Hong Kong. Privacy commissioner responds to public enquiries about the issue of "Employer Collecting Employees' Fingerprint Data for Attendance Purpose". http://www.pcpd.org.hk/english/infocentre/press_20090716.html. Accessed 30 Jan 2010.
Proença, H., S. Filipe, R. Santos, J. Oliveira, and L.A. Alexandre. 2009. The UBIRIS.v2: A database of visible wavelength iris images captured on-the-move and at-a-distance. IEEE Transactions on Pattern Analysis and Machine Intelligence, available online (to appear).
PROnetworks. Face recognition cigarette vending machines. http://www.pronetworks.org/forums/face-recognition-cigarette-vending-machines-t102463.html. Accessed 30 Jan 2010.
Ramus, S.J., and H. Eichenbaum. 2000. Neural correlates of olfactory recognition memory in rat orbitofrontal cortex. *Journal of Neuroscience* 20: 8199–8208.
Ranger, S. 2006. Photos: Iris scanning at the airport. Silicon.com. http://www.silicon.com/management/public-sector/2006/04/13/photos-iris-scanning-at-the-airport-39158086/. Accessed 10 Oct 2009.
Ratha, N., J.H. Connell, and R.M. Bolle. 2001. An analysis of minutiae matching strengths. *Proceedings of the International Conference on Audio and Video-based Biometric Authentication*, Halmstad, Sweden, 223–228.
Reddy, P.V., A. Kumar, S.M.K. Rahman, and T.S. Mundra. 2008. A new antispoofing approach for biometric devices. *IEEE Transactions on Biomedical Circuits and Systems* 2(4): 284–293.
Rhodes, Henry T.F. 1956. *Alphonse Bertillon, Father of scientific detection*. New York: Abelard-Schuman.
Ross, A., K. Nandakumar, and A.K. Jain. 2006. *Handbook of multibiometrics*. New York: Springer.
Ross, A., J. Shah, and A.K. Jain. 2007. From templates to images: Reconstructing fingerprints from minutiae points. *IEEE Transactions on Pattern Analysis and Machine Intelligence* 29(4): 544–560.
Rowe, R.K., U. Uludag, M. Demirkus, et al. 2007. A multispectral whole-hand biometric authentication system. *Proceedings of the Biometric Symposium, Biometric Consort Conference*, Baltimore.
Schiphol. Fast border passage with iris scan. http://www.schiphol.nl/AtSchiphol/PriviumIrisscan/FastBorderPassageWithIrisScan.htm. Accessed 27 Sept 2009.
Serkan, T. 2009. Women uses tape to trick biometric airport fingerprint scan. CrunchGear. http://www.crunchgear.com/2009/01/02/woman-uses-tape-to-trick-biometric-airport-fingerprint-scan. Accessed 30 Jan 2010.

Shamir, L., S. Ling, S. Rahimi, L. Ferrucci, and I.G. Goldberg. 2009. Biometric identification using knee X-rays. *International Journal of Biometrics* 1(3): 365–370.

Singh, K. 2008. Altered fingerprints, Interpol report. INTERPOL. http://www.interpol.int/Public/Forensic/fingerprints/research/alteredfingerprints.pdf. Accessed 30 Jan 2010.

Singh, R., M. Vatsa, A. Ross, et al. 2009. Online learning in biometrics: A case study in face classifier update. *Proceedings of the BTAS'2009*, Washington, DC.

Song, S., K. Ohnuma, Z. Liu, L. Mei, A. Kawada, and T. Monma. 2009. Novel biometrics based on nose pore recognition. *Optical Engineering* 48(5): 57204.

South Wales Police. A history of fingerprinting. http://www.south-wales.police.uk/fe/master.asp?n1=8&n2=253&n3=1028. Accessed 30 Jan 2010.

Standing Committee 37 ISO Standards Development. http://isotc.iso.org/livelink/livelink?func=ll&objId=2262372&objAction=browse&sort=name. Accessed 30 Jan 2010.

Sun, Z., T. Tan, Y. Yang, et al. 2005. Ordinal palmprint representation for personal identification. *Proceedings of CVPR 2005*, San Diego, 279–284

Sun, Z., A. Paulino, J. Feng, Z. Chai, T. Tan, and A. K. Jain. 2010. A study of multibiometric traits of identical twins. *Proceedings of the SPIE Biometrics*, Florida.

Swabey, M.A., S.P. Beeby, A.D. Brown, and J.E. Chad. 2004. Using Otoacoustic emission as biometric. *Proceedings of ICBA*, Hong Kong, LNCS, vol. 3072, pp. 600–606.

Tabor, Z., D. Karpisz, L. Wojnar, and P. Kowalski. 2009. An automatic recognition of the frontal sinus in x-ray images of skull. *IEEE Transactions on Biomedical Engineering* 56(2): 361–368.

Tao, D., X. Li, X. Wu, et al. 2007. Analysis and gabor features for gait recognition. *IEEE Transactions on Pattern Analysis and Machine Intelligence* 29(10): 1700–1715.

Topping, A., V. Kuperschmidt, and A. Gormley. 1998. Method and apparatus for the automated identification of individuals by their beds of their fingernails. U.S. Patent No. 5751835.

Tronci, R., G. Giacinto, and F. Roli. 2007. Dynamic score selection for fusion of multiple biometric matchers. *Proceedings of the 14th IEEE International Conference on Image Analysis and Processing*, ICIAP, Modena, Italy, 15–20.

Tuyls, P., A.H.M. Akkermans, T.A.M. Kevenaar, et al. 2005. Practical biometric authentication with template protection. *Proceedings of the 5th International Conference on Audio- and Video-based Biometric Person Authentication*, Rye Town, NY, 436–446.

Uhl, A., and P. Wild. 2008. Footprint-based biometric verification. *Journal of Electronic Imaging* 17: 11016.

UIDAI. Creating a unique identity number for every resident in India (Working paper – version 1.1). http://uidai.gov.in. Accessed 30 Jan 2010.

UK Border Agency. Iris Recognition Immigration System (IRIS). http://www.ukba.homeoffice.gov.uk/managingborders/technology/iris. Accessed 30 Jan 2010.

Uludag, U., and A.K. Jain. 2006. Securing fingerprint template: Fuzzy vault with helper data. *Proceedings of CVPR Workshop on Privacy Research in Vision*, New York, 163.

University of Cambridge. John Daugman's webpage. http://www.cl.cam.ac.uk/~jgd1000. Accessed 30 Jan 2010.

US Department of Homeland Security. New biometric technology improves security and facilitates U.S. entry process for international travelers. https://www.dhs.gov/xlibrary/assets/usvisit/usvisit_edu_10-fingerprint_consumer_friendly_content_1400_words.pdf. Accessed 30 Jan 2010.

Veeramachaneni, K., L.A. Osadciw, and P.K. Varshney. 2005. An adaptive multimodal biometric management algorithm. *IEEE Transactions on Systems, Man, Cybernetics. Part C* 35(3): 344–356.

Volokh, E. 2007. Religious freedom and DNA gathering. Volokh conspiracy or CNET news http://volokh.com/2007/02/26/religious-freedom-and-dna-gathering Feb. 2007 or http://news.cnet.com/Feds-out-for-hackers-blood/2100-7348_3-6151385.html. Accessed 30 Jan 2010.

Walt Disney World Resort. http://disneyworld.disney.go.com/. Accessed 30 Jan 2010.

Wayman, J.L., 1997. Large-scale civilian biometric systems - issues and feasibility. *Proceedings of the CardTech/SecureTech Government*, Washington DC, http://www.engr.sjsu.edu/biometrics/nbtccw.pdf. Accessed 30 Jan 2010.

Wickins, J. 2007. The ethics of biometrics: The risk of social exclusion from the widespread use of electronic identification. *Science and Engineering Ethics* 13: 45–54.

Woodard, D.L., and P.J. Flynn. 2005. Finger surface as a biometric identifier. *Computer Vision and Image Understanding* 100(3): 357–384.

Yan, P., and K.W. Bowyer. 2007. Biometric recognition using 3D ear shape. *IEEE Transactions on Pattern Analysis and Machine Intelligence* 29(8): 1297–1308.

Zhang, D., W.K. Kong, J. You, et al. 2003. Online palmprint identification. *IEEE Transactions on Pattern Analysis and Machine Intelligence* 25(9): 1041–1050.

Chapter 4
Biometrics, Privacy and Agency

Annemarie Sprokkereef and Paul de Hert

4.1 Introduction

Technological progress in the development of biometric techniques is unmistakable. The technological functionality of available systems is maturing and showing improved capability. With biometric techniques, we refer to the identification or authentication of individuals based on a physical or behavioural characteristic through the use of mathematical and statistical methods. It is obvious that without information technology (especially ever improving network capabilities) and the availability and advances of new sensor technology, the world of biometrics would not show such rapid development. Proliferation of sensor technology has had immediate implications for biometrics technology, which requires this kind of infrastructure to extend the capabilities offered, particularly in terms of accuracy and intrusiveness. It is the emerging intelligent distributed sensor networks that that are paving the way for the second generation of biometrics.

The average minister, official or parliamentarian involved in law making will already face difficulties in assessing the efficiency and implications of the first generation biometric techniques as used in commercial applications currently available on the market. The main reason for this is that the information available is often contradictory. An additional complication is the blurring of boundaries between potential (=future) and current capacity of biometric applications (Petermann and Sauter 2002). This is, and will remain, a source of technical and political confusion (Van der Ploeg 1999). Even more complex, also for experts, is an assessment of the societal impact of the use of

A. Sprokkereef (✉)
Tilburg Institute for Law, Technology, and Society (TILT), The Netherlands
e-mail: a.c.j.sprokkereef@uvt.nl

P. de Hert
Research Group on Law Science Technology & Society (LSTS), Vrije Universiteit Brussel (VUB), Belgium, and Tilburg Institute for Law, Technology, and Society (TILT), The Netherlands
e-mail: paul.de.hert@uvt.nl

E. Mordini and D. Tzovaras (eds.), *Second Generation Biometrics: The Ethical, Legal and Social Context*, The International Library of Ethics, Law and Technology 11,
DOI 10.1007/978-94-007-3892-8_4,

biometrics (Grijpink 2001, 2005; Ashbourn 2005; JRC 2005; Androunikou et al. 2005; Bray 2004; Rundle and Chris 2007; Sprokkereef 2008). A systematic, constantly updated and forward looking analysis and assessment of the societal, economic and legal impact of further growth of the use biometric applications is needed to inform the political process. In this chapter we will mainly restrict ourselves to an analysis of the legal implications of first and second generation biometric applications.

The handling of first generation biometric data- such as finger- and iris-scans – already creates fundamental discussions about the scope of data protection and human rights law. The introduction of soft biometrics, i.e. the use of general traits such as gender, weight, height, age, or ethnicity for automated classification, is even more contested (Van der Ploeg 2002). It has attracted criticism of indiscriminate social sorting, as automated decisions are created that divide people into categories for further processing. What are the legal implications of automated sorting of people on the basis of their behaviour (and/or general traits) into classifications such as for example, Asians and non-Asians, young and old, gay and hetero, and so forth? On the one hand, as machines are taking the decisions, the act of sorting takes on a seemingly neutral dimension. On the other hand, the embedded systems, ambient intelligence, distant sensing and passive biometrics involved require no conscious cooperation from subjects and thus pose a challenge to the traditional concepts used in the fields of data protection and human rights. In short, problematic legal aspects of these developments are covert data capture, lack of transparency and consent. Some second generation biometrics used for authentication or multiple factor assessments such as heart rate, body temperature, brain activity patterns, and pupil dilation also question the validity of current consensus on what constitute personal data for example. These new body data, often collected in embedded and passive fashion, need to be assessed from a legal viewpoint. So where does the traditional approach become problematic? At first sight this is the issue of transparent collection of biometric data by consent. Many second generation biometrics are collected whilst the data subject is unaware. Data is collected from a distance and the collection does not need to be apparent. The cultural change here is that tracking and tracing thus becomes the norm. This is the fundamental change that creates the surveillance society and the question is: where does this leaves the law (SSN 2006)?

This contribution will focus on the challenges that so called second generation biometrics pose for the evolving legal framework on biometrics. In this analysis we will not discuss the current technical state of the art, unless it is directly relevant for a discussion of the legal framework or the concept of biometrics (Grijpink 2008; EBF 2007). We will first assess the current legal framework governing the first generation of biometrics, and then take a closer look at the relationship between the emerging second generation and the law.

4.2 Legal Principles Governing Personal Data

We take as a starting point that all biometric data are personal data protected by data protection legislation (De Hert et al. 2007; Kindt 2007a, b). Before going into detailed analysis, we will first extract some basic principles behind data protection law and formulate some questions that show how these principles might affect the choice or even admissibility of the use of certain types of biometrics.

In accordance with Article 6 of the 95/46/EC Directive that has been implemented in all EU member states, personal data must be collected for specified, explicit and legitimate purposes only and may not be further processed in a way that is incompatible with those purposes. In addition, the data themselves must be adequate, relevant and not excessive in relation to the purpose for which they are collected (principle of purpose). Once the purpose for which data are collected has been established, an assessment of the proportionality of collecting and processing these data can be made. Thus, the main data protection principles are: confidentiality, purpose specification, proportionality and individual participation. If we translate these principles into day-to-day biometrics they can lead to questions such as:

Confidentiality: Has the biometric system concerned been sufficiently tested to warrant a minimum level of confidentiality protection?
Purpose specification: Is there sufficient protection against unauthorised use of the biometric data for purposes beyond the original goal. Can purposes keep being added? To what extent does the collection of biometric data aim to discriminate?
Proportionality: Are the biometric data adequate, relevant and proportional in relation to the purpose for which they are collected? More specifically, is verification being used, when there is no need for identification? Is one biometric more proportional than another?
Individual participation: Are fallback procedures put into place in cases where biometric data, be it raw images or templates, become unreadable? Can individuals opt not to provide their biometric or not to have their biometrics read?

4.3 The European Data Protection Framework and Biometrics

The abovementioned general Directive 95/46/EC (1995) constitutes the main and general legal framework for the processing of personal data. With the adoption of the Directive, the biggest breakthrough was its scope. The Directive brings electronic visual and auditive processing systems explicitly within its remit. The preamble says "*Whereas, given the importance of the developments under way, in the framework of the information society, of the techniques uses to capture, transmit, manipulate, record, store or communicate sound and image data relating to natural persons, this Directive should be applicable to processing involving such data*".[1] Processing of sound and visual data is thus considered as an action to which the Directive applies. Processing is thus very broad: it can be automatic or manual and

[1] Directive 95/46/EC, Preamble, § 14.

can consist of one of the following operations: collection, recording, organisation, storage, adaptation, alteration, retrieval, consultation, use, disclosure by transmission and so forth. The sheer fact of collecting visual (for example face scan) or sound data can therefore already be considered as processing. Without going into too much detail, we will make a quick scan of the provisions of the Directive in relation to biometrics (Hes et al. 1999; Sprokkereef and de Hert 2007).

The Directive thus aims to protect the rights and freedoms of persons with respect to the processing of personal data. It does so by laying down rights for the person, whose data is processed, and guidelines and duties for the controller, or processor determining when this processing is lawful. The rights, duties and guidelines relate to: data quality; making data processing legitimate; special categories of processing; information to be given to the data subject; the data subject's right of access to data; the data subject's right to object to data processing; confidentiality and security of processing; and notification of processing to a supervisory authority.

Article 2 (a) of that Directive defines personal data as 'any information relating to an identified or identifiable natural person'. Although the Directive does not mention biometric data as such, its legal provisions and principles also apply to the processing of biometric data. It has been claimed that the Directive does not apply to biometric data in specific processing circumstances, that is to say when the data can no longer be traced back to a specific identifiable person (Kindt 2007b). Nevertheless, one has to acknowledge that the essence of all biometric systems per se is that the data processed relate to identified or identifiable persons. These systems use personal characteristics to either identify the person to whom these characteristics belong or to verify that the characteristics belong to a person authorised to use the system. In most cases where biometric data are concerned, there is a link to the person at least for the data controller, who must be able to check the final proper functionality of the biometric system and cases of false rejects or false matches. The four key elements to the definition of personal data are "any information", "relating to", "identified or identifiable" and "natural person". According to the *Biovision Best Practice Report* (Albrecht 2003): personal data that relate to the implementation of a biometric at least include: the image or record captured from a censor; any transmitted form of the image or record between sensor and processing systems; the processed data, whether completely transformed to a template or only partially processed by an algorithm; the stored image or record or template; any accompanying data collected at the time of the enrolment; the image or record captured from the sensor during normal operation of the biometric; any transmitted form or image or record at verification or identification; the template obtained from the storage device; any accompanying data obtained at the time of verification or identification; the result of the matching process when linked to particular actions or transmissions; and any updating of the template in response to the identification or verification. This view has not been challenged. It is clear that the general Directive applies to the processing of personal data wholly or partly by automatic means- and to the processing otherwise than by automatic means – of personal data that form part of a filing system or are intended to form part of a filing system. It does not apply to the processing of personal data by a natural person in the course of a purely personal or household activity and to procession of personal data in the course of an activity that falls outside the scope of European law, such as operations concerning public security, defence or State security.

In summary, in a generic biometric system, processes can be broken down in five groupings: data collection, transmission, signal or image processing, matching decision, storage (Albrecht 2003, 13). Therefore, in the meaning of the Directive, all these five processes in biometric systems should be considered as processing of data unless they take place in a personal or household context such as data handling in a family home biometric entry system.

Article 2 (e) of the Directive defines the controller and the processor of data. In this meaning the data controller or processor of biometric data will be the operator who runs the system in most cases. The instances were consent is necessary are stipulated in Article 7 (a). Article 8 (2) states a general prohibition of processing of sensitive data to which some exemptions apply. Sensitive data in terms of biometrics can include medical, ethnic/racial, behavioural information, or data concerning sex life. In general, processing of sensitive biometric information (relating to ethnic origin for example) will need the explicit consent of the data subject. This requirement sits uncomfortably with the automatic sorting into categories of people on the basis of their bodily traits that characterise some second generation biometrics applications. Article 6 (d) states that data must be accurate and, where necessary, kept up to date. In terms of biometrics, this imposes the obligation to use only such biometric systems that have low false match rates so that there is only a small likely hood of processing incorrect data.

Article 20 states that the Member states shall determine processing operations likely to present specific risks to the freedoms of data subjects and shall check that these processing operations are examined prior to the start thereof. An assessment of risks created by biometrics is subject to interpretation by member states. The margin of interpretation regarding this provision has turned out to be wide. Whilst the French authorities regard every biometric system as presenting potential risks, other national data protection authorities have come to different conclusions. The Dutch data protection authority has so far not used the power of *prior checking* as given by this provision. Finally, Article 15 (1) lays down the rules on automated decisions on individuals. Central here is the notion of a human intervention before a final decision is taken. Examples of an automated decision using biometrics could include border crossing and immigration processes in the public sector, and biometric processes at the working place in the private sector. In cases where the biometric application is only used to support an authentication process and the legal decision is not based solely on the biometric process, Article 15 ought not to apply to biometric processing. We will come back to this provision when discussing second generation biometrics below.

4.4 The Article 29 Data Protection Working Party

The Article 29 Data Protection Working Party (hereinafter the WP29) is an independent European advisory body established by the above Directive whose membership consists of representatives of national supervisory bodies. The statements and opinions of WP29 not only have an impact on national judiciaries but also on the national supervisory bodies themselves (De Hert et al. 2008, 23–24). In fact, one of the core objectives of WP29 has been to help Europe's data protection practices move together instead of moving apart. As part of this mission they have undertaken a review of the interpretation of personal data across member states and adopted guidance to help clarify the position.

In August 2003, the WP29 provided specific guidelines for the processing of biometric data in a working document on biometrics (WP29 2003/2004). These guidelines are highly relevant for biometric identity management systems in general, whether used in the public sphere or for private commercial purposes. The use of biometrics in applications controlled by governments, such as in passports, travel documents and ID cards and in large scale databases such as a Visa Information System (VIS) and the Schengen Information System (SIS II), has been the subject of intense scrutiny after several opinions of the WP29. These opinions have highlighted the risks of the implementation of biometrics in these applications in its current form. WP29 has further reflected on the meaning of biometric data in an opinion on the concept of personal data (WP29 2007). In this opinion, the functionality of biometric data to establish a link with an individual and to function as identifier was stressed. The working party therefore just ran short of regarding all biometric data as sensitive data under the directive. The advantages offered by using biometrics as a key identifier are contested. As all biometric applications and especially large scale systems are still in the roll out phase, the WP29, followed by many technical experts, finds it a dangerous strategy to place complete reliance on the security and reliability of biometrics as a key identifier (Sprokkereef and De Hert 2007; Grijpink 2008; Kindt 2007b).

4.5 Data Protection Agencies

The national Data Protection Agencies (the DPAs) have all embarked on the task of the interpretation of the data protection legislation applied to biometrics for use in the private sector. In most countries national data protection legislation based on the Directive does not explicitly mention biometrics. In the majority of cases, DPAs have reviewed the processing of biometric data upon request for a preliminary opinion by the data controller or upon notification of the processing. In France, it has become mandatory since 2004 for controllers to request such opinion. The controller needs this opinion, in fact an authorization, before the start of the processing of biometric data.[2] France is hereby one of the few countries that has acted proactively

[2]Later on, in 2006, the CNIL has issued some 'unique authorizations' which permit controllers, if they comply with all requirements, to file a declaration of conformity.

to the emerging trend of the use of biometric data by imposing such prior authorization. This has created an enormous workload for the CNIL, resulting in the creation of the concept of 'unique authorisations' to help ease the task in 2006. The CNIL is not alone in being too understaffed a data protection authority to be able to carry out its tasks (Thomas 2008).

The DPAs have many more competences to exercise. These competences include, according to the Directive, endowment with investigative powers, such as access to the (biometric) data processed by a controller and powers to collect all the information necessary. In addition, powers of intervention, including the competence to order the erasure of data or imposing temporary or definitive bans on the use of (biometric) data, and the power to engage in legal proceedings against controllers if they do not respect the data protection provisions. The DPAs can also hear claims of individuals who state that their rights and freedoms with regard to the processing of personal data have been infringed or hear claims of organisations representing such individuals. Appeals against the decisions of the DPAs are in principle possible before the national courts of the country where the DPA is established. Appeals should conform to the existing procedure for appeal against such (administrative) decisions. In the United Kingdom, for example, the 'Information Tribunal' has been set up (formerly the 'Data Protection Tribunal') to determine appeals against notices and decisions served by the Information Commissioner (Turle 2007).

4.6 Understanding the Privacy and Data Protection Challenges of Biometric Data Processing

So far we have said nothing about the threats biometrics create to human rights and values. We have limited ourselves to the observation that European data protection law applies, we have discussed some of the implications and applications of this set of rules and principles made by the WP29 and DPA's. But what is really the issue with biometrics?

A brief description of some general trends in relation to new technologies helps to prepare the answer to this question. The increased use of biometrics in all kind of settings clearly does not develop in a vacuum. Technological advances in biometric technology open up possibilities, and these are taken up based on a supply and demand that is the result of political and societal developments. The latter affect the introduction of the technology in a range of different settings. Some of the current and relevant trends that can be identified are:

- The increasing volume of data that is gathered, processed and exchanged;
- Linking of data banks and data sets both by government agencies and by commercial parties;
- Function creep (as a process by which data are used for different purposes than originally collected for);
- Increased tendency to keep data on file for possible future use (rather than discard data);

- Mounting pressure on individuals to disclose personal information and allow data linkage both by government agencies and by business organisations (the latter mainly through internet);
- proliferation of types of hardware that hold large data sets (USB sticks, small gadgets and portable computers, data holding phones and so forth:) that pose new security risks.

The growth in network capabilities offered by information technology and the availability and advances of new sensor technology combined create the technical possibilities for these trends to re-enforce each other. New possibilities in the information exchange between new technologies and ICT have an attraction difficult to resist. The tendency to use data for other purposes than originally collected for can thus be observed in the private- but is even more striking in the public domain. The opening up the EURODAC site to police and other law enforcement agencies is the most quoted European example of this. The bulk of the literature on the European wide introduction of biometric technologies in the public sector identifies the core problem is that government demand is focusing on surveillance, control and fraud detection more than on data collection minimisation, and security- and risk- mitigation. The consensus is that in the public sector, function creep is a logical development when the emphasis is on surveillance and control and therefore on tracking, identifying and controlling individuals.

Biometric applications are seen as useful tools that can play a role in the new systems of data management that result from these trends. In general, expectations placed on biometrics in the short term seem to be high. On the contrary, the expectations of the impact and precise role of biometrics in the long term appear rather vague and not very well documented. In this light, it is important to keep in mind the technical imperfection of biometrics, which has also been underlined by the European Data Protection Supervisor (EDPS) in its opinion on the proposal for VIS (EDPS 2005). Firstly, 5% of individuals are estimated not to be able to enroll because of having no readable fingerprints or no fingerprints at all. This would mean with regard to the use of VIS, which is expected to include data on 20 million applicants in 2007, that one million persons cannot be checked by the normal procedure. Secondly, with regard to biometrics, an error rate of 0.5–1% is considered normal, which would mean a False Rejection Rate of 0.5–1% with regard to measures and checks based on VIS.

The expectations in connection with biometrics are overestimated. Indeed, biometrics could lead to too much trust in the effectiveness of electronic solutions. Biometrics is based on probabilities: false positives and – negatives are unavoidable. If only 1% of a targeted group of 100,000 people a day suffers from a false negative, this would cause every day 1,000 people to be ‘automatically’ (but wrongfully) stopped. That current studies on the longer term reliability of biometrical data show that for example finger scans change throughout time, illustrates this risk (Wayman 2006, 14).

Furthermore, most of the information we know about the reliability, accuracy and efficiency of biometrics, is provided by the vendors of biometrics, and large scale assessments are not yet available (EBF 2007). The 'Biometric summary table' in a report of 2004 of the Organisation for Economic Co-operation and Development (OECD 2004) shows that the biometric technology – as it is – shows varying rates of reliability. Moreover, no biometric technology seems to be in line with all data protection principles and user acceptance at the same time (data quality principle, transparency principle, data security principle). Whereas the reliability of fingerprint scanning is only 'possibly' very high, the user acceptance is medium to low. Whereas the accuracy of facial recognition is medium to high, the stability and the transparency are low.

The use of biometrics will not exclude identity theft or forgery. Although biometrics prevent so-called 'identity substitution' to a certain degree, the fraudulent issuance of a genuine passport cannot be prevented. This lack of security of new e-passports, currently introduced in a number of EU Member States was first illustrated by a German computer security expert. He demonstrated how personal information stored into the documents could be copied and transferred to another device, including fake passports. According to experts, despite several improvements since, the German passport is still lacking security, particularly with respect to confidentiality (Berthold 2009, 55). Recent works have verified in a formal manner that security leaks exist in basic access control (BAC) and even in (extended access control) EAC. Most of these leaks are known since years and have been demonstrated in experiments. It is likely that new weaknesses appear in the data handling implementation of electronic passports (Sprokkereef and Koops 2009). The trend that can be noted in other EU policies on biometrics too is that expectations of near perfect reliability are gradually dropped over time (Hornung 2005; Gasson et al. 2007). As biometric technology is in the roll out phase (EBF 2007; de Leeuw 2007; Lodge and Sprokkereef 2009), this means that the high expectations in the short term will probably be adjusted rather sooner than later (Brussee et al. 2008).

4.7 The Human Right to Data Protection and Privacy

Let us now turn to the values of privacy and data protection enshrined in European legal texts. The human right to data protection is explicitly recognised in Article 8 of the Charter of Fundamental Rights of the European Union. Whereas the right to data protection is covered in Article 8 of the Charter, the right to privacy is covered in Article 7. This highlights the difference between privacy and data protection, and underlines the need for both rights to coexist. There are, after all, circumstances when the right to privacy applies while the right to data protection does not (and vice versa) (De Hert and Gutwirth 2006; Gutwirth and De Hert 2008).

Article 7 of the Charter of Fundamental Rights of the European Union *and* Article 8 of the European Convention on Human Rights (ECHR*)* provides for the fundamental right to privacy. Article 8 ECHR states: '1. everyone has the right to respect for his private and family life, his home and his correspondence. 2. there shall be no interference by a public authority with the exercise of this right except such as is in accordance with the law and is necessary in a democratic society in the interests of national security, public safety or the economic well-being of the country, for the prevention of disorder or crime, for the protection of health or morals, or for the protection of the rights and freedoms of others.' In Article 7 of the Charter of Fundamental Rights of the EU, it is stated: 'Everybody has the right to respect for his or her family life, home and communications'.

In different judgments, the European Court of Human Rights (ECtHR) has applied Article 8 of the ECHR on the basis that it provides protection against the systematic collection and storage of personal data. Article 8 was invoked in the Rotaru vs. Romania (ECtHR 2000) In June 2006, the ECtHR ruled that the continued storage of personal data (in this case, for a period of more than 30 years) by Swedish security and intelligence agencies was disproportional, and represented a breach of the applicants' right to privacy (ECtHR 2006).

In a judgment on the interpretation of Directive 95/46/EC on the protection of individuals with regard to the processing of personal data and on the free movement of such data the European Court of Justice (ECJ) has confirmed that the criteria and limitations set out in Article 8 apply when assessing whether the processing of personal data conforms to Community law (ECJ 2003).

Article 8 imposes strict limitations on interference in an individual's private life by public authorities – any law that interferes with the private lives of EU citizens requires further justification. The 'necessity criterion' is of considerable importance here. The second paragraph of Article 8 stipulates that state interference must be 'necessary in a democratic society:

- in the interests of national security, public safety or the economic well-being of the country,
- for the prevention of disorder or crime,
- for the protection of health or morals, or
- for the protection of the rights and freedoms of others'.

The necessity criterion relates to the proportionality principle applied under data protection law. In other words, 'personal data must be adequate, relevant and not excessive in relation to the purposes for which they are collected and/or further processed'. Non-compliance with the proportionality principle implies the simultaneous infringement of the necessity criterion (assuming the right to privacy under Article 8 of the ECHR applies of course). This is confirmed by the European Court of Human Rights: 'the notion of necessity implies a 'pressing social need'. To be more specific, the measure(s) employed must be "*proportionate to the legitimate aim pursued*" (ECtHR 1986). If too many or irrelevant data are processed in relation to the processing purpose, the processing can be considered illegitimate.

The 'necessity' criterion – indispensable to interfere with an individual's private sphere – plays a prime role in the discussions on the legitimacy of biometrics. Critics point at the technical imperfections of biometrics discussed above and cast doubts on the reliability, accuracy and efficiency of biometrics. In principle, biometrics makes the processing of data easier. Biometrics are stronger identifiers as they identify individuals and link them to existing data. Since they are unique identifiers they can potentially be used as a key to link data in several databases. The privacy argument then goes that the use of such a powerful tool is not (always) warranted and that the danger of function creep and identity theft is of such a nature that it would make the choice for biometrics disproportional. The proliferation of biometrics makes individual biometric data more accessible. It also makes the biometric data more linkable to other personal data, thus making the technology privacy invasive per se. Instances of function creep, such as the use of biometric data collected for immigration purposes used in the context of unrelated criminal investigations (E.C.J. 2008), might occur in the private sector also. To give examples: biometric applications first introduced with the purpose of fast and efficient entry of employees or customers can be used for time registration or customer profiling at a later stage. Here the information given to customers and legal rights to be informed and the right to correct come into play.

The foregoing can be summarised as follows: because biometrics are powerful, they cannot be deployed in all cases and because they are still imperfect, deployment should be limited. This cautious approach towards the use of biometrics is fuelled by two questions that remain unanswered: Are biometrics sensitive data in the sense of data protection law and does the obligation to be subjected to biometrical identification not conflict with feelings of (bodily) dignity of people?

Firstly, there is the issue whether biometrical data are sensitive data. Article 8 of the EC Directive 95/46 on Data Protection principally prohibits the processing of sensitive data. Sensitive data are "personal data revealing racial or ethnic origin, political opinions, religious or philosophical beliefs, trade-union membership, and data concerning health or sex life." The use of biometrics can involve the processing of sensitive data in the sense of Article 8. Biometrical data of disabled people may relate to their medical condition and correlations could for example be made between papillary patterns and diseases such as leukaemia and breast cancer The WP29 opinion on Council Regulation 2252/2004 (WP29 2003, 68) states: "In the case of storing fingerprints attention will have to be paid in so far as various correlations between certain papillary patterns and corresponding diseases are discussed. As for instance certain papillary patterns are said to depend on the nutrition of the mother (and thus of the foetus) during the 3rd month of the pregnancy. Leukaemia and breast cancer seem to be statistically correlated with certain papillary patterns. Any direct or precise correlations in these cases are not known…". Face recognition can reveal racial or ethnic origin (WP29 2003). The processing of biometrical data may thus reveal – more or less immediately – sensitive information about an individual. This possibility goes far beyond the purpose for which biometrical identification is supposed to be used.

It seems clear that the *taking* of fingerprints and photos may involve the processing of sensitive data. Possibly, case law (of the European Court of Human Rights) will make a distinction between different biometrics (e.g. non-sensitive fingerprint vs. sensitive facial image). Further, it is not clear if the algorithms and machine-readable templates that contain the information are always to be considered as sensitive personal data. Several stages in the processing of biometrical data can be identified. The first stage is the capture or measurement of the human characteristic and the creation of a template. In this stage the 'raw' or unprocessed template sometimes contains information which can directly be interpreted in terms of e.g. race or state of health. Examples are facial images showing skin colour or certain signs of illnesses. These initial templates can in those cases be classified as sensitive data. Subsequent steps often follow in the processing, in which the original data are being manipulated. Whether these processed data still classify as sensitive data is questionable but very relevant for the application of data protection law to second generation biometrics (Hes et al. 1999).

Secondly, there is the question about human dignity. Taking, measuring, and processing of biometrical data may also harm a person's personal feeling or experience of dignity. The fact that people feel uncomfortable with close observations (they are obliged to look into a lens, they are obliged to put fingers on holders used by other people etc....) has already been observed as a possible feeling of intrusion of dignity (Wayman 2006, 15).

That taking facial images is related to this observance, may also be derived from the fact that for example the Quality Assurance (QA) software – used to examine the properties of the applicant's photo for a passport or travel document – can reject a photo thereby *explaining why* (Friedrich and Seidel 2006, 5). Exceptions to the photo requirements are possible for handicapped citizens and for certain religious reasons, but this may at the same time confront people with themselves as being an exception; and force them to reveal their religion.

The use of 'your' body as an identification tool for others might likewise infringe what is called our informational privacy. In other words, as stated by Anton Alterman: "*The degree to which the body is objectified by the process, suggest[s] that biometric identification alienates a part of the embodied self. The body becomes an object whose identity is instantly determinable by purely mechanical means, and subject to external controls on that basis; while those means themselves are removed from the control of the subject. The representations are infinitely reproducible by their owner, but are not even accessible to the subject whose body they represent. The embodied person now bears, more or less, a label with a bar code, and is in this respect alienated from her own body as well as from the technology used to recognize it. If having an iris scan on file is not quite like being incarcerated in the world at large, being made known to mechanical systems wherever they may be is still a tangible loss of privacy that is not precisely paralleled by any other kind of information technology.*" [Emphasis added] (Alterman 2003, 146)

4.8 Some Useful Distinctions for the Privacy and Data Protection Debate

With these unanswered questions, it can come as a surprise to note that the privacy debate around biometrics is in some regards deadlocked. We believe that one way out is to look more closely at the technology itself. Technical developments and shifts in the way the technology is used are of such a nature that they impact on the proportionality question at the core of both the privacy and the data protection debate. In that context one can point at new exciting technologies that have been developed both to improve the possibilities created by encryption and in overcoming the problems of false positives because of noise (Tuyls et al. 2007). More fundamental for the discussion are the following distinctions regarding the use of biometrics.

The most important distinction always made is that between the goals of identification and verification. This distinction separates biometric systems aimed at bringing about automated identification of a particular person and those aimed at verification of a claim made by a person who presents him or herself. As the identification function requires a one to many comparison whilst the verification function requires a one to one comparison, privacy risks involved in the use of the biometric technology vary considerably from application to application. In principle, the verification function permits the biometric characteristic to be stored locally, even under the full control of the individual, so that the risk that the biometric data are used for other purposes is limited. This does not mean that all biometric systems that could be restricted to serving the verification function have actually been designed to maximise the control of the individual over his or her biological data. Most biometric applications used in the semi private domain in the Netherlands for example have been introduced for verification purposes (Sprokkereef and De Hert 2009). In most instances the objective is to grant authorised persons access to physical or digital spaces. Control over personal data in these situations needs to be clarified, and the legal conditions determining the handling of data as well as the enforcement of applicable law are issues that need scrutiny.

The usefulness of the notion of control over biometric data when assessing privacy will be discussed below. What suffices here is to establish that the identification function cannot be performed without the use of a (central or de-centralized) database. Therefore, in the case of biometric identification procedures, biometric data are never under the strict and full control of the individual. PETs or privacy friendly options are another distinctive category of products that already exist on the market (Sprokkereef and Koops 2009). When verification is the goal of the system, the processor can opt for decentralisation of the template storage and verification. By decentralisation of both the template storage and verification process, the biometrical data are processed in an environment controlled by the individual or an environment from which no connection to a central database can be made.

When identification is the goal these options do not exist. This makes these systems more dangerous in terms of loss of control. Privacy options do however exist with regard to another privacy aspect, viz. security. In case of central template storage and verification, mathematical manipulation (encryption algorithms or hash-function) can ensure encryption of databases so that it is not possible to relate the biometric data to other data stored in different databases, at different locations. In the case of EURODAC, the PET aspect is the HIT-No HIT facility in the first instance, but of course in the case of a HIT, biometrics are de-encrypted so that they can lead back to the other personal details of the person involved.

The way in which data are stored centrally of course matters in terms of privacy protection. The use of biometric encryption in combination with the use of a "protected safe" that is only accessible to a small number of people, greatly enhances the privacy protection of individuals enrolled in a system with central storage. The biometric applications using these techniques are still in development or in the early phases of roll out (Cavoukian and Stoianov 2007; Tuyls et al. 2007). It is also important to make a distinction between the protection of biometric data stored within systems and the use of biometrics by operators to safeguard authorised access to other type of data stored in a system. Equally relevant is the distinction between biometric applications used in the private or semi-public domain and those used in the public domain. Often this boils down to whether providing biometric samples is obligatory or only voluntary for the individual concerned. With first generation biometrics, applications in the private or semi-public domain are used on a voluntary basis. This means that the individuals asked to provide their characteristic could refuse this and use another facility, for example another swimming pool not using a biometric entry system or the individual can make use of a similar service without the biometric at the same institution (for example voice recognition when managing your bank account). When there is an alternative way to obtain the services offered then the element of choice, consent and control is in the hand of the individual. Here information is a key factor. The users of biometric systems often fail to realise the implications of offering their characteristics for storage on a database, and the loss of individual control over their characteristics once this has happened. Rights, such as the right to correct, are seldom exercised and other ex ante measures remain unused.

When there is no alternative available, and this is the case with most public sector introductions of biometrics such as the biometric passport, the biometric characteristic has to be presented. A risk that then arises is that biometric based systems may be used by other persons and for other purposes than foreseen (Hornung 2007; Friedrich and Seidel 2006). In 2005, the 27th International Conference of Data Protection and Privacy Commissioners adopted a Resolution in which it expressed its awareness '*of the fact that the private sector is also increasingly processing biometric data mostly on a voluntary basis*'. The Conference called for '*1. effective safeguards to be implemented <u>at an early stage</u> to limit the risks inherent to the nature of biometrics; 2. the <u>strict distinction</u> between biometric data collected and stored for public purposes (e.g. border control) on the basis of legal obligations and for <u>contractual purposes</u> on the basis of consent; 3. the <u>technical restriction</u> of the use of biometrics in passports and identity cards to verification purposes comparing the data in the document with the data provided by the holder when presenting the document*' (emphases added) (Datacommisioners 2005).

Then there is the importance in the legal implications of storing raw data v the use of templates. Under raw data we understand the unmodified output of the sensor device. This can be the image of a fingerprint, a face, an iris, or a sound from a microphone. Some pre-processing is allowed in the definition, provided that no information or redundancy is added or dropped. Template data are shadows made from raw data. Template data are those data which are compared in the matcher unit. Normally, templates will only contain information necessary for comparison

(Bromba 2006). From a privacy perspective working with templates offers the distinct advantage that if they get stolen or hacked you do not lose your digital fingerprint. There is the possibility of making a new template based on other algorithms. However, when raw data gets stolen, the subject cannot produce other and distinct raw data. Identity theft of raw data is therefore a more serious problem. Some hold that storage of the raw template only occurs in very basic biometric systems the handling of raw data and will soon become a relic of the past. However these systems are still on the market and they are in European passports. Privacy activists are often reassured by the statement that biometric raw data cannot be reconstructed from stored biometric templates. Recent literature however shows that there is strong evidence that raw data can be reconstructed from template data (at least partially). Furthermore, misuse of templates does not necessarily need a reconstruction of raw data (Bromba 2006).

If these distinctions, nuances and developments are taken into account, the privacy balancing might well produce a different outcome in every concrete case. It might even be the case that the use of biometrics in a privacy enhancing way turns out to be the better option. Although there is an abundance of literature on privacy enhancing technology, much of it does not pay any attention to biometric technology (Hes et al. 2000; Tavani and Moor 2001; Levi et al. 2004; Koorn et al. 2004; Philips 2004; Adams 2006; Borking 2008a, b; European Commission 2007). Based on the small body of literature that deals with biometrics, (Hes et al. 1999; Grijpink 2001; Zorkadis and Donos 2004; Andronikou 2007; Sprokkereef and De Hert 2009) we can detail the following illustrations of what constitute privacy enhancing and potential privacy invasive features of biometric applications.

The one-off use of fingerprints in medical screening, for example, is a biometric application that enhances privacy (PET or privacy enhancing technology). This makes having to use patient names to match with their diagnostic results unnecessary. There is a double advantage to the one off use of biometrics in this instance: patients can remain anonymous and there is greater reassurance that data are released to the correct person. Another obvious example is the ex post measure already mentioned *biometric* authentication to restrict operator use in a database. This use of biometrics makes operators more accountable for any use/misuse of data. A more generic example is the match on card-sensor on card: biometric authentication without the biometric characteristics leaving devices owned by the individual (Neuwirt 2001). Biometric cryptography is a privacy enhancing technical solution that integrates an individual's biometric characteristics in a revocable or non-revocable cryptographic key. This method now forms an integral part of all but the cheapest biometric applications on the commercial market. When the key is revocable, this introduces issues relating to "function creep" and law enforcement. At the same time, there are possibilities for a three-way check to come to an architectural design that restricts the number of people having access to data. In addition, there are applications that offer two-way verification and therefore integrate the "trust and verify" security principle. Another privacy enhancing possibility would lie in the certification of the privacy compliance of biometrical identification products; this is therefore not a PET but a certification of the biometric application as a PET.

When we concentrate on biometrics as a privacy invading technology an obvious example would be the storage if information in a manner where medical or racial information can be inferred. Equally intrusive is the use of an individual's biometric data to link their identity between different applications/databases. Another example is the use of biometrics for surveillance or sorting, where no permission is asked to take (moving) images of people for example of face recognition systems. Most second generation biometrics fall under this category (see below). Then there is central storage of raw biometric data, central storage of biometric templates and so forth aimed at identification. Then there are biometric systems aiming for interoperability, maybe even using a biometric as a unique identifiable key (instead of another personal detail such as the name (alphabetical identifier) or a number (numerical identifier)). These systems are built on as much identification as possible and link data that would otherwise go unconnected. Also very privacy intrusive is a privacy invading choice for biometric identification. To be more precise: the use of a biometric system for identification, where verification would already have met the objectives of the application.

4.9 Biometrics and the Second Generation

So what are the important elements of second generation biometrics and will they give rise to a new set of legal issues to be analysed and discussed? We identify two developments in biometrics that together form the main step away from the first generation applications of the technology. The first is the emergence of new biometric traits, the so-called soft biometrics and physiological biometrics, and the second is the shift to embedded biometric systems, with elements such as distant sensing and 'passive' biometrics. These distinct developments are the basic changes that might catapult us into the world of ambient intelligence and ubiquitous computing. Then, the already complex legal assessment of biometric data handling will be taken to a different level altogether and pose serious challenges to existing legal approaches (basically based on data protection law). The dream of second generation of biometrics is a person's identification on the basis of that person's dynamical behaviour. In fact, the attempt is not made to identify a person, no: the objective is to read the person's mind. So instead of enrolling and identifying or verifying a person, second generation biometrics is aimed at a categorisation of individuals. The threats caused by this de-personalisation are many fold. Of course, unjustified selection according to profile will result in discrimination. Stigmatisation will take place and will involve allocation to a group on the basis of relatively random profiles which will impact the future of a person. Confrontation of individuals with unwanted information is another side effect that is very likely to occur. Profiling may result in a limited information supply: according to perceived but incomplete profile. Similarity in profiles will cause de-individualisation. Finally, there will be unknown effects in linking dispersed information.

4.10 Concerns

One of the most fundamental challenges in the protection of persona
is related to the incremental change from visible to invisible data co
us assume first that the individual subject knows that he is subject to biometric processing of the second generation. There is then mainly a tension between the processing of second generation biometric data and the individual participation principle. How to check and verify if the biometrical data are still accurate? The obvious risk that the systems (and not only personal data) may be used by other persons and for other purposes than foreseen is difficult to minimize, without the traditional possibilities for individual participation. How to exercise the right to have the data corrected or the right to object to certain types of data processing? Does an individual for example know that the biometrical identifiers are still working? Here a number of transparency tools can be developed that give the individual more insight into who is taking which decisions on the basis of data collected. The current lack of possibilities to enforce individual participation pales into insignificance, when assessing the applicability of data protection law in situations where the subject is unaware of the invisible data collection. Therefore, our main and immediate legal concern regarding second generation biometrics is the applicability of data protection regulation in those situations and the specific use of the data for profiling.

First, there is the applicability of data protection regulation. If no attempt is made to identify a person, can we define the data concerned as personal data? If not, what guarantees remain against unwarranted and unfit social categorisation?

Secondly, there is the issue of profiling. It is not clear whether and when profiling falls directly under the rights and obligations of the EC Directive 95/46 (Hildebrandt and Backhouse 2008; Hildebrandt and Gutwirth 2008). The Directive may allow statistical processing or profiling of personal data, once the data are made anonymous. Recital 26 states "*whereas the principles of protection shall not apply to data rendered anonymous in such a way that the data subject is no longer identifiable.*" Article 6.1.b is very specific: "*further processing of the data for historical, statistical or scientific purposes is not considered as incompatible provided that appropriate safeguards are provided by the Member States whereas these safeguards must in particular rule out the use of the data in support of measures or decisions regarding any particular individual.*" Article 15 of the Data Protection Directive principally prohibits that a person is subject to automated decisions which produce legal effects concerning him, or significantly affects him, and which are based solely on automated processing of data intended to evaluate certain personal aspects relating to him. This Article thus gives individuals the right to object to decisions affecting them when decisions are made solely on their profiles (Gutwirth and De Hert 2008, 283). This provision however, is accompanied by numerous exceptions that do not set strong and clear-cut limits to targeted profiling actions. The results of data profiling can be applied afterwards to data subjects without them knowing that the profiles are applied to them, for example: people can be individually stopped or checked at a border control because they fall under a certain profile. How is it guaranteed that he data subject is informed that such automated individual decisions are applied to him? Are there any guarantees that the data subject can exercise the right to obtain from the controller knowledge of the logic involved in such automatic processing operations? Will all authorised agents acting upon these automated decisions 'know' this logic involved and be able to communicate this logic to the data subject? As Gutwirth and De Hert have shown (2008), profiling should not only be addressed with a mix of modes of regulation, but this mix must also construct an appropriate articulation of opacity and transparency tools.

In conclusion, the use of second generation biometrics will have to lead to a reassessment of the traditional data protection approach that only data relating to identified or identifiable persons have to be protected. In fact, existing European legal mechanisms cannot guarantee effective protection against profiling. This has already led to a call for widening the protection currently granted through the regulation of 'unsollicited communications' via the new notion of 'unsollicited adjustments'. The notion of 'unsollicited adjustments' would close a legal loophole allowing a situation in which objects that seemingly have a neutral guiding function, in practice secretly track individuals to surreptitiously adapt their performance based on undisclosed criteria (Gonzalez et al. 2010). Similarly, gaps in current data protection, in the case of second generation biometrics applied in real life situations, can lead to forms of profiling that leave some of the rights for individual, such as the right to have data corrected, or the right of access to the data, unprotected. Approaching new phenomena such as profiling with heavy prohibitions may block progress or lead to a situation where the prohibitions are not respected (Gutwirth and De Hert 2008). There does not need to be a choice between the opacity approach of data protection (prescriptive rules) and a data transparency approach (making data handling visible and data handlers accountable). In the regulation of profiling, opacity and transparency tools can each have their own role to play. In a normative weighing of privacy and other interests, some intrusions will turn out just to be too threatening for fundamental rights whilst others will be accepted and submitted to the legal conditions of transparency and accountability.

References

Adams, C. 2006. A classification for privacy technologies. *University of Ottawa Law and Technology Journal (UOLTJ)* 3(1): 35–52.

Albrecht, A. 2003. BIOVISION: D 7.4 Privacy Best Practices in Deployment of Biometric Systems. BIOVISION ROADMAP, oai.cwi.nl/oai/asset/4057/04057D.pdf. Accessed 14 Sept 2010.

Alterman, A. 2003. A piece of yourself: Ethical issues in biometric identification. *Ethics and Information Technology (Kluwer)* 5: 139–150.

Androunikou, V., D. Demetis, and T. Varvarigou. 2005. Biometric implementations and the implications for security and privacy. *Journal of the Future of Identity in the Information Society* 1(1): 20–35.

Ashbourn, J. 2005. The social implications of the wide scale implementation of biometric and related technologies. Background paper for the Euroscience Open Forum ESOF (2006), Munich. http://www.statewatch.org/news/2006/jul/biometrics-and-identity-management.pdf. Accessed 14 Sept 2010.

Berthold, S. 2009. Epass. 5.3. In *D3.16: Biometrics: PET or PIT?* ed. A. Sprokkereef and B.J. Koops. Brussels: FIDIS.

Borking, J. 2008a. Organizational Motives for Adopting Privacy Enhancing Technologies. *Data Protection Review*, Madrid: DPA.

Borking, J. 2008b. The Business Case for PET and the EuroPrise Seal. Europrise deliverable.

Bray, P. 2004. Ethical aspects of facial recognition systems in public places. *Journal of Information Communication and Ethics in Society* 2: 97–109.

Bromba, M. 2006. *On the Reconstruction of Biometric Raw Data from Template Data*. Via http://www.bromba.com/knowhow/temppriv.htm. Accessed 14 Sept 2010.

Brussee, R., L. Heerink, R.E. Leenes, J. Nouwt, M.E. Pekárek, A.C.J. Sprokkereef, and W. Teeuw. 2008. *Persoonsinformatie of Identiteit? Identiteitsvaststelling en Elektronische Dossiers in het Licht van Maatschappelijke en Technologische Ontwikkelingen. Telematica Instituut.* Report TI/RS/2008/034: 1–98. Enschede: Telematica Instituut.

Cavoukian, A., and A. Stoianov. 2007. *Biometric Encryption: A Positive-Sum Technology that Achieves Strong Authentication, Security and Privacy.* Information and Privacy Commissioner's Office, Ontario.

Charter of Fundamental Rights of the European Union. 2000. *Official Journal* C 364 (December).

Data Protection Commissioners. 2005. 27th International Conference of Data Protection and Privacy Commissioners, Resolution on the use of biometrics in passports, identity cards and travel documents, Montreux 16 September 2005. http://www.edps.eu.int/legislation/05-09-16_resolution_biometrics_EN.pdf. Accessed 14 Sept 2010.

De Hert, P., and S. Gutwirth. 2006. Privacy, data protection and law enforcement. Opacity of the individual and transparency of the power. In *Privacy and the criminal law*, ed. E. Claes, A. Duff, and S. Gutwirth, 61–104. Antwerp/Oxford: Intersentia.

De Hert, P., W. Scheurs, and E. Brouwer. 2007. Machine-readable identity documents with biometric data in the EU - part III - Overview of the legal framework. *Keesing Journal of Documents and Identity* 22: 23–26.

De Hert, P., S. Gutwirth, A. Moscibroda, D. Wright, and G. González Fuster. 2008. *Legal Safeguards for Privacy and Data Protection.* Working paper series REFGOV-FR-19. http://refgov.cpdr.ucl.ac.be/?go=publications. Accessed 14 Sept 2010.

de Leeuw, E. 2007. Biometrie en Nationaal Identiteitsmanagement. *Privacy and Informatie* 2(10): 50–56.

Directive 95/46/EC of the European Parliament and of the Council of 24 October 1995 on the protection of individuals with regard to the processing of personal data and on the free movement of such data. *Official Journal European Communities Legislation* L 281 (November).

EBF (European Biometrics Forum). 2007. *Security and Privacy in Large Scale Biometric Systems*: Seville: JRC/ITPS. http://is.jrc.es/documents/SecurityPrivacyFinalReport.pdf. Accessed 14 Sept 2010.

E.C.J. 2003. 20 May 2003 Österreichischer Rundfunk and others, joint cases, C-138-01, C-139/01 and C-465/00.

E.C.J. 2008. 16 December 2008 Heinz Huber V FRG, C 524/06, *Official Journal* C44/5 of 21.2.2009.

ECtHR. 1986. Judgment of 24 November 1986 (Gillow vs. The United Kingdom).

ECtHR. 2000. Rotaru vs. Romania, 4 May 2000, appl. no. 28341/95 *Reports* 2000-V, §§ 43–44.

ECtHR. 2006. Segerstedt-Wiberg and Others v. Sweden, 6 June 2006, Appl. no. 62332/00.

European Commission. 2007. *A Fine Balance 2007: Privacy Enhancing Technologies; How to Create a Trusted Information Society.* Conference Summary. ftp://ftp.cordis.europa.eu/pub/fp7/ict/docs/security/20080228-pet-final-report_en.pdf. Accessed 14 Sept 2010.

European Data Protection Supervisor (EDPS). 2005. Opinion on VIS, Brussels. http://www.edps.europa.eu/12_en_opinions.htm. Accessed 14 Sept 2010.

Friedrich, E., and U. Seidel. 2006. The introduction of the German e-passport. Biometric passport offers firstclass balance between security and privacy. *Keesing Journal of Documents and Identity* 16: 2006.

Gasson, M. et al., eds. 2007. *Fidis deliverable D.3.2.: A study on PKI and biometrics*, www.fidis.net. Accessed 14 Sept 2010.

González Fuster, G., S. Gutwirth, and P. de Hert. 2010. From unsollicited communications to unsollicited adjustments. Redefining a key mechanism for privacy protection. In *Data protection in a profiled world*, ed. S. Gutwirth, Y. Poullet, and P. De Hert. Berlin: Springer.
Grijpink, J. 2001. Biometrics and privacy. *Computer Law and Security Report* 17(3): 154–160.
Grijpink, J. 2005. Two *Barriers* to realizing the benefits of biometrics. *Computer Law and Security Report* 21(3): 249–256.
Grijpink, J. 2008. Biometrie, Veiligheid en Privacy. *Privacy en Informatie* 11: 10–14.
Gutwirth, S., and P. De Hert. 2008. Regulating profiling in a democratic constitutional state. In *Profiling the European citizen. Cross-disciplinary perspectives*, ed. M. Hildebrandt and S. Gutwirth, 271–292. Berlin: Springer Press.
Hes, R., T.F.M. Hooghiemstra, and J.J. Borking. 1999. *At face value, on biometrical identification and privacy*, Achtergrond Studies en Verkenningen, vol. 15, 1–70. The Hague: Registratiekamer.
Hes, R., et al. 2000. *Privacy-enhancing technologies: The path to anonymity*, Achtergrond Studies en Verkenningen, vol. 11, 1–60. The Hague: Registratiekamer (Revised Edition).
Hildebrandt, M., and J. Backhouse, eds. 2008. FIDIS Deliverable D7.2.: *Descriptive Analysis and Inventory of Profiling Practices.* http://www.fidis.net/resources/deliverables/profiling/int-d72000/. Accessed 14 Sept 2010.
Hildebrandt, M., and S. Gutwirth, eds. 2008. FIDIS Deliverable D7.4.: *Implications of Profiling on Democracy and the Rule of Law.* http://www.fidis.net/resources/deliverables/profiling/int-d74000/. Accessed 14 Sept 2010.
Hornung, G. 2005. *Die digitale Identität. Rechtsprobleme von Chipkartenausweisen: Digitaler Personalausweis, elektronische Gesundheitskarte, JobCard-Verfahren.* Reihe "Der elektronische Rechtsverkehr", ed. Roßnagel A., and TeleTrusT Deutschland e.V.,10, Baden-Baden: Nomos Verlagsgesellschaft.
Hornung, G. 2007. The European regulation on biometric passports: Legislative procedures, political interactions, legal framework and technical safeguards. *SCRIPT ED* 4(3): 246–262.
JRC (Joint Research Centre). 2005. *Biometrics at the Frontiers: Assessing the Impact on Society.* Technical Report Series, Seville: Institute for Prospective Technological Studies (IPTS).
Kindt, E. 2007a. Biometric applications and the data protection legislation (the legal review and the proportionality test). *Datenschutz and Datensicherheit (DuD)* 31: 166–170.
Kindt, E. 2007b. FIDIS (Future of Identity in the Information Society) *Deliverable 3.10: Biometrics in Identity Management.*
Koorn, R., et al. 2004. *Privacy enhancing technologies. Witboek voor Beslissers.* The Hague: Ministerie van Binnenlandse Zaken.
Levi, M., et al. 2004. Technologies, security, and privacy in the post-9/11 European Information Society. *Journal of Law and Society* 31(2): 194–200.
Lodge, J., and A. Sprokkereef. 2009. *Accountable and transparent E-Security- the Case of British (In) Security, Borders and Biometrics*, Challenge. http://www.libertysecurity.org/article2488.html. Accessed 14 Sept 2010.
Neuwirt, K. 2001. *Report on the protection of personal data with regard to the use of smart cards.* Strassbourg: Council of Europe.
Organisation for Economic Co-operation and Development (O.E.C.D). 2004. Background material on biometrics and enhanced network systems for the security of international travel Working Party on Information Security and Privacy. http://www.oecd.org/dataoecd/16/18/34661198.pdf. Accessed 14 Sept 2010.
Philips, D. 2004. Privacy policy and PETs. *New Media and Society* 6(6): 691–706.
Petermann, Th., and A. Sauter. 2002. *Biometrische Identifikationssysteme Sachstandsbericht*, TAB Working report nr 76. http://www.tab.fzk.de/de/projekt/zusammenfassung/ab76.pdf. Accessed 25 Jan 2010.
Rundle, M., and C. Chris. 2007. Ethical Implications of Emerging Technologies: A Survey (UNESCO, Information for All – IFAP). UNESCO, Communication and Information Sector: 1–90.

Sprokkereef, A. 2008. Data protection and the use of biometric data in the EU; in IFIP international federation for information processing. In *The future of identity in the information society*, vol. 262, ed. S. Fischer Huebner, P. Duquenoy, A. Zaccato, and L. Martucci, 277–284. Boston: Springer.

Sprokkereef, A.C.J., and P.J.A. de Hert. 2007. Ethical practice in the use of biometric identifiers within the EU. *Law, Science and Policy* 3(2): 177–201.

Sprokkereef, A.C.J., and P.J.A. de Hert, 2009. *The use of privacy enhancing aspects of biometrics: Biometrics as PET (privacy enhancing technology) in the Dutch Private and Semi-Public Domain*. Tilburg: Tilburg Institute for Law, Technology and Society. http://arno.uvt.nl/show.cgi?fid=93109. Accessed 14 Sept 2010.

Sprokkereef, A., and B.J. Koops, eds. 2009. D3.16: *Biometrics: PET or PIT?*, Brussels: FIDIS, August 2009, 68 pp. http://www.fidis.net/fileadmin/fidis/deliverables/new_deliverables2/fidis-WP3-del3.16-biometrics-PET-or-PIT.PDF. Accessed 14 Sept 2010.

SSN (Surveillance Studies Network). 2006. *A Report on the Surveillance Society* - For the Information Commissioner by the Surveillance Studies Network, London: Information Comissioner (Full Report). http://www.ico.gov.uk/upload/documents/library/data_protection/practical_application/surveillance_society_full_report_2006.pdf

Tavani, H., and J. Moor. 2001. Privacy protection, control of information, and privacy-enhancing technologies. *Computers and Society* 31(1): 6–11.

Thomas, R. 2008. The UK Information Commissioner: On funding in evidence to the House of Commons Justice Committee on the Protection of Personal Data Report, H of C Justice Committee report: *Protection of Private Data*, HC 154 January 2008.

Turle, M. 2007. Freedom of information and data protection law: A conflict or a reconciliation? *Computer Law and Security Report* 23: 514–522.

Tuyls, P., et al. (eds.). 2007. *On private biometrics, secure key storage and anti-counterfeiting*. Boston: Springer.

van der Ploeg, I. 1999. The illegal body: 'Eurodac' and the politics of biometric identification. *Ethics and Information Technology* 1(4): 295–302.

van der Ploeg, I. 2002. Biometrics and the body as information, normative issues of the socio-technical coding of the body (chapter 3). In *Surveillance as social sorting: Privacy, risk, and digital discrimination*, 57–73. New York: Routledge.

Wayman, J. 2006. Linking persons to documents with biometrics. Biometric systems from the 1970s to date. *Keesing Journal of Documents & Identity* (16): 15ff.

WP29 (Article 29 Working Party). 2003. *Working document on biometrics 12168/02*, 1.8.2003 and *11224/04*.

WP29 (Article 29 Working Party). 2007. *Opinion 4/2007 on the concept of personal data*, 20 June 2007.

Zorkadis, V., and P. Donos. 2004. On biometrics-based authentication from a privacy-protection perspective – Deriving privacy-enhancing requirements. *Information Management and Computer Security* 12: 125–137.

Part II
Emerging Biometrics and Technology Trends

Chapter 5
Gait and Anthropometric Profile Biometrics: A Step Forward

Dimosthenis Ioannidis, Dimitrios Tzovaras, Gabriele Dalle Mura, Marcello Ferro, Gaetano Valenza, Alessandro Tognetti, and Giovanni Pioggia

5.1 Introduction

As of today, biometrics systems are gaining significant attention due to their ability to protect humans and resources from potential non-legitimate user attacks in high security environments (e.g. airports, access control rooms, etc.). It is well known that humans have used body and other characteristics such as face, gait, etc. for recognizing each other (Gloor 1980). The last decades several biometric systems have been developed and established their applicability in controlled environments such as fingerprints, retina and iris, and facial characteristics (Abate et al. 2007; Bowyer et al. 2008; Jain et al. 2004). These technologies have demonstrated reliable user authentication in very restricted environments and the applicability of biometrics to a wider range of surveillance areas stimulated the research community to the

D. Ioannidis (✉) • D. Tzovaras
Informatics and Telematics Institute, 6th km Charilaou-Thermi Road,
P.O. Box 361, Thermi-Thessaloniki 57001, Greece
e-mail: djoannid@iti.gr; dimitrios.tzovaras@iti.gr

G. Dalle Mura • G. Valenza • A. Tognetti
Interdepartmental Research Centre "E. Piaggio", Faculty of Engineering,
University of Pisa, Via Diotisalvi 2, 56126 Pisa, Italy
e-mail: gabriele.dallemura@iet.unipi.it; gaetano.valenza@iet.unipi.it; a.tognetti@ing.unipi.it

M. Ferro
"Antonio Zampolli" Institute for Computational Linguistics (ILC) National Research Council (CNR), Via G. Moruzzi 1, 56124 Pisa, Italy
e-mail: marcello.ferro@ilc.cnr.it

G. Pioggia
Institute of Clinical Physiology (IFC), National Research Council (CNR),
Via G. Moruzzi 1, 56124 Pisa, Italy
e-mail: giovanni.pioggia@ifc.cnr.it

E. Mordini and D. Tzovaras (eds.), *Second Generation Biometrics: The Ethical, Legal and Social Context*, The International Library of Ethics, Law and Technology 11,
DOI 10.1007/978-94-007-3892-8_5,

design and development of new and emerging biometrics such as gait (Boulgouris and Chi 2007) and characteristics based on user anthropometric profile (e.g. full or partially body measurements) (Ferro et al. 2009). Current work on these approaches demonstrated their authentication potential and their ability to allow the non-stop authentication of the humans in high security environments. However, since these biometric techniques (activity-related signals and body measurements) are mature, more experimental frameworks shall be designed and evaluated for human verification in order to fully deploy them in large-scale security applications. In the following sections, two unimodal biometric traits are presented based on the analysis of body measurements either using dynamic signals (gait) or analyzing body in fixed seat environment (body measurements using a sensing seat sensor).

5.2 On the Potential of Body Measurements for User Authentication

5.2.1 Authentication Potential of Gait as a Biometric

The last 10 years, gait as a biometric has received significant attention due to increase in the importance of surveillance and security in public and private areas. Latest research activities in multi-biometric environments have evolved the use of gait as a promising modality for identification and authentication purposes. The current state of the art is that "databases of over 100 subjects imaged walking outdoors or indoors can be recognised with well over 90% identification rate and factors which affect gait were understood, there was capability to handle application environment and understanding of the measure's potency for recognition purposes" (Gloor 1980). However, recognition rates with change of view angle, clothing, shoe, surface, illumination, and pose usually decreased performance, thus making the human gait recognition a challenging and emerging biometric trait (Boulgouris et al. 2005).

Recent studies on the gait recognition potential are focused mainly in two directions: view-invariant gait analysis (Jean et al. 2009; Bodor et al. 2009; Bouchrika et al. 2009) and novel algorithms for the extraction and fusion of static and kinematic parameters of human locomotion (Chen et al. 2009; Bouchrika and Nixon 2008). In most cases, gait recognition is comprised from two main phases: a feature extraction phase, where motion information is obtained and recognition phase, where a classification technique is applied to the obtain motion patterns. The crux of the gait recognition lies in perfecting the first phase. It is challenging to specify gait features that are sufficiently discriminable and can be reliably extracted from video. The methods utilised for this feature extraction can be broadly classified as being either model-free (appearance-based) or model-based. Appearance-based methods focus on the spatiotemporal information contained in the silhouette images. Model-based methods construct human model to obtain explicit features describing gait dynamics, such as stride dimensions and joint kinematics.

A recent study using model-based approach (five-link biped model) reported recognition rates of 100% in the CMU MoBo data set (25 subjects) (Zhang et al. 2007). When examining a greater data set, Bouchrika and Nixon (2007) reached a correct classification rate of 92%, again, by means of model-based approach. In their study the motion templates describing the motion of the joints as derived by gait analysis, were parameterised using the elliptic Fourier descriptors. The mean error for the extracted joints compared to manual data (10 persons) was 1.36% of height. However, in both model-based approaches the camera was capturing the side-view of the subjects. Promising results for a frontal-view gait recognition have been demonstrated by Goffredo et al. (2008) by means of an appearance-based approach. The proposed method for the front-view gait analysis is based on two consecutive steps: the gait cycle detection and the gait volume description. Without any knowledge of the camera parameters the authors found a mean percentage of recognition rate equal to 96.3%, when examining three public available databases.

Concluding, both model- and appearance-based approaches have increased their performance in gait recognition the last years. When examining gait recognition under several certain view angles (especial frontal or near frontal view) appearance-based approaches seem to outperform the model based approaches. However when the issue is an approach, which should be independent from viewpoint, i.e. enrolment and test are taking place at different view angles, model based methods seem to be more appropriate. It is possible that a fusion of model- and appearance-based approaches can contribute to higher rates, whereas the static and dynamic cues of gait are extracted using compact representations with robust performance in dynamic changing environments.

5.2.2 Authentication Potential of Body Measurements as a Biometric

Several companies are currently working to realize comfortable interactive seats. These systems, some of them already on the market, use different technologies and materials, but they share the use of sensors that measure the pressure exerted on the seat by the subject. Actually, the existing sensing seats are not able to perform the human authentication task and no result on this topic, even if in a preliminary stage, was found in literature review. The BPMS by Tekscan measures the pressure distribution of a human body on support surfaces such as seats, mattresses, cushions, and backrests and it is used for automotive driver seat design, hospital and home seat design, comfort analysis. The study developed by Tan et al. (2001) is focused on sensing chair using pressure sensors placed over the seat pan and back rest of the chair for real-time capturing of contact information between the chair and its occupant in order to implement static posture classification. A kind of technology, developed in the textile domain, is Softswitch (The Softswitch Company 2008: it is based on pressure sensors that can be integrated in fabrics. Softswitch combines conductive textile materials and a quantum tunnelling composite (QTC) with unique pressure controllable switching properties. The research of seating comfort in the transportation industry is still an open problem; for solving this question some studies (The Johnson Controls Company 2008) have been effected to evaluate the advantages and the disadvantages of automotive seat. It is possible to note that the presented systems are more focused on comfort monitoring, pressure mapping, air-bag activation and event-related tasks. However, the information supplied by these systems may be used to extract features useful for the authentication task, as for instance pressure profile. UNIPI, starting from this consideration, works to realize a system for subject authentication based on sensing seat. The first prototype of this system is part of the HUMABIO (Human Monitoring and Authentication Using Biodynamic Indicators and Behavioral Analysis) project for multi-modal human authentication. HUMABIO is a EC co-funded Specific Targeted Research Project (STREP) where new types of biometrics were combined with state of the art sensor technologies in order to enhance security in a wide spectrum of applications like transportation safety and continuous authentication in safety critical environments like laboratories, airports or other buildings. In this project, the enrolment and the authentication procedures were carried out with the cooperation of the user, according to the instructions supplied by the system. The mentioned prototype is able to supply a one-dimensional deformation profile, and, after a feature extraction process, the system is able to perform the human authentication task. The ACTIBIO project aims to perform a continuous authentication, without interfering with the user actions and according to the detection of predefined events. According to this purpose, the previous SensingSeat prototype was upgraded and a new control system was developed to handle the issues regarding the continuous authentication as well as the event notifications of the ACTIBIO core system.

5.3 Gait Biometric Technology

5.3.1 Proposed Approach and Motivation

The main purpose and contributions of this paper are summarized as follows:

- A novel gait recognition system is proposed based on the use of 2D and new 3D appearance-based features of the image silhouette sequence.
- Three novel feature extraction techniques are presented: the two of them are based on the generalized Radon Transform, namely the Radial Integration Transform (RIT) and the Circular Integration Transform (CIT), whilst the third descriptor is based on the weighted Krawtchouk moments. The former are utilized in order to provide an analytical representation of the static and dynamic cues of the human body shape using a few coefficients, and the latter, are well known for their discriminating capability and compactness.

- The paper also introduces the use of range data, captured by a stereo camera, for gait signal analysis. Depth related data are assigned to the binary image silhouette sequences using an innovative transform: the 3D Geodesic Silhouette Distribution Transform. This transform is utilized to encode information about the position of the human body segments on the image plain and their 3D distribution on the hull that depicts the visible human body.

The proposed algorithms were tested and evaluated in two main different databases, namely the "Gait Challenge" of USF and in the proprietary gait database of the HUMABIO EU IST project consisting of 75 persons. Extensive experiments have been carried out in these databases and the proposed methods were found to be robust in existence of noise and with increased recognition accuracy, when compared to other similar state-of-the-art algorithms.

5.3.2 Silhouette Extraction and Pre-processing Steps

5.3.2.1 Background Estimation and Binary Silhouette Extraction

The first step in a human gait recognition system that uses appearance-based techniques is the extraction of the walking subject's silhouette from the input color image sequence. In this paper, a sequential number of steps are introduced in order to provide the final binary silhouettes, as illustrated in Fig. 5.1.

Initially, the background is estimated using a temporal median filter on the image sequence, assuming that the background is static and the foreground is moving. In the next step, the silhouettes, denoted as B_k^{Sil}, are extracted by comparing each frame of the sequence with the background. The areas where the difference of their intensity from the background image is larger than a predefined threshold are considered as silhouette areas. The silhouette images that are extracted by this method are noisy. Therefore morphological filtering, based on anti-extensive connected operators (Salembier and Marqués 1999) is applied in order to denoise the binary silhouettes. Finally, potential shadows are removed by analyzing the sequence in the HSV color space (Cucchiara et al. 2001), as illustrated in Fig. 5.1d.

5.3.2.2 Silhouette Enhancement Using Range Data

A novel technique is introduced that exploits range data, if they are available. At this stage, each gait sequence is composed of k preprocessed binary silhouettes $\tilde{B}_k^{Sil}$. Initially, the triangulated version of the 3D silhouette that also includes depth information is generated. Then, using Delaunay triangulation on the available range data, the 3D hull (Moustakas et al. 2007), for each image of the gait sequence is estimated and finally using the Geodesic Transform (Ioannidis et al. 2007) the 3D Geodesic silhouette is extracted, denoted as $\hat{G}_k^{Sil}(x, y)$. Figure 5.2 depicts extracted (a) binary silhouettes and (b) their corresponding 3D geodesic distributed silhouettes.

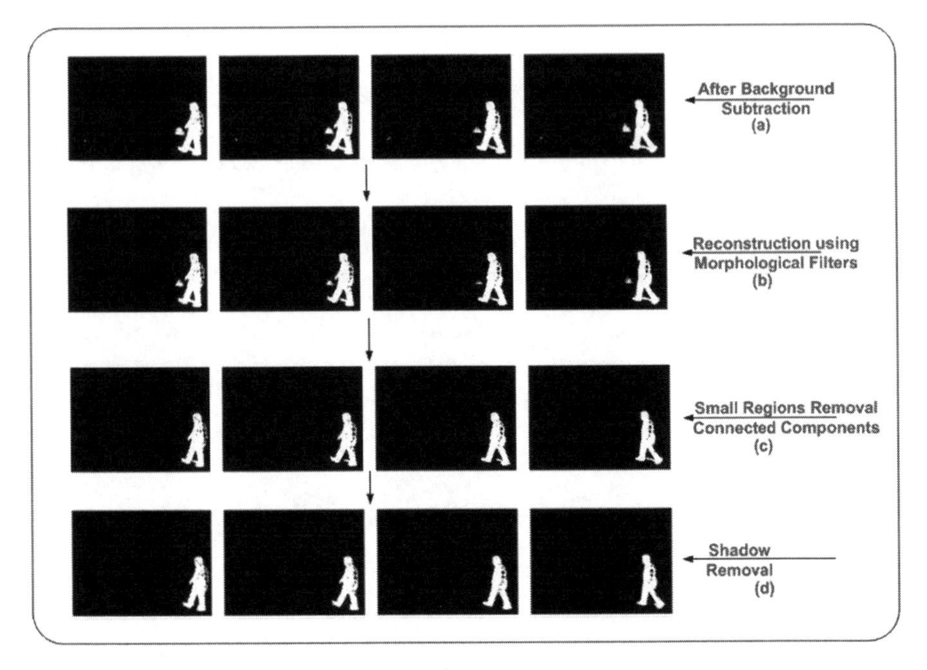

Fig. 5.1 Denoising the initial silhouette images (**a**) using morphological filters (**b**), connected component labelling (**c**) and shadow suppression in terms of HSV colour space (**d**)

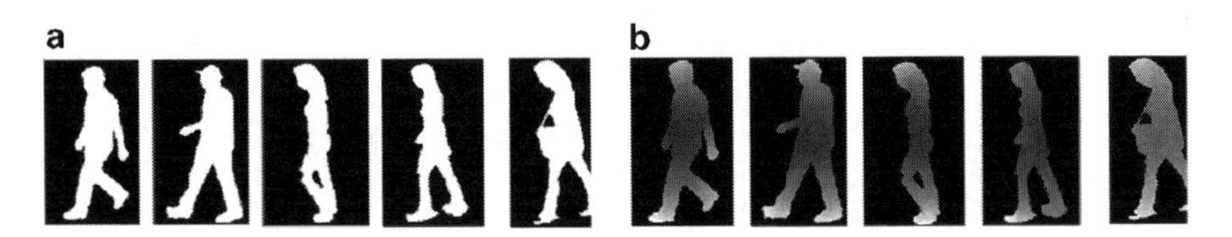

Fig. 5.2 Illustration of silhouette representation used by the proposed system, (**a**) binary silhouette, (**b**) 3D geodesic distributed silhouette

In the final step of the preprocessing stage, the denoised binary ($\tilde{B}^{Sil}$) or 3D silhouette sequence ($\hat{G}^{Sil}$) are scaled and aligned to the center of the frame in each frame (Sarkar et al. 2005).

5.3.3 Feature Extraction Phase

In this paper, three appearance-based techniques are employed in order to extract the most discriminative characteristics of the human locomotion. In all cases, the input to this phase of the gait system is assumed to be either the binary silhouettes ($\tilde{B}_k^{Sil}$) or the 3D-Geodesic image sequence ($\hat{G}_k^{Sil}$).

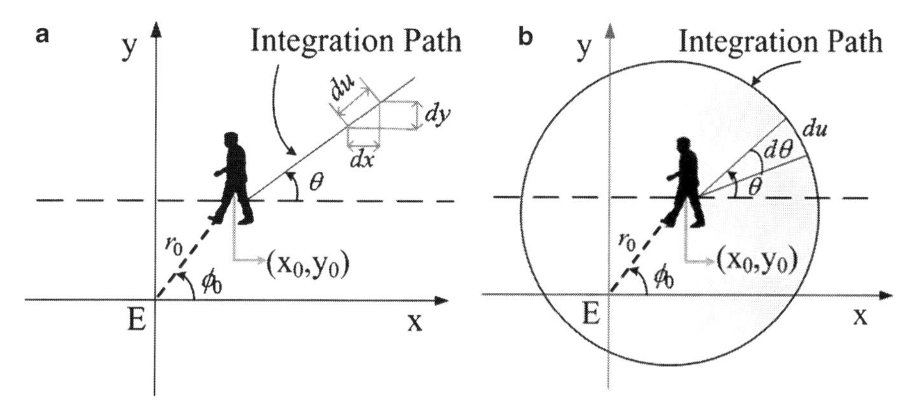

Fig. 5.3 Applying the RIT (**a**) and CIT (**b**) transforms on a silhouette image

5.3.3.1 Generalized Radon Transformations

Two one-dimensional Radon transformations are introduced for feature extraction, namely the Radial Integration Transform (RIT) and the Circular Integration Transform (CIT), which are proven to provide a full analytical representation of the human silhouette (Daras et al. 2006). In particular, the RIT of a function f(x,y) is defined as the integral of f(x,y) in the direction of a straight line starting from the point (x_0, y_o) and forming angle θ with the horizontal axis x (Fig. 5.3). In the proposed approach, the discrete form (Daras et al. 2006) of the RIT transform is used:

$$RIT(t\Delta\theta) = \frac{1}{J}\sum_{j=1}^{J} Sil(x_0 + j\Delta u\cdot\cos(t\Delta\theta), y_0 + j\Delta u\cdot\sin(t\Delta\theta)) \tag{5.1}$$

where $t = 1,\ldots,T$, Δu and $\Delta\theta$ are the constant step sizes of the distance (u) and angle (θ), J is the number of silhouette pixels that coincides with the line that has orientation θ and are positioned between the center of the silhouette and the end of the silhouette in that direction, Sil represents the correspondent binary or 3D silhouette image, and finally $T = 360^o / \Delta\theta$.

In a similar manner, CIT is defined as the integral of a function f(x,y) along a circle curve h(ρ) with center (x_0, y_0) and radius ρ and its discrete form that is utilized by the proposed gait system is given by:

$$CIT(k\Delta\rho) = \frac{1}{\mathrm{T}}\sum_{t=1}^{T} Sil(x_0 + k\Delta\rho\cdot\cos(t\Delta\theta), y_0 + k\Delta\rho\cdot\sin(t\Delta\theta) \tag{5.2}$$

where k = 1,…,K, $\Delta\rho$ and $\Delta\theta$ are the constant step sizes of the radius and angle variables, K $\Delta\rho$ is the radius of the smallest circle that encloses the binary or 3D silhouette image Sil, and finally $T = 360^o / \Delta\theta$.

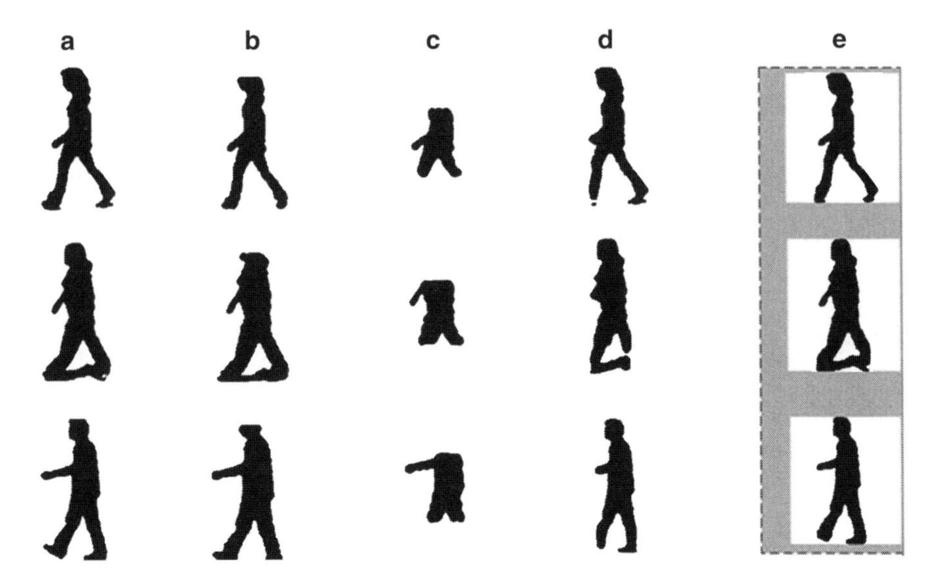

Fig. 5.4 Reconstruction of silhouette images using Krawtchouk moments for different moment order values (N, M), (**a**) Original Silhouette (W × H = 188 × 200), (**b**) N = W/10, M = H/4, (**c**) N = W/10, M = H/16, (**d**) N = W/30, M = H/2 and (**e**) N = W/15, M = H/3

5.3.3.2 Orthogonal Discrete Transform Using Krawtchouk Moments

In this work, a new set of orthogonal moments is proposed based on the discrete classical weighted Krawtchouk polynomials (Mademlis et al. 2006). These moments are proposed due their capability to extract local shape characteristics of images and in addition their orthogonality ensures minimal information redundancy. The Krawtchouk moments Qnm of order (n + m) are estimated using the weighted Krawtchouk polynomials for a silhouette image (binary or 3D) with intensity function $Sil(x,y)$ as follows:

$$Q_{nm} = \sum_{x=0}^{N-1}\sum_{y=0}^{M-1} \bar{K}_n(x; p1, N-1) * \bar{K}_m(y; p2, M-1) \cdot Sil(x,y) \tag{5.3}$$

$$\bar{K}_n(x; p, N) = K_n(x; p, N)\sqrt{\frac{w(x; p, N)}{\rho(n; p, N)}} \tag{5.4}$$

where $\bar{K}_n$, $\bar{K}_m$ are the weighted Krawtchouk polynomials, and (N−1) × (M−1) represents the pixel size of the silhouette image. Figure 5.4 shows a graphical representation of the reconstructed silhouette images using different orders of N (for width) and M (for height).

Table 5.1 Comparative verification rates Pv of the proposed optimal weighted features (RCK-G) and the baseline algorithm (Sarkar et al. 2005) at a false rejection rate of 1%, and 10% using z-Norm scores and the binary silhouette transform, due to the lack of range data in the USF database

	ZN (z-Norm)		ZN	
	P_v (%) at P_{FRR} 1%		P_v (%) at P_{FRR} 10%	
	RCK-G	USF	RCK-G	USF
A	**92**	86	**99**	96
B	**88**	76	**97**	90
C	**72**	59	93	**80**
D	**46**	42	78	**70**
E	41	52	**78**	60
F	25	41	**64**	60
G	32	36	**67**	45

5.3.4 *Signature Matching*

The following notations are used in this section: the term gallery is used to refer to the set of reference sequences, whereas the test or unknown sequences to be verified or identified are termed probe sequence. An important step in the recognition system, formally before the matching stage, is gait cycle detection of the gallery/probe sequence. In this paper, the gait cycle is detected using a similar approach to Boulgouris et al. (2004), whereas the signature matching is based on the method described analytically in Ioannidis et al. (2007). Specifically, for each classifier of the proposed system, a distance score is estimated between the probe and the gallery $D_T(\Pr obe, Gallery)$. Finally, the final distance is calculated based on the weighted algorithm (RCK-G) that is presented analytically in Ioannidis et al. (2007).

5.3.5 *Experimental Results and Conclusions*

The proposed methods were evaluated on two different databases: (a) the publicly available HumanID "Gait Challenge" dataset (Sarkar et al. 2005), and (b) the proprietary large indoor HUMABIO dataset (Ioannidis et al. 2007). For evaluation of the proposed approach in a verification scenario, Rate Operating Characteristic curves (ROC) are used that illustrate the probability PV of positively recognizing an authorized person for different values of the false acceptance rate P_{FAR}.

Verification results on USF dataset for RCK-G algorithm are reported in Table 5.1 in comparison with the baseline algorithm (Sarkar et al. 2005).

As seen, the proposed method based on the silhouette sequences and using the weighted classifiers generally outperforms the baseline method. Using the normalized distances-scores, the verification performance is improved, e.g. for a false rejection rate of 10% the verification rate is above 64% for all experiments A-G.

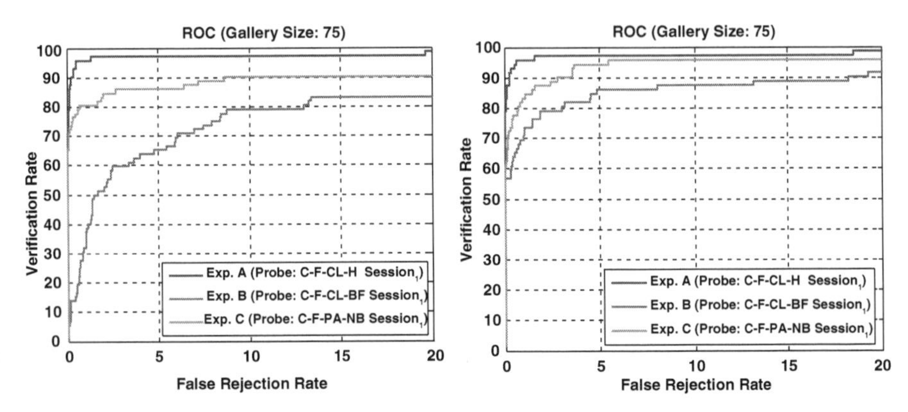

Fig. 5.5 Rate Operating Characteristic (ROC) curves for the HUMABIO gait database using the weighted classifiers (RCK-G) and normalized scores (zNorm) based on the Binary Silhouette Transform (*left*) and the Geodesic Silhouette Distribution Transform (*right*)

Verification results are also reported for the HUMABIO gait database. The ROC using the z-Norm scores of the weighted feature algorithm (RCK-G) is shown in Fig. 5.5. Verification rates are increased when 3D silhouette sequences were used instead of the binary silhouettes. For example, for a false rejection rate of 5% the verification rate is increased by 6% for shoe (experiment C) condition, when Geodesic silhouette distribution transform is used.

In this section, a novel feature-based gait recognition framework was presented that uses the 2.5D information of the captured sequence captured by a stereo camera. This information is initially transformed into a 3D hull and then the 3D protrusion transform is proposed to generate the "geodesic" silhouette. Three novel feature extractor algorithms are combined together using a weighted algorithm in order to extract the static and dynamic cues of the human gait shape either in the binary silhouette or in the geodesic silhouette distribution. Experimental results demonstrate the efficiency of the proposed method when compared to state of the art approaches.

5.4 An Innovative Sensing Seat for Human Authentication

5.4.1 Sensing Seat Technology

The UNIPI module is an unobtrusive and versatile sensing seat system for human authentication that can be employed in different scenarios such as truck and car pilots, airplane pilots, plant and office personnel, and, in general, environments where the security is mandatory and a soft seat is available. It is an anthropometric system based on pressure sensors integrated in seats, in order to enhance the security and reliability of the other biometric system but also increase its applicability to scenarios where the physiological profile of an individual cannot be obtained.

The sensing seat is realized by a seat coated by a removable Lycra sensing cover. The sensing cover is able to respond to simultaneous deformations in different directions by means of a piezoresistive network which consists of a mixtures of polymers deepened with coal directly printed onto the fabric. The strain sensors developed by the University of Pisa are realized by means of Conductive Elastomers (CE) composites (Lorussi et al. 2004, 2005). CE composites show piezoresistive properties when a deformation is applied and can be easily integrated into fabric or other flexible substrate to be employed as strain sensors. The used CE is based on a WACKER Ltd (Elastosil LR 3162 A/B) product (The Wacker Company 2008). It consists in a mixture of graphite and silicon rubber. WACKER Ltd guarantees the non-toxicity of the product that, after the vulcanization, can be employed in medical and pharmaceutical applications. It can be smeared on flexible and elastic substrate or arranged in films applicable on elastic supports. Sensors were realized by directly smearing the CE on a Lycra®-cotton fabric previously covered by an adhesive mask. The mask is designed according to the shape and the dimension desired for the sensors. The production phase is structured as follows: after depositing the material, the mask is removed and the treated fabric is placed in an oven at about 130°C. During this phase the cross-linking of the solution speeds up and, in about 10 min, the sensing fabric is ready to be employed. It is important to underline that the integration of CE materials does not change the mechanical properties of the underlying texture, thus maintaining a good comfort for the user. For a correct determination of sensor positions and orientations, experimental trials were performed in order to validate the proposed design. To obtain a specific sensor topology over the fabric, an adhesive mask representing the drawing of sensors and connections was realized. The mask is designed starting from the desired sensor positions and orientations traced on a three-dimensional model of the human body. Then, the mask is realized by cutting a sheet of adhesive paper with a laser milling machine.

5.4.1.1 Static and Dynamic Characterization of Conductive Elastomeric Sensor

The piezoresistive properties of CE composites, in literature, are statically described by using percolation theory. In our work, the attention has been focused on the behavior of the electrical resistance of CE during the transient time and other non-linear phenomena occurring after a deformation. In our application, CE has been integrated into fabric and employed as strain sensors. The main objective of the CE characterization has been to determine the relationship between the electric resistance R(t) of a CE sample and its actual length L(t), where t is the time.

For instance, in terms of quasi-static characterization (Lorussi et al. 2005), a CE sample of 5 cm in length and 1.7 cm in width presents an non-stretched electrical resistance of about 3KOhm per cm, and its gauge factor (GF) is about 2.9. In order to obtain the gauge factor, it is necessary to construct the static calibration curve shown in Fig. 5.6.

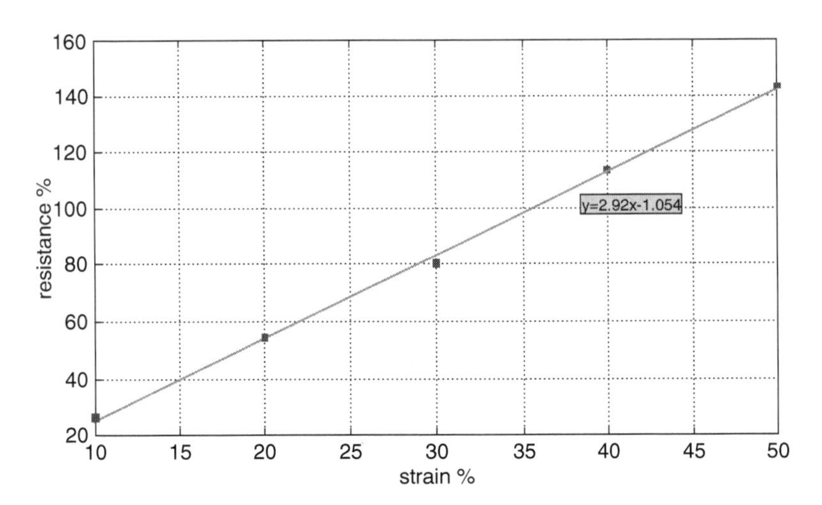

Fig. 5.6 The static calibration curve

The curve is realized by stretching the CE sample with the step in deformation having different strain (Fig. 5.7) and at the same time acquiring the final resistance value (Scilingo et al. 2003). The GF represents the angular coefficient of the static calibration curve.

Sensor response shows a peak in correspondence to every mechanical transition. Sensor responses during constant pressure time intervals may be approximated by decreasing exponential, assuming the local minimum as the steady-state value. The longer the pressure time interval, the more the above mentioned approximation is accurate. In order to remove the contribution of high order exponential, the first order time constants were extracted by means of a window filter. This choice allowed quantization errors introduced by the acquisition device in response to rapid transitions to be avoided and sensor steady state deformation, related to slower frequency components, to be maintained. Taking into account the first-order components of the sensor response (resistance variation) to a rectangular stimulation (applied deformation), the equivalent circuit represented in Fig. 5.8 can be derived.

The power supply V is the electrical equivalent of the imposed deformation. The switch T1 (initially open) is closed and opened in correspondence to the beginning and the end of the imposed deformation respectively. The switch T2 (initially open) is closed when T1 is opened again. Following a simple analysis of this circuit, it is easy to recognize that the variation of the charging and discharging currents of the capacitance in consecutive phases of stimulation are equivalent to the variation of the resistance of the sensor during its deformation and the following release respectively. The circuit parameters R1, R2, R3 and C can be derived by using the features, extracted from reference experimental signals, listed in Table 5.1. The features values listed above were extracted from ten cycles of a reference experimental signal and were used to derive the circuit parameters.

According to these ten cycles of stimulation, the solution of this equivalent circuit provided the results reported in Fig. 5.9.

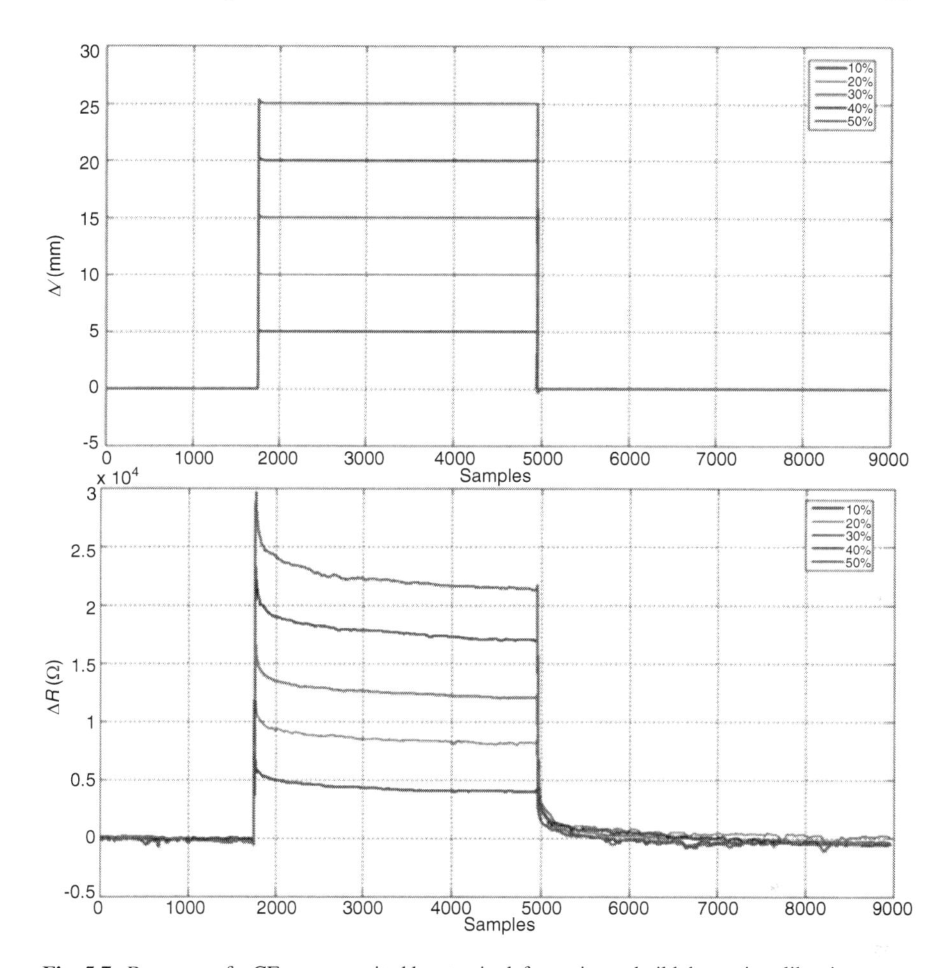

Fig. 5.7 Response of a CE sensor excited by step in deformation to build the static calibration curve

Figure 5.10, which has been reported as an example of this analysis, shows the output of a sample stretched with trapezoidal ramps in deformation having different velocities $\dot{L}(t)$ (where $L(t)$ is the length of the sample).

The main remarks on sensor behavior are summarized in the following:

- Both in the case of deformations which increase or decrease very quickly the length of the specimen and in the case of deformations which reduce it, two local maxima greater than both the starting and the regime value are shown.
- If the relationship between $R(t)$ and $L(t)$ were linear, one of the extreme described in the previous point would be a minimum.
- The height of the overshoot peaks increases with the strength velocity $\dot{L}(t)$.
- The relaxing transient time, which lasts up to several minutes, is too long to suitably code the human movements.

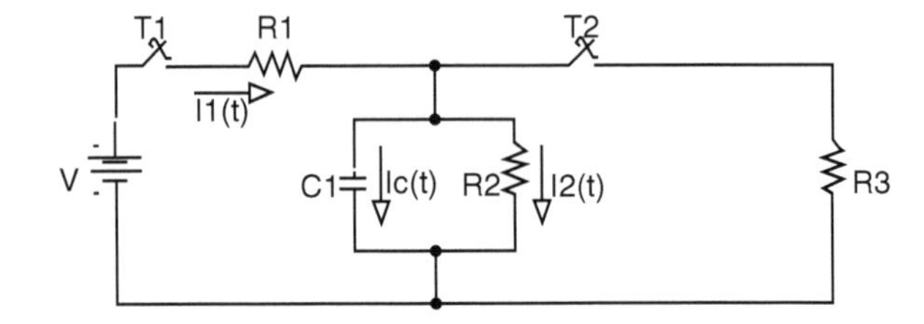

Fig. 5.8 Electric model of a conductive elastomeric strain sensor

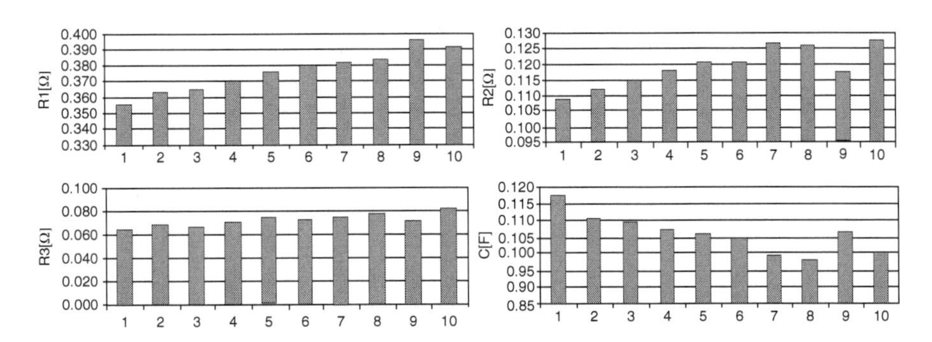

Fig. 5.9 Values of the parameters of the equivalent electric model extracted from ten cycles of a reference experimental signal

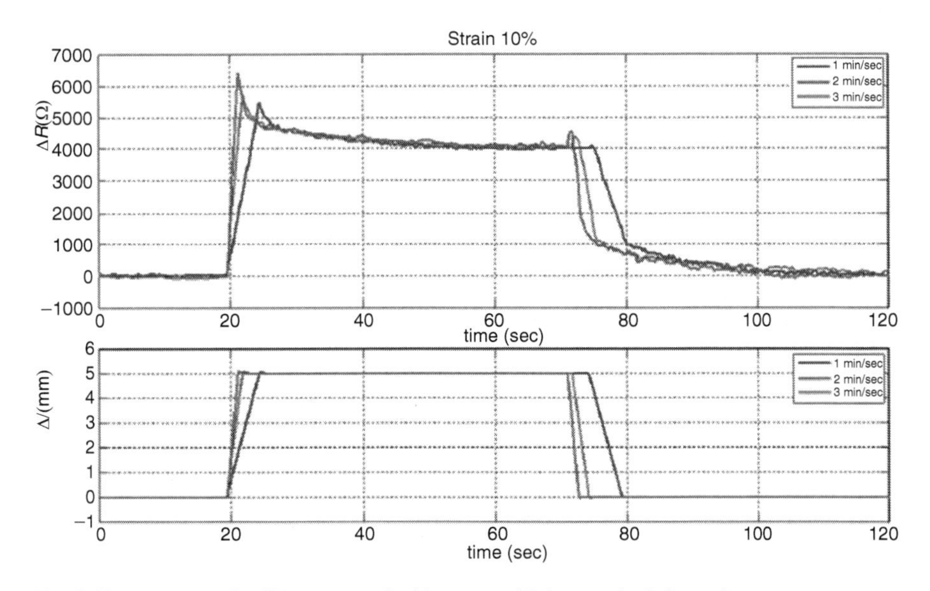

Fig. 5.10 Response of a CE sensor excited by trapezoidal ramps in deformation

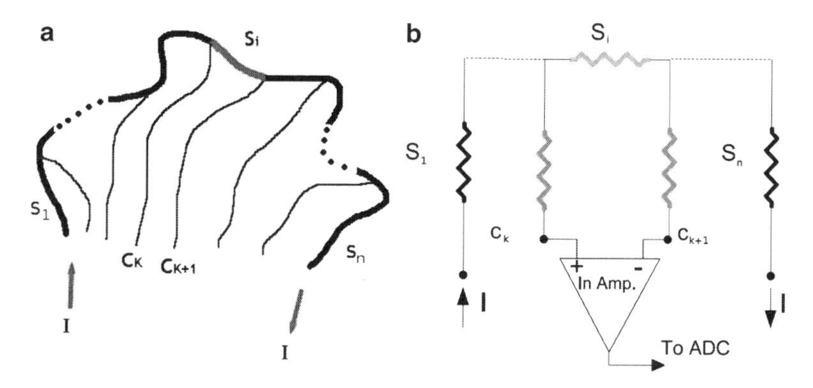

Fig. 5.11 Sensing cover electrical model (**a**) and electronic acquisition front-end (**b**)

5.4.1.2 Electrical Model and Acquisition Electronic Design

The sensors and their connections to the acquisition unit was realized by means of the same materials, in this way no metallic cables are needed on the seat. This is an advantage in terms of user comfort as we maintain the fabric original elasticity. Moreover, by using this approach, the electrical contacts on the CE material can be placed in areas where the fabric deformations and stresses are reduced. Figure 5.11a, b represent the generic configuration of a sensing cover and the electronics acquisition front-end respectively. The sensors are connected in series thus forming a single sensor line (larger line of Fig. 5.11a) while the connections (represented by the thin lines of Fig. 5.11b) intersect the sensor line in the appropriate points.

Since connections and sensors are made by the same material, both of them change their electrical resistance when the user moves. For this reason, the front-end of the acquisition unit had to be designed in order to compensate the connection resistance variations. To obtain this result, the sensor line is supplied with a constant current I and the voltage falling across two consecutive connections are acquired using high input impedance amplifiers (i.e., instrumentation amplifiers), as shown in Fig. 5.11b. Considering the example of sensor S_i, if the amplifier is connected between C_k and C_{k+1}, only a little amount of current flows through the connection lines compared to the current that flows through the sensor line (and in the sensor S_i). In this way, if the current I is well dimensioned, the voltage read by the amplifier is almost equal to the voltage fall across the sensor (that is proportional to the sensor S_i electrical resistance). Taking into account the above described strategy, the analog front-end of the electronic unit included a number of instrumentation amplifiers equal to the sensors number. The data coming from the front-end, are low pass filtered, digitalized and acquired in a PC by means of a general purpose card or transmitted by a dedicated electronic interface. Moreover, the power consumption is near zero resulting in a completely safe system. The fabric equipped with distributed and redundant unobtrusive strain sensors guarantees to address plasticity, low dimension, lightness, and low cost. Since the strain sensors can be directly printed on the

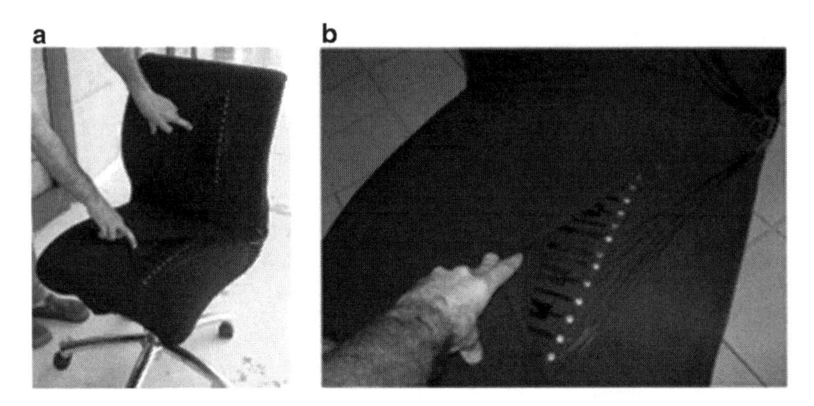

Fig. 5.12 The Sensing Seat system prototype: (**a**) The office seat equipped with the sensing cover; (**b**) details of the sensor connections on the bottom side of the sensing cover

fabric, specific cover layouts may be designed to coat different seat shapes obtaining a good adherence to the seat. As a result, the sensing seat system does not interfere with the mechanical structure of the seat and it is designed as an extension of the seat itself (Fig. 5.12).

5.4.2 Sensing Seat Experimental Results

5.4.2.1 Recording Protocol and Data Analysis

Several topology layouts were taken into account (series, parallel and quadrupole network of sensors) and finally the best compromise between the technical complexity and the classification performance of the system was found using the series network. The sensing cover prototype is equipped with 32 strain sensors: 16 sensors in the bottom side and 16 in the upper side. The existing sensing cover prototype was tailored to a real office seat. Different layouts could be developed to handle different types of seats (e.g. office seats, car seats). It should also be remarked that the seat must be soft enough to guarantee the sensors to be adequately stretched as a human subject is seated. Moreover, as it will be explained below, since the signals supplied by the sensors depend on the positioning and the initial stretching of the cover, data are consistent only after the cover is mounted over the specific seat. In fact, the enrolment signatures are not valid if the cover is dismounted and mounted again even on the same seat. In a normal scenario, once the sensing cover has been mounted, it should not be removed. However, in order to overcome this inconvenience, UNIPI studied some solutions. The best way is a kind of automatic recalibration. When the sensing cover is removed and mounted again, unavoidably the initial deformation of the cover is different, and consequently the electrical signals too. The recalibration phase needs only two steps: the signals in no-seated and in full-seated condition. With these information, using an interpolation function, it is possible to map the status of the sensors from the previous configuration into the current one. In this way, only using a scale factor, it is possible to adapt the previous signature and no new enrolment phase is necessary. Some methods to extract the features were tested, and the one, among the others, having the best performance in term of classification, was the method using as features the minimum, maximum and mean values, and this one was implemented. In this way during an action the 32 signals from the sensorized cover are acquired and stored, and the start and stop action time are taken in account. The initial and final 5% of the entire time period are not considered for the analysis in order to bypass the transitory signal phenomena. The remaining part of signal is elaborated in order to extract the minimum value, the maximum value and the mean value. This process is done over all the 32 signals. At this point, this matrix of values recorded during the enrolment phase forms the biometric signature of a particular subject performing a particular action, and all the signatures are stored into a database. This process is done over all the subjects and over all the actions. During the authentication phase, the features are extracted with the same strategy and then they are compared with the stored signature using a classifier. Some classifiers were tested in order to evaluate the recognition phase. On the base of the results, the classifier based on the Euclidean vector distance was chosen. No specific protocol is required to extract the features, since the recording protocol is very adaptable. Only the subject is asked to start from full-seated position and to return in the same position after to have performed the action. Since these elements are linked to the deformation of the sensors due to the subject pressure, each signature represents the deformation of the seat. As a result, the voltage vectors available for each measurement and for each predefined position are related to the pressure exerted by the subject on the seat (Ford Global Technologies Inc. 2001; Hilliard 2002; Federspiel 2004).

The ACTIBIO recordings gave the opportunity to have real data to be used to evaluate the system authentication performances. Relevant data for the SensingSeat system were collected in a fixed seat office scenario, giving to the users the opportunity to act according to a specified protocol. The actions have been chosen in order to simulate in the best way a real office scenario. The main activities are answer to phone, typing, using the mouse, writing with a pencil, taking a glass. The list of the actions for the events triggering are shown below (Table 5.2).

As previously explained, the data have been analysed with some algorithms in the way to test and chose the that one with best performance. The classifier VDC, based on Euclidean vector distance, was trained on 80% of the available examples for each action and subject, while the remaining 20% was used as the test set. The results in term of FAR and FRR for each action and for each subject are shown in the figures below. The data were analyzed in order to classify each subject, with respect to the all the others, performing each action (Figs. 5.13 and 5.14).

The results show a mean FAR=0.9% +/−3.1% and a mean FRR equal to 0.1% +/−0.4%. Even if the results are encouraging, more tests must be conducted since the number of repetitions per action and subject is currently too small to assess the system reliability. The EER is calculated for each specific activities; in the following

Table 5.2 The actions used for the event-related authentication analysis

Action ID	Action description
0	Phone conversation
1	Phone conversation (light)
2	Phone reached the ear
3	Phone left from ear
4	Interacting with mouse
5	Interacting with mouse (light)
6	Write typing
7	Write typing (light)
8	Writing with pencil
9	Writing with pencil (light)
10	Talking to the microphone panel
11	Talking the microphone panel (light)
12	Pressing buttons in the office panel
13	Pressing buttons in the office panel (light)
14	Drinking from glass
15	Drinking from glass (light)
16	Filling glass with water
17	Yawning
18	Raising hands
19	User is seated
20	Watching video
21	No activity

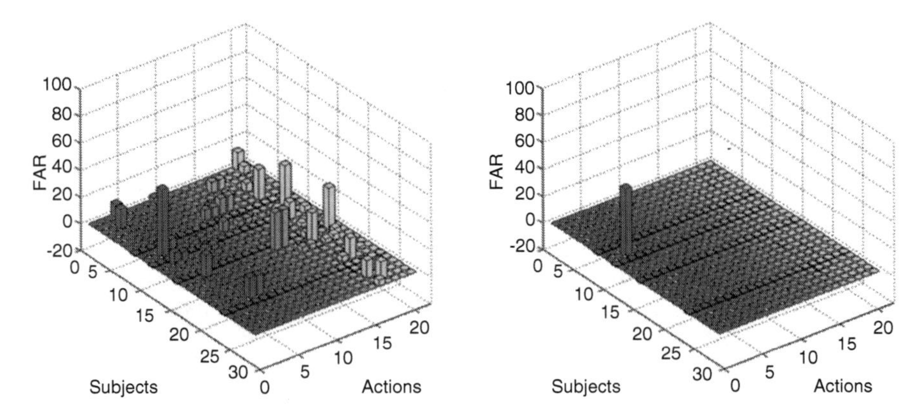

Fig. 5.13 SensingSeat authentication error: FAR and FRR (24 subjects, 21 actions, 6 repetitions per subject per action)

figures, for simplicity, the "Phone reached the ear" and "Phone conversation" actions are shown, considering for instance the subject 13. The EER is calculated for the considered subject performing the specific action with respect to all the other subjects. This parameter permits to see how the system is able to recognize a subject during a specific event, and to see how it is able to distinguish the considered subject respect to all the others (Fig. 5.15).

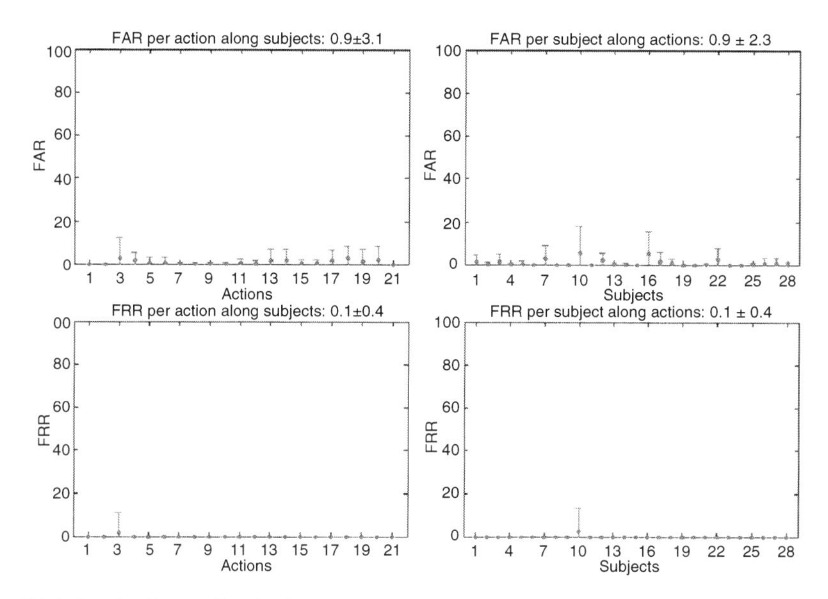

Fig. 5.14 SensingSeat authentication error: mean and standard deviation of FAR and FRR along subjects (*left*) and actions (*right*); 24 subjects, 21 actions, 6 repetitions per subject per action

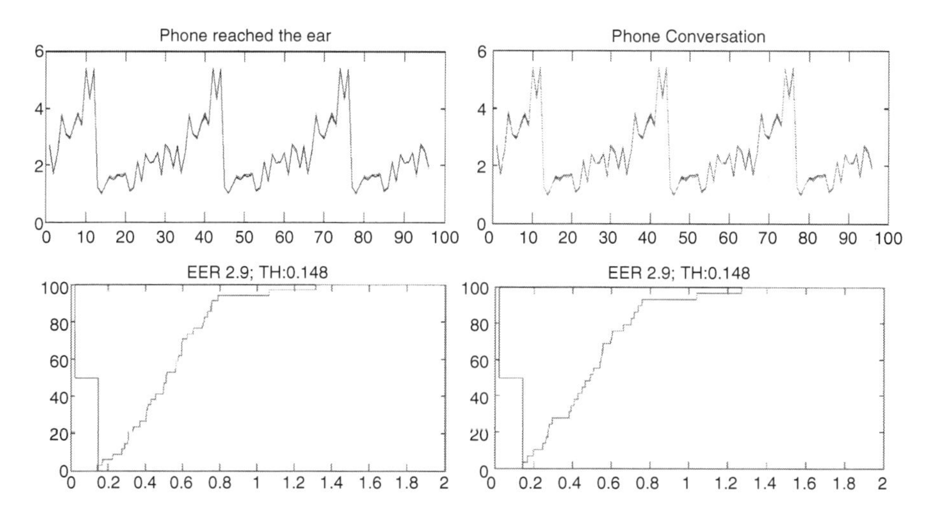

Fig. 5.15 Subject ID 13: comparison of two actions: "Phone reached the ear" (*left*) and "Phone conversation" (*right*); the features are shown at the *top*; the FRR and FAR vs. threshold curve is shown at the *bottom*

Considering the global performance of the system, the best useful parameter is the EER calculated for each action considering all the subject. In the following table the results are reported (Table 5.3).

Table 5.3 EER calculated for every actions considering all the subjects

Action ID	Action description	EER per action
0	Phone conversation	1.268
1	Phone conversation (light)	1.087
2	Phone reached the ear	2.536
3	Phone left from ear	5.435
4	Interacting with mouse	0.867
5	Interacting with mouse (light)	0.945
6	Write typing	1.087
7	Write typing (light)	1.087
8	Writing with pencil	2.355
9	Writing with pencil (light)	0.925
10	Talking to the microphone panel	4.167
11	Talking the microphone panel (light)	4.167
12	Pressing buttons in the office panel	1.345
13	Pressing buttons in the office panel (light)	1.449
14	Drinking from glass	4.167
15	Drinking from glass (light)	4.167
16	Filling glass with water	2.213
17	Yawning	3.456
18	Raising hands	8.333
19	User is seated	4.456
20	Watching video	6.703
21	No activity	3.442

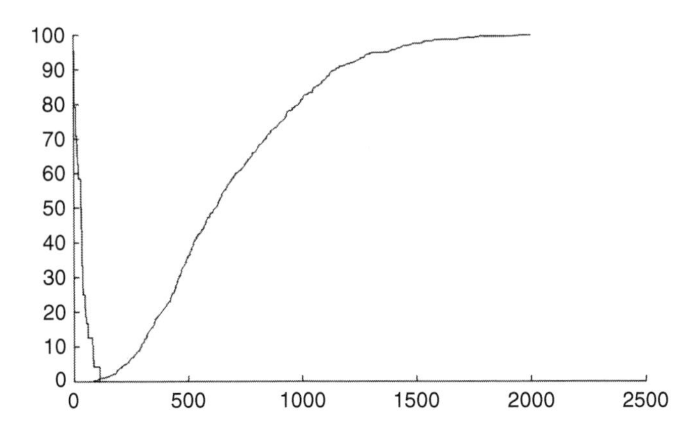

Fig. 5.16 EER calculated for "Interacting with the mouse" considering all the subjects. The value of EER is 0.867

With respect to the previous summarizing table, it is possible to show the comparison between the actions with the best and the worst score, respectively "Interacting with the mouse" and "Raising hand" (Figs. 5.16 and 5.17).

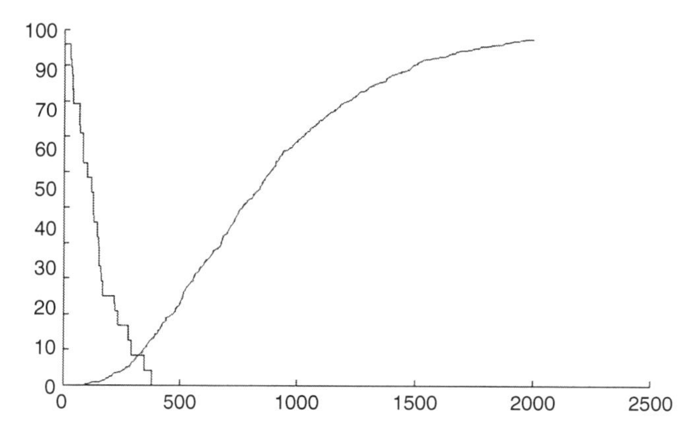

Fig. 5.17 EER calculated for "Raising hand" considering all the subjects. The value of EER is 8.333

5.4.2.2 Conclusions

In this section the development of a novel sensing seat system based on an unobtrusive piezoresistive sensor array is described. The main result is a positive assessment of the use of the reported sensing seat in the authentication task, showing the robustness of the system in terms of biometric rates. Another relevant result is the assessment of the strain sensor technology and of the classification modules based on personal and event-related classifiers. The proposed system is still under development even if the actual prototype was successfully tested within unimodal and multimodal environments. The main advantage in respect of the previous Sensing Seat prototype is represented by the absence of the cooperation of the human subject during the monitoring stage. The open issues include the performance study in extreme environmental conditions (e.g. very low and very high environmental temperature scenarios). Moreover, the strain sensor stability over time as well as its chemical properties must be investigated thoroughly in order to study the sensor degeneration over time (i.e. sensor aging). Additionally, in order to make the system really unobtrusive, the objects inside the clothes and the pockets (e.g. keys, wallet) should be treated as a point of disturbance to increase the final user convenience. All the above mentioned topics will be taken into account in future developments.

5.5 Concluding Remarks

Biometrics measure physical or behavioral characteristics of an individual in order to recognize or authenticate their identity. Usually fingerprints, hand or palm geometry, retina, iris, or facial characteristics are used as physical biometric variables. Signature, voice (which also has a physical component), keystroke pattern and gait are included in behavioral characteristics.

Although some technologies have gained more acceptance than others, it is beyond doubt that the field of access control and biometrics as a whole shows great potential for use in end user segments, such as airports, stadiums, defence installations but also the industry and corporate workplaces where security and privacy are required.

In this chapter, two emerging biometric technologies, namely Gait and Anthropometric profile, have been presented that exploit the human static and dynamic body characteristics for human recognition. Even if these technologies are mature, their high authentication accuracy has been demonstrated and their deployment in existing or new biometric security solutions has been indicated in order to: (i) improve the reliability and accuracy of the multimodal biometric frameworks, (ii) provide new means of unobtrusive subject authentication based on activity-related signals and (iii) to ignite further research on emerging and second generation of biometrics that exploit high recognition performance and take into account user privacy and convenience.

References

Abate, A.F., M. Nappi, D. Riccio, and G. Sabatino. 2007. 2D and 3D face recognition: Survey. *Pattern Recognition Letters* 28(14): 1885–1906.

Bodor, R., A. Drenner, D. Fehr, O. Masoud, and N. Papanikolopoulos. 2009. View-independent human motion classification using image-based reconstruction. *Image and Vision Computing* 27(8): 1194–1206.

Bouchrika, I., and M. Nixon. 2007. Model-based feature extraction for gait analysis and recognition. *Computer vision/computer graphics collaboration techniques*, 150–160.

Bouchrika, I., and M. Nixon. 2008. Gait recognition by dynamic cues. In *19th IEEE International Conference on Pattern Recognition, 2008*, Tampa, FL, USA.

Bouchrika, I., M. Goffredo, J. Carter, and M. Nixon. 2009. Covariate analysis for view-point independent gait recognition. In *The 3rd IAPR/IEEE International Conference on Biometrics*, Italy.

Boulgouris, N.V., Chi, Z.X., 2007 "Gait recognition using radon transform and linear discriminant analysis," *IEEE Transactions on Image Processing* 16(3): 731–740, March 2007.

Boulgouris, N.V., K.N. Plataniotis, and D. Hatzinakos. 2004. Gait recognition using dynamic time warping. In *IEEE 6th Workshop on Multimedia Signal Processing*, 263–266, September 29–October 1, 2004.

Boulgouris, N.V., D. Hatzinakos, and K.N. Plataniotis. 2005. Gait recognition: A challenging signal processing technology for biometric identification. *IEEE Signal Processing Magazine* 22(6): 78–90.

Bowyer, K.W., K. Hollingsworth, and P.J. Flynn. 2008. Image understanding for iris biometrics: A survey. *Computer Vision and Image Understanding* 110(2): 281–307.

Chen, C., J. Liang, H. Zhao, H. Hu, and J. Tian. 2009. Frame difference energy image for gait recognition with incomplete silhouettes. *Pattern Recognition Letters* 30(11): 977–984.

Cucchiara, R., C. Grana, M. Piccardi, A. Prati, and S. Sirotti. 2001. Improving shadow suppression in moving object detection with HSV color information. In *2001 IEEE Intelligent Transportation Systems Proceedings*, 334–339.

Daras, P., D. Zarpalas, D. Tzovaras, and M.G. Strintzis. 2006. Efficient 3-D model search and retrieval using generalized 3-D radon transforms. *IEEE Transactions on Multimedia* 8(1): 101–114.

Federspiel L. Sensor mat for a vehicle seat. I.E.E. International Electronics & Engineering S.a.r.l. Patent, US6794590, Sept. 2004.

Ferro, M., G. Pioggia, A. Tognetti, N. Carbonaro, and D. De Rossi. 2009. A sensing seat for human authentication. *Transactions on Information Forensics and Security* 4(3): 451–459.
Ford Global Technologies Inc. Vehicle air bag deployment dependent on sensing seat and pedal positions. Patent, priorities: [US09681903 Jun. 22, 2001], UKC Headings: G4N Int Cl7 B60R 21/01, B60R 21/16, GB2377536 (GB0212617.5), May 2002.
Gloor, P.A. 1980. Bertillon's method and anthropological research; A new use for old anthropometric files. *Journal of the Forensic Science Society* 20(2): 99–101. ISSN 0015–7368.
Goffredo, M., J.N. Carter, and M.S. Nixon. 2008. Front-view gait recognition. In *IEEE Second International Conference on Biometrics: Theory, Applications and Systems, BTAS.*
Hilliard G.G. Seat cushion. Patent, GB0228513.8, Dec. 2002.
Ioannidis, D., D. Tzovaras, I.G. Damousis, S. Argyropoulos, and K. Moustakas. 2007. Gait recognition using compact feature extraction transforms and depth information. *IEEE Transaction Information Forensics and Security* 2: 623–630.
Jain, A.K., A. Ross, and S. Prabhakar. 2004. An introduction to biometric recognition. *IEEE Transactions on Circuits and Systems for Video Technology* 14(1): 4–20.
Jean, F., Alexandra Branzan Albu, and Robert Bergevin. 2009. Towards view-invariant gait modeling: Computing view-normalized body part trajectories. *Pattern Recognition* 42(11): 2936–2949.
Lorussi, F., W. Rocchia, E.P. Scilingo, A. Tognetti, and D. De Rossi. 2004. Wearable redundant fabric-based sensors arrays for reconstruction of Body segment posture. *IEEE Sensors Journal* 4(6): 807–818.
Lorussi, F., E.P. Scilingo, M. Tesconi, A. Tognetti, and D. De Rossi. 2005. Strain sensing fabric for hand posture and gesture monitoring. *IEEE Transactions on Information Technology in Biomedicine* 9(3): 372–381.
Mademlis, A., A. Axenopoulos, P. Daras, D. Tzovaras, and M.G. Strintzis. 2006. 3D content-based search based on 3D Krawtchouk moments. In *3DPVT 2006*, University of North Carolina, Chapel Hill, NC, USA, June 2006.
Moustakas, K., D. Tzovaras, and M.G. Strintzis. 2007. SQ-Map: Efficient layered collision detection and haptic rendering. *IEEE Transactions on Visualization and Computer Graphics* 13(1): 80–93.
Salembier, P., and F. Marqués. 1999. Region-based representations of image and video: Segmentation tools for multimedia services. *IEEE Transactions on Circuits and Systems for Video Technology* 9(8): 1147–1169.
Sarkar, S., P.J. Phillips, Z. Liu, I.R. Vega, P. Grother, and K.W. Bowyer. 2005. The human ID gait challenge problem: Data sets, performance, and analysis. *IEEE Transaction Pattern Analysis and Machine Intelligence* 27(2): 162–177.
Scilingo, E.P., F. Lorussi, A. Mazzoldi, and D. De Rossi. 2003. Sensing fabrics for wearable kinaesthetic-like systems. *IEEE Sensors Journal* 3(4): 460–467.
Tan, H.Z., L.A. Slivovsky, and A. Pentland. 2001. A sensing chair using pressure distribution sensors. *IEEE/ASME Transactions on Mechatronics* 6(3): 261–268.
The Johnson Controls Company. http://www.johnsoncontrols.com/. Accessed on 17 Mar 2008.
The Softswitch Company. http://www.softswitch.co.uk/. Accessed on 17 Mar 2008.
The Tekscan Company. http://www.tekscan.com. Accessed on 17 Mar 2008.
The Wacker Company. The grades and properties of Elastosil LR liquid silicon rubber. http://www.wacker.com/internet/webcache/de_DE/_Downloads/EL_LR_Eigensch_en.pdf. Accessed on 17 Mar 2008.
Zhang, R., C. Vogler, and D. Metaxas. 2007. Human gait recognition at sagittal plane. *Image and Vision Computing* 25: 321–330.

Chapter 6
Activity and Event Related Biometrics

Anastasios Drosou and Dimitrios Tzovaras

Abbreviations

APF	Annealed particle filter
ATM	Automatic teller machine
CIT	Circular integration radon transform
DNA	Deoxyribo nucleic acid
GMM	Gaussian mixture model
HHMM	Hierarchical hidden Markov models
HMM	Hidden Markov model
MHI	Motion history image
PF	Particle filter
RIT	Radial integration radon transform

6.1 Introduction

Contrary to old fashioned methods, such as ID cards ("what somebody possesses") or passwords ("what somebody remembers"), biometrics render it possible to confirm or establish an individual's identity based on "who somebody is". Depending on the application context, biometrics are nowadays mobilized to give answers to

A. Drosou (✉)
Department of Electrical and Electronic Engineering, Imperial College London,
SW7 2AZ, London, UK
e-mail: a.drosou09@imperial.ac.uk

D. Tzovaras
Informatics and Telematics Institute, 6th km Charilaou-Thermi Road,
P.O. Box 361, Thermi-Thessaloniki 57001, Greece
e-mail: dimitrios.tzovaras@iti.gr

E. Mordini and D. Tzovaras (eds.), *Second Generation Biometrics: The Ethical, Legal and Social Context*, The International Library of Ethics, Law and Technology 11,
DOI 10.1007/978-94-007-3892-8_6,

either the verification (*Is the user Mr. X?*) or the identification problem (*Who is the user?*).

Established biometric systems relate mostly with static, physiological human biometric characteristics. Static biometrics include fingerprint, DNA, face, iris and or retina and hand or palm geometry recognition.

A general shortcoming of these biometric traits is their obtrusive process of obtaining the biometric feature. The subject has to stop, go through a specific measurement procedure, which can be very uncomfortable, wait for a period of time and get clearance after authentication is positive. Moreover, the discrimination to groups of people whose biometrics cannot be recorded well for the creation of the database reference, for example people whose fingerprints do not print well or they even miss the required limb or feature. These people are de facto excluded by the system. In that respect, the research on new biometrics that use features that exist in every human, thus rendering them applicable to the greatest possible percentage of the population has become very important.

As explained above, although some technologies have gained more acceptance than others, it is beyond doubt that the field of access control and biometrics as a whole show great potential for use in end user segments and covering areas such as airports, stadiums, defence installations but also the industry and corporate workplaces where security and privacy are required.

Alternatively to static physiological biometrics, the following sections will focus on activity-related (behavioral) biometric signals. Emerging biometrics can potentially allow the non-stop (on-the-move) authentication or even identification in an unobtrusive and transparent manner to the subject and become part of an ambient intelligence environment.

Specifically, the remaining part of the chapter is structured as follows; In Sect. 6.2 the concept of the activity-related biometrics is introduced. In Sect. 6.3 activities, which have exhibited high authentication potential are presented. The classification of various activities in normal and abnormal ones is elaborated in Sect. 6.4. Two methods for the tracking of anthropometric characteristics during are suggested in Sect. 6.5, while the importance for the extraction of invariant features from the tracked data is stated in Sect. 6.6. Next, in Sect. 6.7 we deal with the authentication potential exhibited by the method used. The ethical issues that arise from the insertion of biometrics in commercial products are thoroughly examined and discussed in Sect. 6.8. The current chapter ends (Sect. 6.9) with a general discussion about further enhancement of biometric technology and their possible use in other areas than security.

6.2 The Concept of Activity Related Biometrics

Typical physiological biometric (Jain et al. 2004b) technologies for recognition like fingerprints, palm geometry, retina and/or iris, and facial characteristics demonstrate among others, a very restricted applicability to controlled environments.

Moreover, static physical characteristics can be digitally duplicated, i.e. the face could be copied using a photograph, a voice print using a voice recording and the fingerprint using various forging methods. In addition, static biometrics could be intolerant of changes in physiology such as daily voice changes or appearance changes.

These drawbacks in the person recognition problem could be overcome with the mobilization of behavioural (activity-related) biometric characteristics (Jain et al. 2004b), using shape based activity signals (gestures, gait, full body and limb motion, etc.) of individuals as a means to recognize or authenticate their identity. Behavioural and physiological dynamic indicators (as a response to specific stimuli) could address these issues and enhance the reliability and robustness of biometric authentication systems when used in conjunction with the usual biometric techniques. The nature of these physiological features allows the continuous authentication of a person (in the controlled environment), thus presenting a greater challenge to the potential attacker. Further they could potentially allow the non-stop (on-the-move) authentication or even identification (Boulgouris and Chi 2007; Kale et al. 2002) which is unobtrusive and transparent to the subject and become part of an ambient intelligence environment.

Previous work on human identification using activity-related signals can be mainly divided in two main categories: (a) sensor-based recognition (Junker et al. 2004) and (b) vision-based recognition. Recently, research trends have been moving towards the second category, due to the obtrusiveness of sensor-based recognition approaches. Additionally, recent work and efforts on human recognition have shown that the human behavior (e.g. extraction of facial dynamics features) and motion (e.g. human body shape dynamics during gait), when considering activity-related signals, provide the potential of continuous authentication for discriminating people (Ioannidis et al. 2007; Boulgouris and Chi 2007; Kale et al. 2002).

Moreover, shape identification using behavioral activity signals has recently started to attract the attention of the research community. Behavioral biometrics are related to specific actions and the way that each person executes them. The most known example of activity-related biometrics is gait recognition (Boulgouris and Chi 2007). Earlier, in Kale et al. (2002) and Bobick and Davis (2001) person recognition has been carried out using shape-based activity signals, while in Bobick and Johnson (2001), a method for human identification using static, activity-specific parameters was presented.

The new concept discussed in the current chapter, suggests analyzing the response of the user to specific stimuli generated by the environment, while the user is performing everyday tasks, i.e., without following a specific protocol. An illuminating example of such an activity is a phone conversation, held in an office environment, while the user is seated at his desk. The expected response of the user to the ringing of the phone (stimuli) is to raise the latter and bring it to his/her ear before starting talking. Similarly, at the end of the discussion, the user is expected to place the earphone back on its base. Both of these activities are considered to contain biometric information, since each human is used to answer the phone in a different behavioral way (e.g. using different hands, moving the head towards the phone, etc.).

Although state-of-the-art solutions show inferior performance compared to static biometrics (fingerprints, iris, etc.), this drawback could be eliminated by the inferential integration of different independent activities and different modalities. There are plenty, event triggered, activities in everyday life (Sect. 6.3), which can be very revelational about our behavioral biometric characteristics. Generally speaking, any activity could be revealing about our behavioural biometry and these information can be useful if we only manage to detect the starting and ending of the certain activity (Sect. 6.4).

Further, the robustness and the performance of a solely behavioural biometric system can be significantly improved by the fusion of soft biometrics recognition capacity in conjunction with the one of conventional, physiological biometrics like fingerprints, iris, etc. Specifically, soft biometrics, such as gender, height, age, weight etc., are believed to be able to significantly improve the performance of a biometric system in conjunction with conventional static biometrics (Jain et al. 2004a), yet their exploitation remains an open issue. Microphones for voice recognition, sound based sensors for monitoring activities or other modalities could be considered as well.

The novelty of such an approach lies in the fact that the measurements that will be used for authentication will correspond to the response of the person to specific events being however, fully unobtrusive, comfortable and also fully integrated in an Ambient Intelligence infrastructure. The objective of our study is the description of a biometric system, able to authenticate a subject without the latter noticing any sensor and without having to wait or stand in any pre-defined posture. This is exactly the concept called 'on-the-move biometry' (Matey et al. 2006).

6.3 Activities with Increased Authentication Potential

As analyzed above, the limitations from which conventional biometric resources suffer (cooperation, obtrusion), can be overcome with the implication of a modern biometrics, the activity related ones, in security systems. That way a continuous authentication would be possible, without being the person being disturbed.

This section of the chapter deals with the potential of using the characteristics of human movement during daily activities at work, such as reaching, grasping and using objects, as a signature for authenticating an individual. Even though a motion looks similar for human eyes, form and content of the captured data may be greatly different. Hence, the goal is to find features, whose inter-individual differences are greater enough than their intra-individual differences, in order to extract a characteristic signature. Focus will be given on work-related activities which are taking place in a daily and ritual basis, so that the requirement of higher safety in work environment to be met.

The European funded research project ACTIBIO (Unobtrusive Authentication Using **ACTI**vity Related and Soft **BIO**metrics – FP7) deals exactly with this problem.

Specifically, ACTIBIO aims to present new biometrics based on activity during work. That way a continuous authentication is possible, without being the person disturbed. According to ACTIBIO documentation, regarding activity related biometrics, no other body motions than gait have been reported to be utilized as biometric for authentication purposes.

Activities, taking place in three different pilot scenarios that have been implemented within the framework of ACTIBIO and that simulate typical work environments, will be discussed below with reference to their authentication potential. These scenario sets are following:

- Activities for driver pilot:
  It involves a driver inside a commercial vehicle. The purpose is to authenticate the driver while seated at the wheel of a car. The applications can be seen in security as well as safety areas: security from preventing unauthorized people to drive a certain car and car hijacking can be avoided by continuously authenticating the driver. Safety issues are rather related to monitoring the physical state of the driver and are thus not explained in detail here.
- Activities for office pilot (Fixed Seat User):
  The purpose of this scenario is the unobtrusive authentication (continuous or not) of the subject and his/her permission to access high-level security resources (e.g. data access) in a controlled environment.
- Fixed work place pilot:
  The third installation of the behavioural subsystem is located in an office environment. Individuals will be able to move freely in the area of their workspace. Its purpose is to protect resources from unauthorised access. Therefore, the identity of the authorised employees will be authenticated (continuously or not), while they are performing their daily (simple or complex) activities, as listed in (Table 6.1).

It must be pointed out that the authentication potential rates are not compared to biometrics such as fingerprint, since this way most of them would be characterized by a low authentication potential. On the opposite, the rating has been normalized within the behavioral biometric authentication potential exhibited by all activities tested. A modality is rated with "no", when it has no potential to provide any useful features for authentication for the specific activity. Furthermore, soft biometrics will be utilized rather for classification and detection and not for authentication. Thus, in case of soft biometrics the rating is not referred to its authentication potential, but to their potential to detect certain characteristics, which can be utilized in "culling" identities from the database.

Activities, which can be adequate for authentication purposes in safety environments have been selected on the basis of their frequency, reproducibility, distinctiveness and rituality.

As demonstrated in the tables, there is quite a number of common, everyday activities taking place in work environments, which demonstrate high authentication capacity.

Table 6.1 Table with activities with high authentication potential

Activities for driver pilot	Authentication potential
Pull fasten seat belt	High
Regulate radio/CD player	Potential identifier
Regulate air conditioning	Potential identifier
Check and adjust the Radio	Potential identifier
Put on/off glasses	High
Handle steering wheel	Potential identifier
Handle the gear	Potential identifier
Unfasten seat belt	High
Abnormal activities for driver pilot	Authentication potential
Rapid move of the head and keep it there for a while	High
Activities for office pilot (Fixed Seat User) and activities for fixed work place pilot	**Authentication potential**
Walking in the room	High
Wearing protective clothes, e.g. gloves	Potential identifier
Using keyboard	Potential identifier
Writing (e.g. notes)	High
Phone Conversation	High
Throw away objects in the waste bin	High
Drinking from glass	High
Stretching (head, arms, shoulder, trunk)	Potential identifier
Getting up	Potential identifier
Abnormal activities for office pilot (Fixed Seat User) and abnormal activities for fixed work place pilot	**Authentication potential**
Hands up	High
Pressing buttons too often	Potential identifier
Rapid, explosive action	High
Sudden stand up	Potential identifier

6.4 Normal/Abnormal Activity Recognition/Applications (Office, Workplace, Driver)

Activity recognition is a very crucial issue in activity-related biometric recognition. In order a biometric system to identify or to rank an individual for his behavioral pattern, it has to be fully aware about the activity the latter is performing. So, the main difficulty with human activity recognition lies in the segmentation of the activity of interest in a given image sequence. In order to recognize each action that appears in a given sequence, it is in some sense necessary to identify the portion of the sequence supposedly corresponding to an action. In the cases, where the activity is triggered by an event (e.g. the answering of a phone call follows the ringing of the phone) the segmentation of the sequence of activities to its partial activities is simple. In all other cases, however, the activity has to be detected otherwise.

So, an entity that initiates and completes the action, as well as a set of objects that the action is invoked upon, are required. Based upon the monitoring of the actuator, we can associate different states of him with the activity that is in progress. In this way, some association rules can be established, like high body movement means exercising activity, etc.

The interest in human activity recognition stems from a number of applications that rely on accurate inference of activities that a person is performing. These include, context aware computing to support for cognitively impaired people (Patterson et al. 2004), health monitoring and fitness, and seamless services provisioning based on the location and activity of people. In general, activity recognition methods can be categorized in two main approaches: methods based on various sensors placed on the subject to extract meaningful features and video based methods for human activity detection.

In the recent past, advances in wearable sensing and computing devices have enabled the estimation of a person's activities over extended periods of time. Feature selection modules typically work on high-dimensional, high frequency data coming directly from sensors (such as cameras, microphones, and accelerometers) to identify relatively small numbers of semantically higher-level features, such as objects in images, phonemes in audio streams, and motions in accelerometer data (Bao and Intille 2004). One of the drawbacks of the methods based on multiple sensors and measurements taken all over the body is that they often lead to unwieldy systems with large battery packs. To overcome this problem, a small low-power sensor board that is mounted on a single location on the body was presented in Lester et al. (2005). The relationship between the extracted features and the corresponding activities still remains an open problem. The reasoning may include identifying ongoing activities, detecting anomalies in the execution of activities, and performing actions to help achieve the goal of the activities.

In Zouba et al. (2007), a multimodal approach for the automatic monitoring of everyday activities of elderly people was presented, whereby video analysis from cameras in the apartment, contact sensors on the doors and the windows and pressure sensors on the chairs were combined. Moreover, a real-time video understanding system, which automatically recognizes activities occurring in environments observed through video surveillance cameras, was presented in Fusier et al. (2007). Similarly, a system for recognizing activities in the home setting using a set of small and simple state-change sensors was introduced in Tapia et al. (2004).

A possible answer to the initially stated problem of the segmentation of particular activities in a given image sequence was given by Kawanaka et al. (2006), whereby hierarchical Hidden Markov Models (HHMM) were employed.

A major challenge met also in the ACTIBIO project (see Sect. 6.3) is to determine the types of patterns, events and activities that the system will be trained for and detect during its operation. Computationally challenging is also to recognize the many variations of common activities. A crucial issue is the extraction of the user's usual behavioural patterns, which the system will characterize as "normal". Consequently it will track down behavioural patterns that deviate significantly from the normal patterns and the system will characterize as "abnormal". This will lead to the detection of potential emergencies (e.g. detect falls from body posture patterns).

In connection with the three pilot scenarios presented in Sect. 6.3, vision-based methods for activity recognition can be applied in the following cases:

- Security operator pilot in the company indoor premises (offices/labs/corridors).
- Authentication via activity recognition and control in transactions through "always on" machines.
- Vehicle driver unobtrusive continuous authentication to prevent hijacking.
- Target user groups include:
- Sensitive facilities and infrastructures that require very high security and safety levels.
- Security companies managing control rooms of sensitive infrastructures, critical infrastructures requiring 7/7 protection, control stations of production lines in industries, etc.
- "Always on" networked infrastructures, such as the control points of machines in industrial environments, the ATM points of transactions in banks, public services one-stop shops, etc.
- Vehicle drivers (dangerous goods vehicles, trucks, taxis, even passenger cars).
- Healthcare carriers and elderly population wishing to live independently at home.

For these scenarios, a combination of a statistical and a structural approach should be considered. The statistical approach refers to the description of the object of study in terms of quantitative features extracted from the data (mean, standard deviation, etc.) and classification based on statistical methods such as similarity. The structural approach, on the other hand, deals with the description of the object by a set of primitives identified within the data and their interrelationships (relational graph) and classification by parsing the relational graph by syntactic grammars, one for each group of classification.

An efficient approach could also be the implementation of a hybrid concept as hierarchical (to decomposition of the decision process into several successive stages) or parallel (to reinforcement of the decision at each stage) arrangements. However, the selection of the most appropriate methods is largely constrained by the youth of the area of research, which is the cause of a lack of both a priori knowledge and realistic collection of data.

In any case, the most dominating approach within the framework of ACTIBIO project is the method proposed in by Bobick and Davis in (2001). A Motion History Image (MHI), composed by the cumulation of the differences of successive images from a sequence, forms a static image template. Thus, normalized to the user's body size, prototype **M**otion **H**istory **I**mages (center of Fig. 6.1), are stored in the database. These prototype MHIs are compared to a "live" MHI, which is updated in every frame during the recording of the individual, in terms of their 1D RIT (Radial Integration Radon Transform) and CIT (Circular Integration Radon Transform). An activity is only then recognized, when the Eucledian Distance and the Correlation Factor between these transformed signals exceed the similarity thresholds, set for each activity.

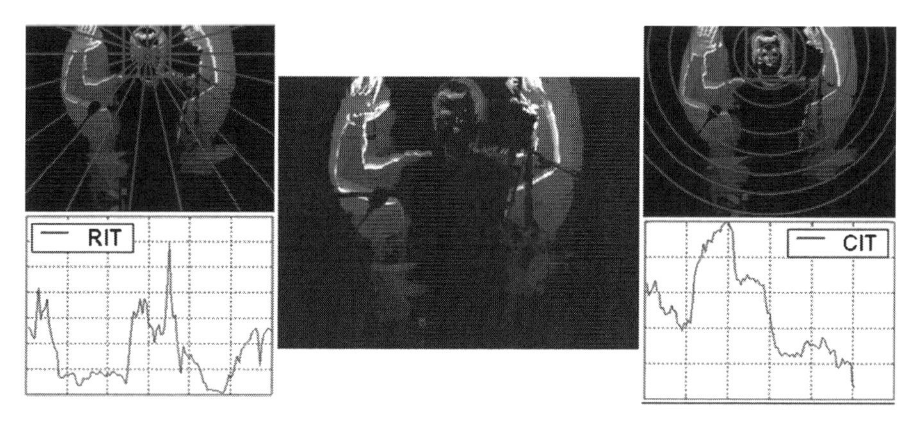

Fig. 6.1 Motion history image from "Raising Hands" activity and its radon transformations (*left*: RIT; *right*: CIT)

6.5 Tracking of Anthropometric Characteristics

The use of biometrics assumes two requirements; that the biometric data can be efficiently and accurately, (a) collected and (b) verified (Wickins 2007). The analysis of the dynamic information of the body motion and the implementation of novel algorithms that would distinguish people based on their gesture and shape characteristics, when performing work-related activities is a requirement for a robust behavioural biometric system. Thus, methodologies that will optimize both the collection and verification phase of activity related biometric identification, will be analyzed in this section. Specifically, focus will be cast on trunk and upper limbs motion during walking, sitting or standing.

The old fashioned way of having sensors attached to the human in order to collect biometric data has been long been outdated, due to obtrusiveness, inflexibility and inefficiency in terms of power supply reasons. For the least obtrusive tracking of anthropometric characteristics, trends are turning into the visual tracking of the body. State of the art algorithms, nowadays, are able to perform full tracking of any part of the body and thus manage to provide us with a big variety of anthropometric characteristics. The successive deployment of the latter over time, can reveal behavioural biometric information of the subject, since it includes not only physical characteristics of him but also the way he handles them.

Two methods for the of the human body:

- Skeleton Models (Sect. 6.5.1),
- Limb Tracking (Sect. 6.5.2).

6.5.1 Human Body Tracking Based on Skeleton Models

Human Body Tracking, also referred by several authors as Pose Estimation, is the inference of the configuration at a precise time instant of the underlying kinematic structure of the human body. Since it implies the characterization of a high-dimensional configuration using multiple-view cues, it is a challenging and often ill-posed problem. However, body tracking has a high number of applications in several fields, such as computer animation, medical tools, smart interfaces, etc. Based on the use of a human body model, Human Body Tracking algorithms can be classified as follows:

- Model-free: These approaches do not use explicit prior body models, thus directly mapping images to poses. There are two main branches within this category. The first one is known as probabilistic assembly of parts and aims at detecting individual human body parts and then joining them to extract the pose information. The second group comprises the example-based methods that automatically learn the mapping from image to pose from a set of training images.
- Model-based: These algorithms take advantage of the usage of a skeletal model that is fleshed out to approximate the human body shape. Within these approaches, analysis-by-synthesis techniques have become the dominant technology. Analysis-by-synthesis framework comprises two basic steps. First, a set of hypothesis is generated according to the rules imposed by the body model. Second, this set is evaluated with some evidence in order to infer the best hypothesis.

The use of prior knowledge about the human body is likely to increase the efficiency of pose estimation. The reason is found when prior knowledge is seen as a constraint on the motion within the boundaries of what is physically possible. Body models make possible the incorporation of such constraints, thus constituting the basis of efficient schemes for body tracking. Consequently, the majority of the best performing algorithms and research activities are developed based on model-based approaches. Among the more promising technologies that take advantage from the knowledge provided by the model, the aforementioned analysis-by-synthesis constitutes one of the most important groups of algorithms in pose estimation and tracking. In this family of strategies we can find Particle Filters (PFs) that have been the seminal idea for many tracking algorithms. One of the main issues concerning PFs is its inefficiency in high-dimensional spaces, such as pose space. To tackle such a problem several methods have been proposed. Annealed Particle Filter (APF) (Deutscher et al. 2000) is considered as one of the most successful among them.

Thus, a possible approach for human body tracking that can be employed in the three scenarios described in Sect. 6.3 is following. An articulated model depicting a virtual human skeleton is the core element of the human modeling framework. This skeleton is parameterized by a set of angles corresponding to the values of all the possible rotations, a 3D reference position with respect the world coordinates and limb sizes. Both the angles and the reference position are time-varying variables that define the state space and the states themselves.

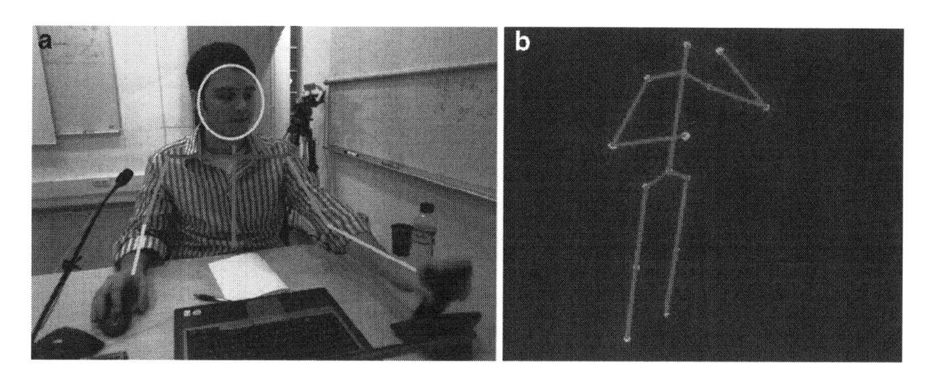

Fig. 6.2 (**a**) Full virtual skeleton; (**b**) skeleton model upper-body tracking

The articulated model points are computed using the kinematic chain framework and the exponential map formulation (Park 1994). To be able to give the human body positions corresponding to the estimated angles and reference positions, a default initial configuration must be precisely defined. To make possible the matching between poses and images, we need to flesh out this virtual skeleton (a). To this end, the head and the limbs are modeled as conical sections and cylinders and the torso is modeled by a cuboid. Planar approximations are used to efficiently project the volume surfaces onto images. Besides, circular regions on the image are defined, where face and hands are expected to be found, according to the projected body model (Fig. 6.2b).

6.5.2 Limb Tracking

The second approach that can extract useful activity-related biometric characteristics from an individual, performing his work in any of the suggested pilot scenarios, focuses on the Limb tracking. Specifically, the detection of the head and hand points throughout the performed activity can achieved with the successive filtering out of non-important regions of the captured scene (Fig. 6.3a) as described in the following paragraphs.

6.5.2.1 Face

First, the tracking of the face is implemented in the current approach by a combination of a face detection algorithm, a skin color classifier and a face tracking algorithm.

The face detection is the main method that confirms the existence of one or more persons in the region under surveillance. The face detection algorithm, which has been implemented in our framework, is based on the use of haar-like feature types and a cascade-architecture boosting algorithm for classification, (i.e. AdaBoost) as described in Viola and Jones (2001) and in Freund and Schapire (1999). In case of

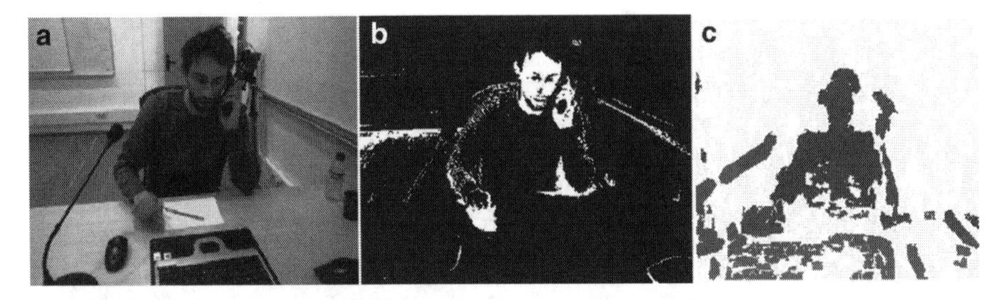

Fig. 6.3 (**a**) Limb upper-body tracking; (**b**) biometric signature from phone-conversation activity

multiple face detection, the front-most face (blob) is retained, utilizing the depth information provided by the stereo camera, while all others are discarded. Detected facial features allow the extraction of the location and the size of human face in arbitrary images, while anything else is ignored.

As an enhancement to the algorithm described above, the output of the latter is evaluated according to a skin classifier 6.5.2.2. In case the skin color restrictions are met, the detected face is passed over to an object tracker.

The face tracker, integrated in our work relies on the idea of kernel-based tracking (i.e. the use of the Epanechnikov kernel, as a weighting function), which has been proposed in Ramesh and Meer (2000) and later exploited and developed forward by various researchers.

6.5.2.2 Skin Color Classifier

A first approach towards the locating of the palms is achieved by the detection of all skin colored pixels on each image. Theoretically, the only skin colored pixels that should remain in a regular image in the end are the ones of the face and the ones of the hands of all people in the image (Fig. 6.4). Each skin color modeling method defines the metric for the discrimination between skin pixels and non-skin pixels. This metric measures the distance (in general sense) between each pixel's color and a pre-defined skin tone.

The method incorporated in our work has been based on Gomez and Morales (2002). The decision rules followed, realize skin cluster boundaries of two color-spaces, namely the RGB (Skarbek and Koschan 1994) and the HSV (Poynton 1997), which render a very rapid classifier with high recognition rates.

6.5.2.3 Using Depth to Enhance Upper-Body Signature Extraction

An optimized L1-norm approach is utilized in order to estimate the disparity images captured by the stereo *Bumblebee* camera of Point Gray Inc. (Scharstein and Szeliski 2002). The depth images acquired (Fig. 6.4c) are gray-scale images, whereas the furthest objects are marked with darker colors while the closest ones with brighter colors.

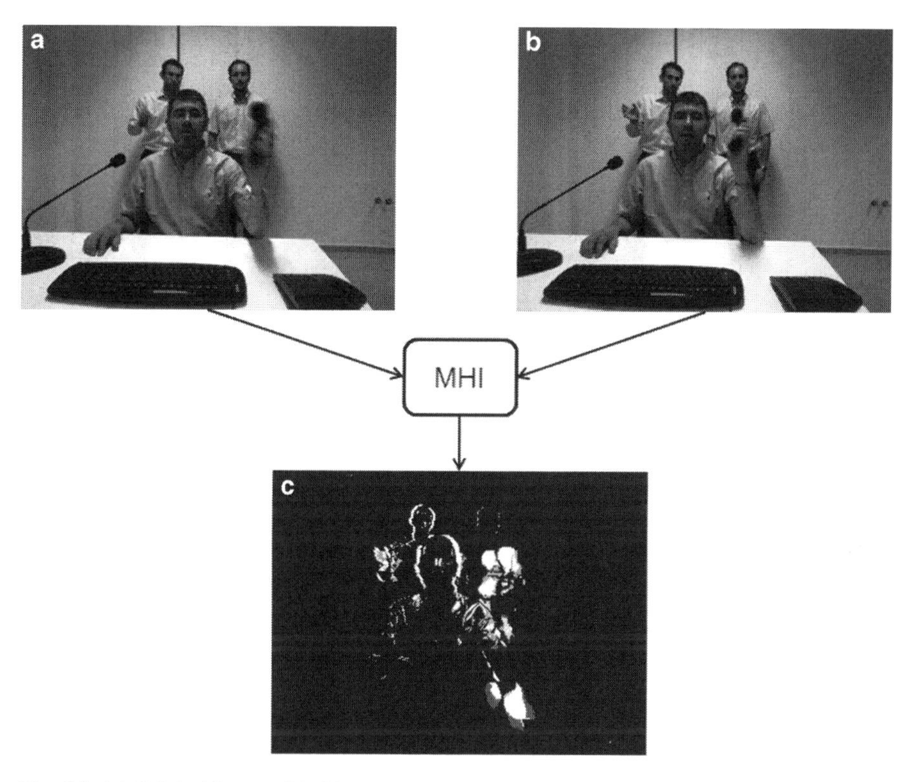

Fig. 6.4 (**a**) Original image; (**b**) skin colored filtered image; (**c**) depth disparity image

Given that the face detection was successful, the depth value of the head can be acquired. Any object with a depth value greater than the one of the head can now be discarded and thus excluded from our observation area. On the other hand, all objects in the foreground, including the user's hands, remain active. A major contribution of this filtering is the exclusion of all other skin colored items in the background, including other persons or surfaces in skin tones (wooden floor, shades, etc.) and noise.

6.5.2.4 Using Motion History Images to Enhance Upper-Body Signature Extraction

Another characteristic, which can be exploited for isolating the hands and the head of a human, is that they are moving parts of the image, in opposition to the furniture, the walls, the seats and all other static objects in the observed area. Thus, a method for extracting the moving parts in the scenery from a sequence of images could be of interest. For this reason, the concept of Motion History Images (MHI) (Bobick and Davis 2001) was mobilized.

The MHI is a static image template, where pixel intensity is a function of the recency of motion in a sequence (recently moving pixels are brighter). Recognition is accomplished in a feature-based statistical framework. Its basic capability is to

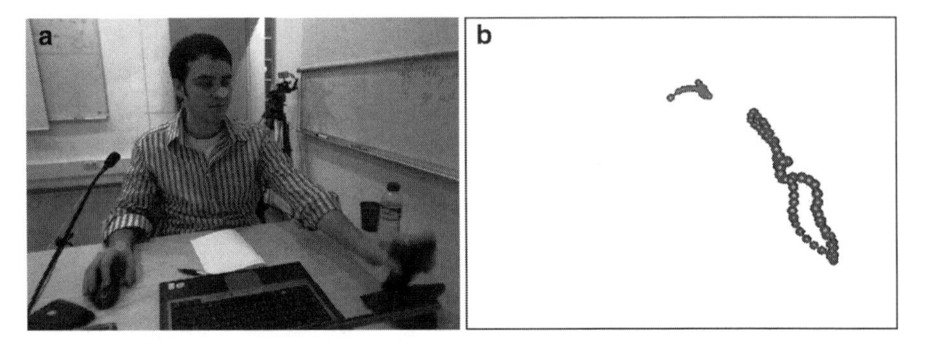

Fig. 6.5 Motion image mask generation

represent how (as opposed to where) motion in the image behaves within the regions that the camera captures.

Its basic idea relies on the construction of a vector-image, which can be matched against stored representations of known movements. A MHI H_T is a multi-component image representation of movement that is used as a temporal template based upon the observed motion. MHIs can be produced by a simple replacement of successive images and a decay operator (Eq. 6.1).

$$H_T(x,y,t) = \begin{cases} t, & \text{if } D(x,y,t) = 1 \\ max(0, H_T(x,y,t-1)-1), & \text{otherwise} \end{cases} \tag{6.1}$$

The use of MHIs has been further motivated by the fact that an observer can easily and instantly recognize moves even in extremely low resolution imagery with no strong features or information about the three-dimensional structure of the scene. It is thus obvious, that significant motion properties of actions and the motion itself can be used for activity detection in an image. In our case, if we set $t = 2$, we restrict the motion history depicted on the image to one frame in the past, which only concerns the most recent motion recorded (Fig. 6.5). In other words, a mask image is created, where all pixels are black except for the ones, where movement last detected.

Last, after having the face at least once successfully detected (red spot), the left/right hand location is marked with a blue/green spot on its location, when motion and skin color and the depth criteria are met (Fig. 6.3a). The tracking of the whole activity is only then complete, when the last image of the annotated sequence has been processed.

6.6 Feature Extraction

Once the tracking process is over, the respective raw trajectory data is acquired. In this section the location points of all tracked body parts (head and hands) is discussed. The block diagram of this refinement approach can be seen in Fig. 6.6. The refined, filtered raw data will result to an optimized signature (Wu and Li 2009), which will carry all the biometric information needed for a user to be identified.

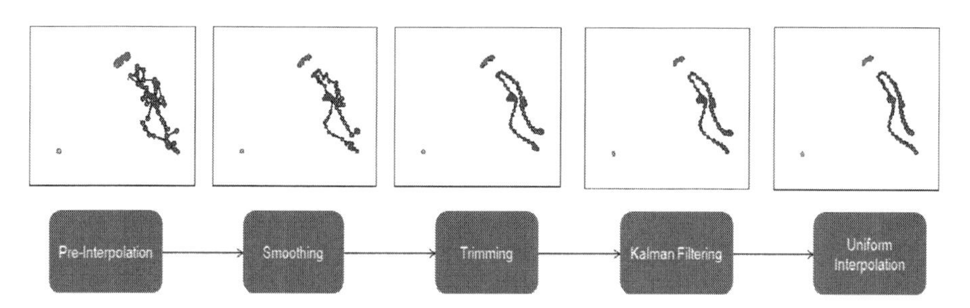

Fig. 6.6 Successive signature filtering

6.6.1 Filtering Processes

During the performance of the activity, there are some cases whereby the user stays still. Absence of motion as well as some inaccuracies in tracking may result in the absence of hand tracking. Within the first pre-interpolation step, the missing spots in the tracked points are interpolated between the last -if any- and the next -if any- valid tracked position.

The second step involves the smoothing of the successive tracked points according to the cumulative Moving Average method. In this step short-term fluctuations, just like small fluctuations or perturbations of the exact spot of the hands due to light flickering or rapid differentiations in the capturing frame rate of the camera, are discarded, while longer-term movements, which show the actual movement of the head or hands, are highlighted.

Further, the relative distances between the head and the hands are checked and the signals are trimmed accordingly. More specifically, the distance between the user's head and his hands is not expected to be bigger than a certain (normalized) value in the 3D space, due to anthropometric restrictions and due to continuity of the physical movements. Thus, in the unusual case of a wrongly tracked point, this point is substituted by the correct one or is simply discarded.

Next, each signal undergoes a Kalman filtering process (Welch and Bishop 1995). The Kalman filter is a very powerful recursive estimator that is executed in two stages (Predict and Update).

Last but not least, a uniform resampling algorithm targeting interpolation is applied on the signals. The uniform interpolation ensures a uniform spatial distribution of the points in the final signature. The sampled points of the raw signature are rearranged, in such a way that a minimum and a maximum distance between two neighboring points is preserved. When necessary, virtual tracked points are added or removed from the signature. The result is an optimized, clean signature with a slightly different signature data set, without loss of the initial motion information (Wu and Li 2009). With this, the proposed signature concerns more about the description for a continuous trajectory rather than a sequence of discretely sampled points.

6.7 Authentication Using Activity Related Biometrics

The last step of a biometric recognition process includes the final decision, whether the individual is accepted as client to the system, or excluded as impostor. In order to build a complete biometric profile or signature for each individual and classify it, the biometric signatures of the latter are combined in statistical models such as HMMs (Hidden Markov Models) or GMMs (Gaussian Mixture Models) (Wu and Li 2009).

HMMs are already used in many areas, such as speech recognition and bio-informatics, for modeling variable-length sequences and dealing with temporal variability within similar sequences (Rabiner 1990). Therefore, HMMs are suggested in gesture and sign recognition to successfully handle changes in the speed of the performed sign or slight changes in the spatial domain. Within the enrollment and authentication stages of ACTIBIO project, a Hidden Markov Models (HMMs) is proposed, as the means for modeling and classifying each signature according to the maximum likelihood criterion (Rabiner 1990).

Specifically, in order to form a cluster of signatures to train a $HMM_{n,k}$ model for any activity n of subject k, biometric signatures extracted from the same subject, performing the same activity three times, are used. Two vectors (left/right hand and head times x, y and z position in space) that correspond to the biometric signature, are contained in each behavioral signature (activity-related feature vectors).

Each feature vector set $(e_k(l_{head}, l_{hand}))$ is then used as the observation vector for the training of the HMM of Subject k. The training of the HMM is performed using the Baum-Welch algorithm. The classification module receives all the features calculated in the hand- and head-analysis modules as input. After the training, each cluster of behavioral biometric signatures of a subject is modeled by a five-state, left-to-right model, fully connected HMM (i.e. $HMM_{n,k}$).

Similarly, in the authentication/evaluation process, it can be claimed claim that, given a biometric signature, the inserted feature vectors form the observation vectors. Given, further, that each HMM of the registered users is computed, the likelihood score of the observation vectors is calculated according to Rabiner (1990), where the transition probability and observation *pdf*s (probability density function) are used. The biometric signature for an activity is only then recognized as Subject k, if the signature likelihood, returned by the HMM classifier, is bigger than an empirically set threshold.

The final decision can be become more accurate with the combination of the recognition results of more than one activity (multi-activity concept). Specifically, the scores from two activities (i.e. phone Conversation and interaction with the panel) contribute according to a weight-factor to the final score for a given subject (Fig. 6.7). Given that one activity usually demonstrates higher recognition capacity than the others, the final decision should mainly base on the outcome of the first activity, while the recognition results of the rest activities should have a supportive contribution.

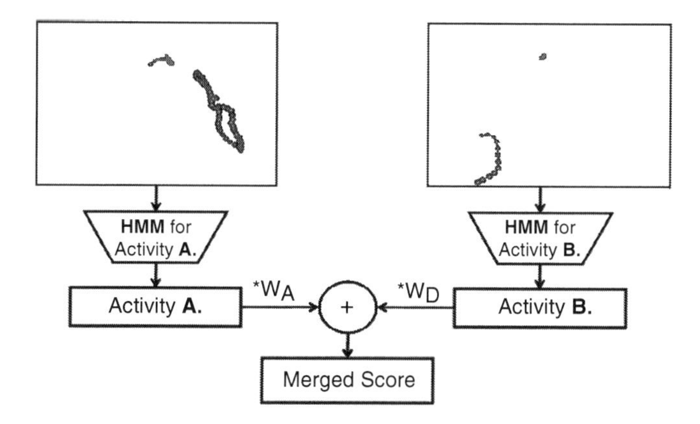

Fig. 6.7 Combination of signature from different activities

6.8 Ethical Implications of Activity Related Biometrics

Although behavioral biometrics address traits that are unstable and with low discriminating content, they can fruitfully used for any kind of recognition, from detection, to authentication, identification, or screening. Personal information that can be elicited by behavioral biometric is huge. Moreover most sensors are not intrusive, i.e., the subject may not even be aware of them. Finally it is to note that usually data cannot be filtered to support only one application, which might raises further ethical and privacy concerns.

Speaking of human action classification, engineers usually refer to a problem of computer vision research. Yet for the purposes of an ethical analysis it is more helpful to refer to the standard neurophysiological classification (Goetz 1998) in four categories: (1) voluntary, (2) semivoluntary, (3) involuntary, and (4) automatic actions.

The most crucial ethical issue is that any of these categories of actions can be misused in such a way that it will reveal other kind of information about the individual, other than the pure biometric characteristics of the latter. Specifically, there are two types of the individual's personal information that are endangered, namely the intentions and emotional states of him/her and his/her medical condition.

Paralanguage and specifically the body language or kinesics (gestures, expressions, and postures) and proxemics (the study of the distance between persons in social interactions distances and other culturally defined uses of space) are the most obvious personal characteristics. Additionally to these come body and hair paint, tattoos, decorative scaring, piercing, branding, rings, perfumes, cosmetic surgery, clothing and other forms of bodily adornment to communicate status, intentions, and other messages that can reveal the cultural or educational background of the individual. All these elements are important clues which reveal personal data implicitly and without the individual's awareness.

Surveillance technologies raise some of the most prominent ethical and political questions of our age. However, given sufficient democratic safeguards, governmental control could benefit all citizens, as their representatives know what is good for society and will not abuse their powers. The Council of Europe's Recommendation R(87)15 on the processing of personal data for police purposes provides the general principle that "new technical means for data processing may only be introduced if all reasonable measures have been taken to ensure that their use complies with the spirit of existing data protection legislation".

6.9 Other Applications of Activity Related Biometrics

Behavioral biometrics are a valuable tool in many security tasks, which require identification or verification of an individual. Behavioral biometrics are often employed because they can be easily collected non-obtrusively and are particularly useful in situations, which do not provide an opportunity for collection of stronger - more reliable biometric data. Moreover, they demonstrate great adaptability to current security installations, since just a couple of cameras is required.

Recognition capacity of a behavioral biometric system can be further improved if combined with other modalities. In Jain et al. (2004a), it is claimed that the performance of a biometric system can be significantly improved by using soft biometrics (e.g. gender, height, age, weight) in conjunction with conventional biometrics like fingerprints, iris, etc. Although up to the present time there have been some research on the extraction and the incorporation of soft biometrics in an identification system, no complete framework has been presented for the joint accurate extraction and deployment of a biometric system. With these, an aliveness detection can be rendered possible.

Behavioral biometrics can form a very welcome application in the field of video gaming. The way a game is played and the strategy that it is used can form the behavioral signature for a player, according to Yampolskiy and Govindaraju (2009). In Wang and Geng (2009), for example, behavior based intrusion detection is extended to a new domain of game networks. It is shown that a behavioral biometric signature can be generated and stored based on the strategy used by an individual to play a Poker game. Once a behavioral signature is generated for a player, it is continuously compared against player's current actions. Any significant deviations in behavior are reported to the game server administrator as potential security breaches. The results are used in multiple ways; for user verification and identification as well as for the generation of synthetic poker data and potential spoofing attempts by the players. Utilizing techniques, developed for behavior based recognition of humans to the identification and verification of intelligent game bots is also proposed.

Last, there is a high potential that medical conditions can be detected by behavioral biometrics. Thus, behavioural biometrics can also be proved useful to medical applications (esp. orthopedic ones). Specifically, behavioural biometrics, using body dynamics, could potentially detect (a) psychiatric conditions, such as dissociative disorders, acute anxiety, panic and major depression, (b) neurological conditions, such as movement disorders, (c) muscolo-skeletal and articular disorders, such as foot and ankle disorders, joint disorders and (d) all those conditions that can generate symptoms similar to those of previous disorders.

6.10 Conclusions

Activity related biometrics is a completely new concept in the field of biometric recognition with great potential in the biometric recognition field and many possible in other fields (e.g. gaming, medicine, etc.) as well. Their main characteristic is the comfort in use and its unobtrusiveness regarding the biometric characteristics extraction from the individuals. They also manage to be applicable to a much bigger part of the population, than classic biometrics.

However, as all the majority of new technologies, behavioural biometrics have also a dark side. If misused, they can render the mean for discriminating people and for abstracting personal information about the individuals physical or emotional shape without his reconciliation. Thus, a series of ethical issues has to be covered by legislation and of course, protected by the scientists and the developers of the biometric system.

References

Bao, L., and S.S. Intille. 2004. *Activity recognition from user-annotated acceleration data*, 1–17. Berlin/Heidelberg: Springer.

Bobick, A., and J. Davis. 2001. The recognition of human movement using temporal templates. *IEEE Transactions on Pattern Analysis and Machine Intelligence* 23(3): 257–267.

Bobick, A.F., and A.Y. Johnson. 2001. Gait recognition using static, activity-specific parameters. *IEEE Computer Society Conference on Computer Vision and Pattern Recognition* 1: 423–430.

Boulgouris, N., and Z. Chi. 2007. Gait recognition using radon transform and linear discriminant analysis. *IEEE Transactions on Image Processing* 16(3): 731–740.

Deutscher, J., A. Blake, and I. Reid. 2000. *Articulated body motion capture by annealed particle filtering*, vol. 2. IEEE Computer Society.

Freund, Y., and R.E. Schapire. 1999. *A short introduction to boosting*. San Francisco: Morgan Kaufmann.

Fusier, F., V. Valentin, F. Bremond, and M. Thonnat. 2007. Video understanding for complex activity recognition. *Machine Vision and Applications Journal* 18: 167–188.

Goetz, C. 1998. Parkinson's disease and movement disorders. *JAMA: The Journal of the American Medical Association* 280: 1796–1797.

Gomez, G., and E.F. Morales. 2002. Automatic feature construction and a simple rule induction algorithm for skin detection. In *Proceedings of the ICML Workshop on Machine Learning in Computer Vision*, 31–38.

Ioannidis, D., D. Tzovaras, I.G. Damousis, S. Argyropoulos, and K. Moustakas. 2007. Gait recognition using compact feature extraction transforms and depth information. *IEEE Transactions on Information Forensics and Security* 2(3): 623–630.

Jain, A.K., S.C. Dass, and K. Nandakumar. 2004a. Soft biometric traits for personal recognition systems. In *Proceedings of International Conference on Biometric Authentication,* Hong Kong, 731–738.

Jain, A.K., A. Ross, and S. Prabhakar. 2004b. An introduction to biometric recognition. *IEEE Transactions on Circuits and Systems for Video Technology* 14(1): 4–20.

Junker, H., P. Lukowicz, and G. Tröster. 2004. User activity related data sets for context recognition. In *Proceedings of the Workshop on Benchmarks and a database for context recognition held in conjunction with Pervasive*, IEEE Computer Society. Vienna, Austria.

Kale, A., N. Cuntoor, and R. Chellappa. 2002. A framework for activity-specific human identification. *IEEE International Conference on Acoustics, Speech, and Signal Processing* 4: IV3660–IV3663.

Kawanaka, D., T. Okatani, and K. Deguchi. 2006. HHMM based recognition of human activity. *IEICE Transactions on Information and Systems* 89(7): 2180–2185.

Lester, J., T. Choudhury, N. Kern, G. Borriello, and B. Hannaford. 2005. A hybrid discriminative/ generative approach for modeling human activities. In *Proceedings of the International Joint Conference on Artificial Intelligence* (*IJCAI*), 766–772.

Matey, J., O. Naroditsky, K. Hanna, R. Kolczynski, D. LoIacono, S. Mangru, M. Tinker, T. Zappia, and W. Zhao. 2006. Iris on the move: Acquisition of images for iris recognition in less constrained environments. *Proceedings of the IEEE* 94(11): 1936–1947.

Park, F. 1994. Computational aspects of the product-of-exponentials formula for robot kinematics. *IEEE Transactions on Automatic Control* 39(3): 643–647.

Patterson, D.J., L. Liao, K. Gajos, M. Collier, N. Livic, K. Olson, S. Wang, D. Fox, and H. Kautz. 2004. Opportunity knocks: A system to provide cognitive assistance with transportation services. In *International Conference on Ubiquitous Computing (UbiComp)*, 433–450. Springer.

Poynton, C. 1997. Frequently asked questions about color, 1–24. http://www.poynton.com/PDFs/ColorFAQ.pdf

Rabiner, L. 1990. A tutorial on hidden Markov models and selected applications in speech recognition. *Readings in Speech Recognition* 53(3): 267–296.

Ramesh, D., and P. Meer. 2000. Real-time tracking of non-rigid objects using mean shift. *IEEE Conference on Computer Vision and Pattern Recognition* 2: 142–149.

Scharstein, D., and R. Szeliski. 2002. A taxonomy and evaluation of dense two-frame stereo correspondence algorithms. *International Journal of Computer Vision* 47: 7–42.

Skarbek, W., and A. Koschan. 1994. *Colour image segmentation-a survey*. Berlin: Technische Universitaet.

Tapia, E.M., S.S. Intille, and K. Larson. 2004. *Activity recognition in the home using simple and ubiquitous sensors*. Berlin/Heidelberg: Springer.

Viola, P., and M. Jones. 2001. Rapid object detection using a boosted cascade of simple. *Proceedings of the IEEE Computer Society Conference on Computer Vision and Pattern Recognition* 1: I511–I518.

Wang, L., and X. Geng. 2009. *Game playing tactic as a behavioral biometric for human identification*, chapter 18, 385–413. Global, IGI.

Welch, G., and G. Bishop. 1995. *An introduction to the Kalman filter*. Chapel Hill: University of North Carolina.

Wickins, J. 2007. The ethics of biometrics: The risk of social exclusion from the widespread use of electronic identification. *Science and Engineering Ethics* 13(1): 45–54.

Wu, S., and Y. Li. 2009. Flexible signature descriptions for adaptive motion trajectory representation, perception and recognition. *Pattern Recognition* 42(1): 194–214.

Yampolskiy, R.V., and V. Govindaraju. 2009. Strategy-based behavioural biometrics: A novel approach to automated identification. *International Journal of Computer Applications in Technology* 35(1): 12.

Zouba, N., F. Bremond, M. Thonnat, and V.T. Vu. 2007. Multi-sensors analysis for everyday activity monitoring. In *4th International Conference: Sciences of Electronic, Technologies of Information and Telecommunications (SETIT2007),* Tunis, Tunisia.

Chapter 7
Electrophysiological Biometrics: Opportunities and Risks

Alejandro Riera, Stephen Dunne, Iván Cester, and Giulio Ruffini

Abbreviations

AR	Autoregression
BCI	Brain computer interface
CC	Cross correlation
CO	Coherence
ECG	Electrocardiogram
EEG	Electroencephalogram
EER	Equal error rate
EMG	Electromyogram
EOG	Electrooculogram
ERP	Event related potential
EU	European Union
FP	Framework program
FPR	False positive rate
FT	Fourier transform
Hz	Hertz
MI	Mutual information
TPR	True positive rate
USB	Universal serial bus

A. Riera (✉) • S. Dunne • I. Cester • G. Ruffini
Starlab Barcelona S.L., Teodor Roviralta 45, 08022 Barcelona, Spain
e-mail: alejandro.riera@starlab.es; stephen.dunne@starlab.es;
ivan.cester@starlab.es; giulio.ruffini@starlab.es

E. Mordini and D. Tzovaras (eds.), *Second Generation Biometrics: The Ethical, Legal and Social Context*, The International Library of Ethics, Law and Technology 11,
DOI 10.1007/978-94-007-3892-8_7, © Springer Science+Business Media B.V. 2012

7.1 Introduction

The market of biometry is growing every year, as a result of the great interest in this field, mainly for security reasons. The use of electrophysiological signals for biometric purposes is a novel approach and offers some advantages compared to more classical biometric modalities such as fingerprinting and retina or voice recognition. For instance, continuous authentication can be performed as long as the subject is wearing the recording electrodes. Moreover, EEG, ECG and EMG (all variants of electrophysiology) can also provide information about the emotional state, sleepiness/fatigue level, the stress level, and continuously monitor the vital signals of the subject, which could also be useful for preventive medicine and telemedicine.

The chapter is organized as follows. The next subsection explains the two main concepts of this text: electrophysiology and biometry. Then the concept of electrophysiological biometrics is provided. This section finishes with a discussion of the advantages of electrophysiological biometrics over more classical biometric modalities. The second section -'Biometric Technology'- is divided in six subsections. 'Artifact rejection/correction' explains our approach to reduce this undesirable noise that corrupts virtually all the electrophysiological recordings. 'EEG', 'ECG', 'EOG' and 'EMG' describe respectively the corresponding biometric modalities we have implemented. The last point of this section, 'BCI', describes a new biometric approach based on the control of a brain computer interface. Next section, 'Multimodal System: Fusion' deals with the fusion of the different modalities in order to extract a more reliable biometric result. The two next sections explain the 'Technology Trends and Opportunities' and the 'Vision for the Future and Risks'. The last section summarizes this chapter.

7.2 Electrophysiological Biometrics

7.2.1 What Is Electrophysiology?

Electrophysiology is the study of the electrical properties of biological cells and tissues. Several techniques have been developed depending on the scale we want to record: from patch-clamp techniques for single cells or even ion channels to electrocardiography or electroencephalography for whole organs such as the heart or the brain, respectively.

Many particular electrophysiological readings have specific names, referring to the origin of the bioelectrical signals:

- Electrocardiography (ECG) – for the heart
- Electroencephalography (EEG) – for the brain
- Electrocorticography (ECoG) – from the cerebral cortex
- Electromyography (EMG) – for the muscles
- Electrooculography (EOG) – for the eyes

- Electroretinography – for the retina
- Electroantennography – for the olfactory receptors in arthropods
- Audiology – for the auditory system

In this chapter we will study the biometric potential of ECG, EEG, EMG and EOG. These electrophysiological modalities can be easily recorded by placing some electrodes in the skin of the subject, making them less obtrusive than for instance ECoG. The electrode configuration and the basic principles of each one of these modalities will be explained in their respective sections, but in order to show the importance of such techniques, it is interesting to mention the following facts:

- As soon as 1872, the first ECG was recorded by Alexander Birmick Muirhead, but it was not until the work of Waller (1887) that the ECG was studied in a more systematic way. Finally, the invention of the string galvanometer by Willem Einthoven (Moukabary 2007) supposed a breakthrough in the study of the ECG. His works were awarded with a Nobel Prize in Medicine in 1924.
- A good timeline history of EEG is provided by Schwartz (1998). The first findings were presented in 1875 by Richard Caton. He recorded the EEG signals of the exposed cerebral hemispheres of rabbits and monkeys. The first EEG recorded to a human is credited to Hans Berger in 1920.
- The first recording of EMG was made in 1890 by Marey, although since 1666 it was known that certain specialized muscles produce electricity. In 1791, Galvanni demonstrated that electricity could initiate muscle contractions (Galvani 1791).
- Finally, the developers of the patch-clamp technique, Erwin Neher and Bert Sakmann (1992), received the Nobel Prize in 1991. Briefly, this technique allows the study of a single or several ion channels present in some types of cells by the use of an electrode and a micropipette.

7.2.2 *What Is Biometry?*

The term biometrics has a Greek origin: it is composed by the words "bios" (life) and "metron" (measure). A biometric identifier is originally defined as the objective measurement of physical characteristics. The term has been used in medicine, biology, agriculture and pharmacy (e.g. in biology, biometrics is a branch that studies biological phenomena and observations by means of statistical analysis).

The term biometrics here refers to automated methods and techniques that analyze human characteristics in order to recognize a person, or distinguish this person from another, based on a physiological or behavioral characteristic.

Another meaning has also been acquired in the last decades, focused on the characteristic to be measured rather than the technique or methodology used (Zhang 2000): "A biometrics is a unique, measurable characteristic or trait of a human being for automatically recognizing or verifying identity."

A biometric trait ideally satisfies the following requirements:

Universal: Each user should have it.

Unique: In order for something to be unique, it has to be the only existing one of its type, have no like or equal, be different from all others. When trying to identify an individual with certainty, it is absolutely essential to find something that is unique/distinctive to that person.

Measurable: In order for recognition to be reliable, the characteristic being used must be relatively static and easily quantifiable.

Permanent: Traits that change significantly with time, age, environment conditions or other variables are of course not suitable for biometrics.

Characteristic or trait: The measurable physical or personal behavioral pattern used to recognize a human being. Currently, identity is often confirmed by something a person has, such as a card or token, or something the person knows, such as a password or a personal identification number. Biometrics involves something a person is or does. These types of characteristics or traits are intrinsic to a person, and can be approximately divided into physiological and behavioral. Physiological characteristics refer to what the person is, or, in other words, they measure physical parameters of a certain part of the body. Some examples are fingerprints, that use skin ridges, face recognition, using the shape and relative positions of face elements, retina scanning, etc. Behavioral characteristics are related to what a person does, or how the person uses the body. Voice or gait recognition, and keystroke dynamics, are examples of this group.

Robust: Intra – class variability should be as small as possible, which means that different captured patterns from the same user should be as close as possible.

Accessible: it should be easy to present to the sensor.

Acceptable: it should be well accepted by the public – non obtrusive and non intrusive.

Hard to circumvent: it should be difficult to alter or reproduce by an impostor who wants to fool the system.

Moreover the recognition system should be automatic, i.e. must work by itself, without direct human intervention. For a biometric technology to be considered automatic, it must recognize or verify a human characteristic in a reasonable time and without a high level of human involvement.

A biometric recognition system has two main operational modes: verification (or authentication) and identification. Recognition refers to no particular operational mode, as we now discuss.

Verification: To verify something is to confirm its truth or establish its correctness. In the field of biometrics, verification is the act of proving the claim made by a person about their identity. A computer system can be designed and trained to compare a biometrics presented by a person against a stored sample previously provided by that person and identified as such. If the two samples match, the system confirms or authenticates the individual as the owner of the biometrics on file.

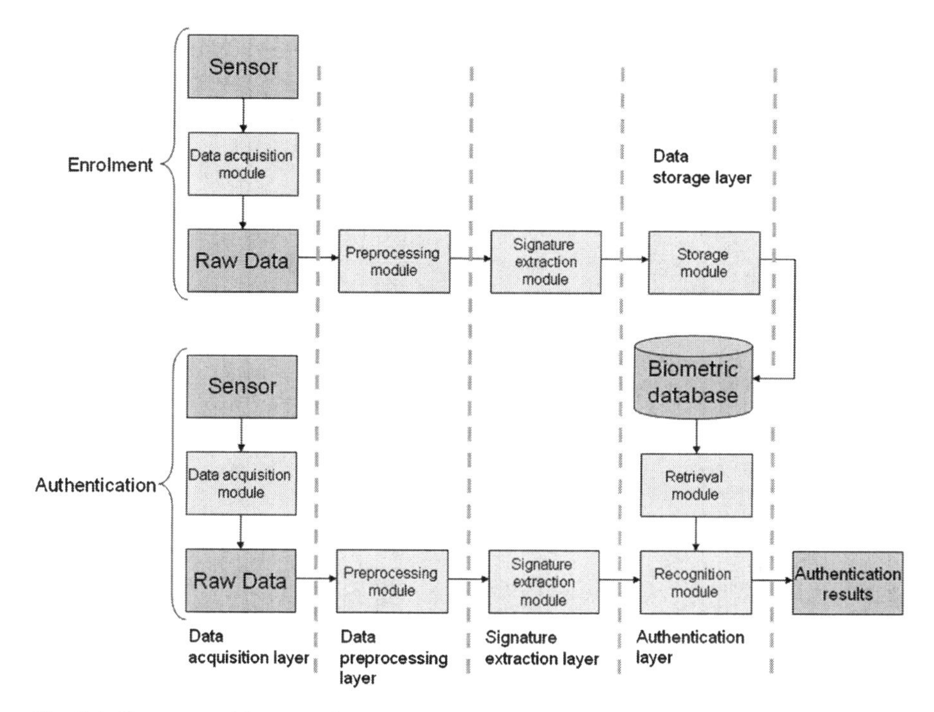

Fig. 7.1 Common architecture of a biometric system

Identification: Identity is the answer to the question about who a person is, or the qualities of a person or group which make them different from others, i.e., being a specific person. Identity can be understood either as the distinct personality of an individual regarded as a persistent entity, or as the individual characteristics by which this person is recognized or known. Identification is the process of associating or linking specific data with a particular person.

Recognition: To recognize someone is to identify them as someone who is known, or to distinguish someone because you have seen, heard or experienced them before (to "know again"). A person cannot recognize someone who is completely unknown to them. A computer system can be designed and trained to recognize a person based on a biometric characteristic, comparing a biometric presented by a person against biometric samples stored in a database If the presented biometric matches a sample on the file, the system then recognizes the person.

Depending on the application context, a biometric system may operate either in authentication mode or identification mode. The probability of having a false true value for the authorization of a subject is higher with identification than with authentication, and thus the later is preferable, especially for high level security requirements. The block diagrams of an authentication system and an identification system are depicted in Fig. 7.1; user enrolment, which is common to both the tasks is also graphically illustrated.

Throughout this chapter, we will use the generic term recognition where we do not wish to make a distinction between authentication and identification. Some other expressions frequently used in biometrics that are worth defining here:

Biometric sample: Biometric samples or data are biometric information presented by the user and captured by the biometric system.

Biometric template: A biometric template is the individual mathematic data set calculated from a biometric sample. Biometric systems need templates for comparison.

Biometric system: A biometric system is an automated system capable of capturing a biometric sample, extracting biometric data, comparing it with other biometric data and deciding whether or not the recognition process has been successful.

Biometric technology: In the present study the term biometric technologies refers to all computer-based methods to recognize human beings using biometric characteristics.

7.2.3 *How Can Electrophysiological Signals Be Used for Biometry?*

Now that the two main concepts of this chapter, biometry and electrophysiology, have been explained, we can link them in order to explain our approach to biometrics. We have used four electrophysiological signals: EEG, ECG, EMG and EOG. Another approach based on a BCI has been also been explored. What are the advantages of using these signals in the field of biometrics? First of all, it is interesting to note that every living person has an active brain and a heart beat, making those signals completely universal. Typical biometric traits, such as fingerprint, voice, and retina, are not universal, and can be subject to physical damage (dry skin, scars, loss of voice, etc.). In fact, it is estimated that 2–3% of the population is missing the feature that is required for the authentication, or that the provided biometric sample is of poor quality.

It has been proven that the EEG and ECG are unique enough to be used for biometric purposes (Marcel and Millán 2007; Mohammadi et al. 2006; Paranjape et al. 2001; Poulos et al. 1998, 1999, 2001, 2002; Riera et al. 2008a, b; Biel et al. 2001; Chang 2005; Israel et al. 2005; Kyoso 2001; Palaniappan and Krishnan 2004). In fact, if we think on the huge number of neurons present in a typical adult brain (10^11) and their number of connections (10^15), we can definitively claim that no 2 brain are identical. A similar argumentation could be done for the heart.

From a more philosophical point of view, we can also think on what the ultimate biometric system will be like. That is, the one that will be impossible to spoof (or almost). This question is clearly related to the issue of where our identity lies. An intuitive answer to this question is that our identity must lie in the brain, or, to be a bit broader, in our Central Nervous System. We could replace our fingertips and hearts or have a face-lift and retain our identities, but replacing somebody's brain will distort our conception of spoofing in a radical way. If we could "clone" somebody's brain, then one could argue that we can no longer say that person is not the real person. At any rate, the reader will agree with the statement that "the ultimate seat of identity lies in the living, dynamic brain", or, at least, in part of it (e.g., in the abstract set of neuronal connections). In recent work we have advanced a great deal in the development of physiologically based biometric systems exploiting EEG (Marcel and Millán 2007; Mohammadi et al. 2006; Paranjape et al. 2001; Poulos et al. 1998, 1999, 2001, 2002; Riera et al. 2008a, b) and ECG (Biel et al. 2001; Chang 2005; Israel et al. 2005; Kyoso 2001; Palaniappan and Krishnan 2004) signals and classification algorithms. The derived systems rely on spontaneously generated electrophysiological signals, and as such they are in some sense weaker to spoofing attacks than they could be. After all, one could record EEG/ECG spontaneous activity and play it back during authentication. This would be hard, but not impossible. If one were free to challenge the impostor asking for different features of their EEG, or to stimulate the subject and study the response of their brains, the biometric system would become much more robust.

On the other hand, physiologically based systems are bound to be more obtrusive than other ones (especially EEG), so they must provide a substantial added value in relation to others (Graff et al. 2007) and minimize the intrusiveness as much as possible by applying wearable electrophysiological recording devices (Ruffini et al. 2006, 2007). Another element of interest is that biometrics technologies will definitely become very important in immersive interactive environments, where we will be able to control voice, body, gestures, etc. How will others know that you are you, and not an avatar controlled by somebody else, or, worse yet, an agent?

7.2.4 *What Are the Comparative Advantages of Electrophysiological Biometrics?*

Electrophysiological biometrics has an advantage over the classical biometric modalities. Normally a biometric system is used in order to access a secure area or in order to unlock a computer, for instance. This scenario is called initial authentication and it is the typical scenario used with fingerprint authentication (there are laptops in the market that incorporate a fingerprint authentication system in order to unlock the computer). On the other hand, by wearing a band set with bioelectrodes, the biometric characteristic can be recorded in real time, and for long periods of time, thus permitting the biometric system to continuously authenticate the user. For instance, once the computer is unlocked, the system is still extracting the biometric features, and this way a impostor could not use such a system, even if it was unlocked previously by the legal user.

In order to make the system accepted by the users, it should be as transparent as possible. ENOBIO is a sensor developed by Starlab Barcelona SL with interesting features: it is wearable, wireless and can work in dry mode, that is, without the need to use conductive gel. It consists of four electrodes and a unit which can be worn as a

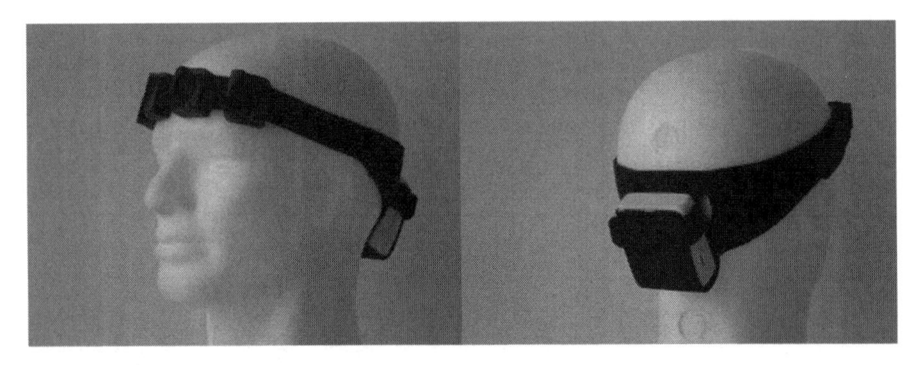

Fig. 7.2 Enobio sensor

head band. The unit has all the electronics and a radio that transmit the recorded data wirelessly to the receiver that is connected to a computer by a USB connection. We can see the electrodes and the recording unit placed with the headband in Fig. 7.2.

In order to record ECG, an electrode can be placed in the left wrist with a longer cable and with the help of a band.

Besides being able to perform a continuous authentication, we can also perform what we call *biometry on the move*. That is being able to authenticate subjects while they are moving about and not performing any specific protocol. Regarding other biometric modalities such as the ones based on fingerprint or iris recognition, the subjects have to place their finger or retina in a specific place for a specific amount of time. This fact is not always well accepted by the users, since they lose time and, specifically for iris, people do not like to place their eyes in front of a camera. With electrophysiological biometrics, the opportunity to record continuously allows the user to be authenticated on the move, thus not loosing time while undertaken the recognition.

7.3 Biometric Technology

7.3.1 Methods for Recording Electrophysiological Biosignals

Before discussing the general methods used for recording the physiological biosignals we give an explanation of artifacts as it is useful for the reader to be briefly informed about these and how they affect work with electrophysiological biosignals.

7.3.1.1 Artifact Rejection/Correction

In order for continuous authentication to take place, we face a very well known problem by electrophysiologists: the artifacts. These are electrical signals originating in places other than the desired one (e.g., electrical activity of the brain).

They can originate, for example, from the electrical contact points on the skin, or from other bio-electrical sources (e.g., muscles). In all recordings, we find artifacts that can come from different sources. The artifacts can be considered as noise that do not contain, in general, useful information. The artifacts can be caused by several factors that we should take into account. It is not straightforward to distinguish among them. In the next list we can see a list of the major categories of artifacts:

- Machine and impedance artifacts: the most common ones relates to problems with the electrode, such as the electrode itself being broken or improperly attached to the subject.
- Presence of 50 Hz artifact (or, e.g., 60 Hz in the USA), either from nearby equipment or the very common ground loop.
- Cardiac artifacts: caused by the heart
- Oculographic artifacts: caused by the movement of the eyes. The retina acts like a dipole, so it should be noted that these artifacts are not caused by the muscles that control the eyes movement, but by the movement of this dipole.
- Myographic artifacts: caused by the electrical activity of the muscles.
- Interference between biosignals: caused by capturing the electrical activity of other tissues than the ones monitored, e.g. the electrical activity of the arm muscles can influence the ECG, or of the face muscles the EEG.

In our case we are interested on the one hand in correcting the artifacts (physiological and others) from the EEG, ECG and EMG in order to have a cleaner signal and on the other hand to record the oculographic artifacts (EOG) in order to use them as a biometric signal. In the case of EMG, bipolar electrodes were placed in each forearm of the subjects so there is no presence of ECG and EOG artifacts. For each one of the EEG and ECG modalities, a different artifact corrector algorithm has been implemented. They will be explained in their respective sections.

7.3.2 EEG

The EEG is the recording of the brain activity by the mean of electrodes placed on the scalp of the subject. The electrical activity is due to the firing of the neurons. Many references can be found in order to find a deeper explanation regarding the EEG, such as Kandel (1981).

The ENOBIO sensor has been used by our team for this approach within the ACTIBIO project. Two electrodes are placed in the forehead of the subject (FP1 and FP2 locations of the 10–20 international system), and referenced to a third electrode placed in the right ear lobe. In the following figure we can see an EEG sample recorded with the ENOBIO device (Fig. 7.3).

Our approach to correct the artifacts is based on the detection of sudden changes in the signal and then we perform a detrending of the signal. Rather than getting in technical detail, we prefer to show the performance of the artifact corrector for each physiological signal with some figures.

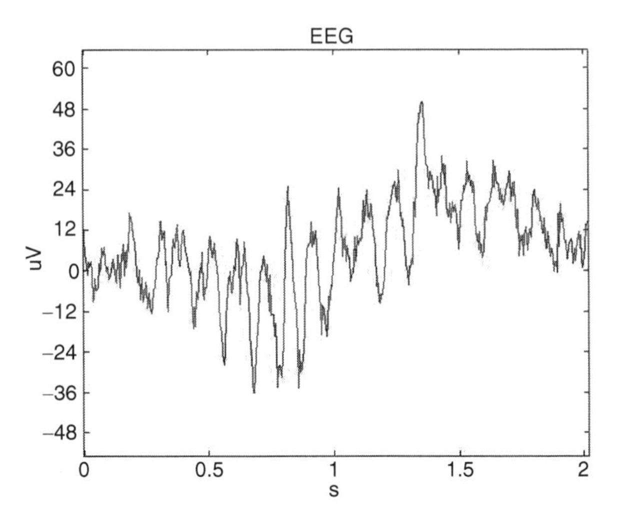

Fig. 7.3 ENOBIO EEG recording sample of 2 s with no pre-processing. The alpha wave (10 Hz characteristic EEG wave) can be seen

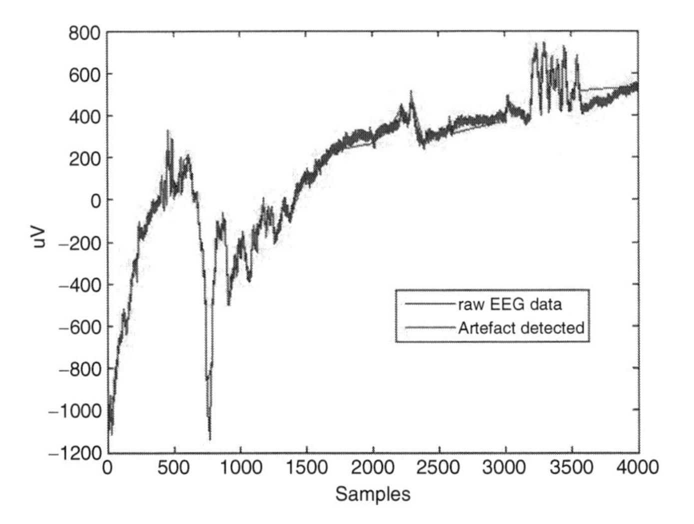

Fig. 7.4 *Light grey* is the artifact estimation that will be subtracted to the raw EEG signal (*black line*). The algorithm works fairly well for this section

In the next figure, we can see a raw EEG signal with movement artifacts (big oscillations at the beginning of the signal) and with blink artifacts (6 peaks at the end of the signal). We see that the big movement artifacts and the blink artifacts are well detected by our artifact correction module (Fig. 7.4).

In the following figure we can see the signal after applying the artifact corrector module. We notice that the movement artifacts (around sample 750) are still present but they are very much reduced. The same applies to the blink artifacts (around sample 3,250). On the other hand, the drifts are completely removed (Fig. 7.5).

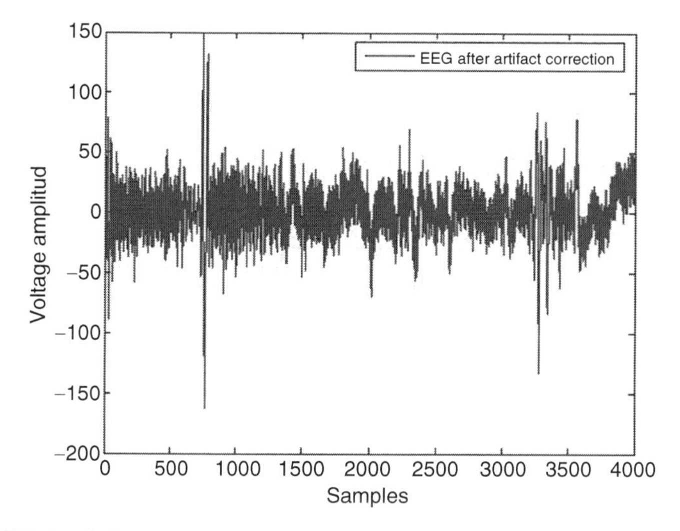

Fig. 7.5 EEG signal after subtracting the artifacts detected by the artifact corrector module

Two protocols have been used. In the first one, the subjects have to stay with their eyes closed, seated on a chair, relaxed and avoiding moving. Doing so, we can minimize the eye movements and blink artifacts, and also the more general movement artifacts. The enrolment consisted of four 3-min takes in the same conditions.

The second protocol is much less restrictive. The subject is free to move and work in front of his or her office table, but in this first study, the subject had to remain seated. The enrolment consisted in this case of four 2-min takes, and the subject had to watch a movie during the recording time.

Several features were extracted. For single channel we used autoregressive coefficients (AR) and Fourier transform (FT) (4 different features since we extract AR and FF for each channel). We also extracted 3 synchronicity features: cross correlation (CC), mutual information (MI) and coherence (CO) (3 features since we extract each one of those for the 2 channels). Fisher Discriminant Analysis was used for the classification, with 4 different discriminant functions. We thus had a total of $(4+3)\times4=28$ possible combinations between channels, features and classifiers.

Using the enrolment takes and performing a cross-fold validation over the enrolment takes, we computed the five best combinations per subject. We called this approach "personalized classifiers", and indeed the performance of the system improves.

Without applying the artifact correction the results are summarized in Table 7.1.

The performances are given in terms of the Equal Error Rate (EER) which is equal to the False Positive Rate (FPR) when the FPR + the True Positive Rate (TPR) are equal to 100%. The EER provides an optimal compromise between the highest possible TAR and the lowest possible FAR.

As we can see the performance of the office takes shows some biometric potential, but it is not very high. The mean of the EER is 36%.

Table 7.1 Classification results of EEG (office take) without applying the artifact correction module

Take	TPR (%)	FPR (EER) (%)
1	64	36
2	63	37
3	65	35

Table 7.2 Classification results of EEG applying the artifact correction module

Take	TPR (%)	FPR (EER) (%)
1	71	29
2	82	18
3	70	30

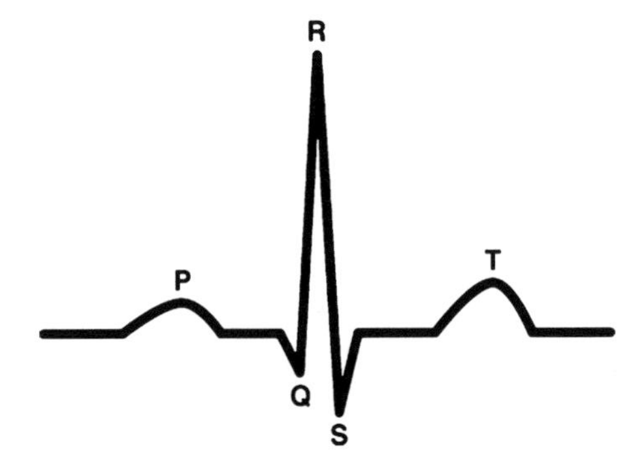

Fig. 7.6 A typical schematic ECG waveform. We can see the different complexes of the ECG

Applying the artifact corrector we see that the results improve considerably as we can see in Table 7.2.

It is worth comparing these results with the ones acquired with the first protocol, where the subjects were asked to stay seated, relaxed and with the eyes closed. In such condition we reached an EER equal to 20%. The performance in this case is remarkably higher than in the office takes, but the recording protocol is much more obtrusive. In the second case, the subject can be freely working or doing his daily tasks while being authenticated, making such an approach much more transparent.

7.3.3 ECG

The ECG is recorded with the help of electrodes placed on the skin of the subject in specific places. It measures the voltage between pairs of electrodes. This voltage is generated by electrical impulses that are at the origin of the contractions of the myocardial muscle fibers. A typical schematic ECG waveform can be seen in the following figure (Fig. 7.6):

In order to record ECG, we place an electrode in the left wrist of the subject, attached with the help of an elastic band. The protocols are the same described in

Table 7.3 Classification results of ECG biometric modality in the office takes

Situation	Take	TPR (%)	FPR (EER) (%)
Office	1	87	13
Office	2	88	12
Office	3	88	12

the EEG part. In fact the ECG and EEG were recorded at the same time, since ENOBIO has 4 recording electrodes (2 for EEG, 1 for ECG and the forth one used as reference). In the following figure we can see an EEG sample recorded with the ENOBIO device.

The enrolment was performed at the same time and in the same conditions than the EEG enrolment (this applies for the 2 described protocols).

As in the second protocol the subjects are moving freely while seated in their office, movement artifacts appear in the ECG signal, and thus an artifact rejection module was developed specifically for this purpose. This module works by performing the following steps:

1. We apply a band pass filter with frequency cut-offs of 0.5 and 35 Hz. That way we remove the drifts present in the signal and we also remove the high frequencies which are of no interest in this case.
2. We remove the high peaks since they correspond to movement artifacts that distort the ECG signal. We apply a simple threshold: all the values higher or lower than 1,000 microVolts are discarded.
3. We apply a peak detector to localize the R peaks in the ECG signal in order to cut each ECG waveform.
4. Now we align all the detected ECG waveform. Note that at this stage, the length of the ECG waveforms is not uniform. We discard the ECG shapes that have at least one point outside of 3 standard deviations of the average of all ECG waveforms.
5. Now all the ECG waveforms have more or less the same shape. The last step is to normalize the length of the ECG waveforms by resampling the part between the P and T complex, which is the one that is more dependant on the Heart Beat Rate.

The ECG we obtain that way have exactly the same length and a homogenous shape. These vectors are the ones we are going to use as features and input in our classifiers. Similarly to what was done with the EEG, we use Fisher Discriminant Analysis for the classification. In this case we have only 1 feature and 4 discriminant functions. The personal classifier approach in this case selects the best discriminant function for each subject.

The results are summarized in the following tables (Tables 7.3 and 7.4):

It is interesting to note that there is a potential in the ECG biometric modality while the subject is walking although the performance is much higher in the offices takes. There are two reasons that explain this difference. On the one hand there are much less movement artifacts in the office takes and on the other hand the office take is longer (around 3 min versus 1 min).

Table 7.4 Classification results of ECG biometric modality in the walking takes

Situation	Take	TPR (%)	FPR (EER) (%)
Walking	1	67	33
Walking	2	64	36

Finally, with the recordings made with the first protocol in which the subjects were relaxed and seated we reached an EER equal to 3%. Again, the performance is remarkably higher than using the office takes, but the recording protocol is much more obtrusive and therefore not ideal or easily accepted by the users. Using the second protocol, in which the subjects are free to work or perform their daily activities, the biometric system becomes transparent for the users.

7.3.4 EOG

The process of measuring eye movements in different environmental contexts is called electrooculography (EOG). The EOG technique is concerned with measuring changes in electrical potential that occur when the eyes move. The EOG has been useful in a wide range of applications from the rapid eye movements measured in sleep studies to the recording of visual fixations during normal perception, visual search, perceptual illusions, and in psychopathology. Studies of reading, eye movements during real and simulated car driving, radar scanning and reading instrument dials under vibrating conditions have been some of the practical tasks examined with eye movement recordings. Eye blinks are easily recorded with EOG procedures and are particularly useful in studies of eyelid conditioning, as a control for possible eye blink contamination in EEG research, and as measures of fatigue, lapses in attention, and stress. There are also the periodic eye blinks that occur throughout the waking day that serve to moisten the eyeball. Still another type of eye blink is that which occurs to a sudden loud stimulus and is considered to be a component of the startle reflex. The startle eye blink is muscular and is related to activity in the muscles that close the lids of the eye. Research on the eye blink component of startle has revealed interesting findings that have implications for both attentional and emotional processes. A deeper overview can be found at (Andreassi 2007).

A preliminary study has been performed to determine the potential of the EOG signal for authentication. The approach we used is based on blinking temporal patterns and in the shape of the blinks. This approach addresses different issues and unknowns; we do not know how dependent of time, mood, illumination, etc… the features will be and thus if the differences between subjects will be big enough to make an authentication robust trough this intra-subject changes.

The hypothesis is that the features we will use have an intra-subject variability smaller than the inter-subject variability. Of course many factors can affect the intra-subject variability such as time of the day, mood, illumination, etc…

We expect that the EOG module, even if it is not reliable enough by itself, can contribute to improve the system performance after the fusion of the different modalities.

This section explains the process followed to extract the features for the authentication process, and some preliminary results of identification and authentication using these features. The whole process can be separated in four main steps:

- Artifact detection and rejection
- Blink detection
- Feature extraction
- Identification/Authentication

7.3.4.1 Artifact Detection and Rejection

In this case, we can not apply the artifact detection described in section the artifact detection/correction, because the blinks would be treated as artifacts, and they would be removed from the signal. In this module, all the biometric information is found in the blinks, thus we want to keep them rather than removing them.

We first apply a band filter between 0.8 and 30 Hz. A threshold is then applied to the resulting signal to localize the parts of the signal over 450 and under −300 uV.

In a second process a wavelet analysis is applied to the data of both channels: we look at frequencies between 1 and 30 Hz. The wavelet used is a Morlet. To calculate the threshold for a specific frequency, we use the mean of the wavelets coefficient for this frequency that we obtained for the parts of the signal that was considered correct using the amplitude method previously explained. The threshold is set at 2.5 times the standard deviation for each separate frequency.

The bad samples of the data are now detected by any of the 2 methods in any of the 2 channels. When a bad sample is detected, there is a rejection of all the data located 25 ms before and after this sample. Then we look for the remaining epochs of good data of more than 10 s of length. From these signals we extract the blinks, from which we extract the features. In order to detect the blinks, we developed an algorithm that detects the blinks in an automatic manner (see Fig. 7.7).

Four different set of features were extracted:

1. Shape of the average over all detected blinks.
2. Mean inter-blink distance.
3. Blink rate.
4. Standard deviation of the blink rate variability.

Features 2, 3 and 4 are based on blinking temporal patterns. Each feature is a number that corresponds to a component of a 3 dimension vector that is then classified. Feature 1 is classified separately and it is an 80 components vector that corresponds to the EOG time series containing the blink.

A preliminary classification was performed applying the K-Nearest neighbor algorithm. The number of subjects used for the test is 23, so the classification rate due to random classification would be of 4.35%. Table 7.5 shows the results of this test.

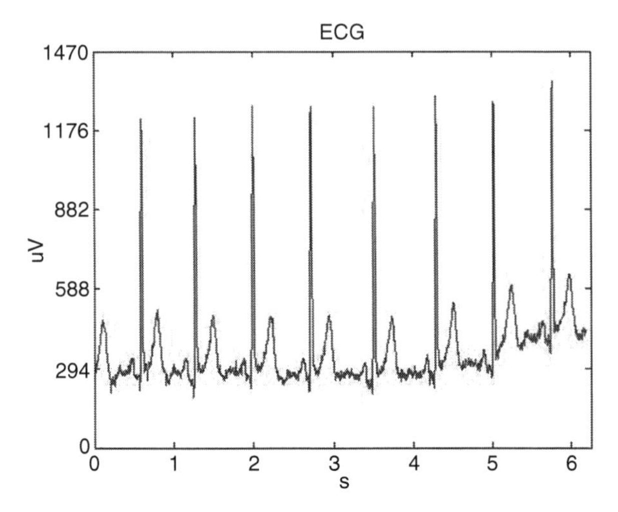

Fig. 7.7 ENOBIO ECG recording sample of approximately 6 s with no pre-processing

Table 7.5 Classification rate of the EOG features using LDA and K- nearest neighbors classifiers

	K-nearest neighbor (%)
Blink shape	24.6
Blinking temporal patterns	12.5

We can conclude that the blink shape and in some extent the Blinking Temporal Pattern shows some biometric potential, since the classification performance is much higher than a random classification. Although by itself it might not be a robust biometric modality, combined with other modalities it might increase the authentication performance. Moreover, with the ENOBIO system the EOG signal can be recorded at the same time that the EEG and ECG signals, making it convenient for a later fusion of those modalities.

7.3.5 EMG

Electromyography is the technique for measuring and recording electrical potentials that are associated with contractions of muscle fibers. The EMG is often used in the clinic to study muscular disorders. Very thin needle electrodes can be inserted into muscle tissue, and recordings can be made from limited muscle regions or even from single motor units. The EMG can also be recorded from the skin surface, because some portion of the action potentials produced in muscle fibers is transmitted to the skin. The closer the muscle tissue is to the skin surface, and the stronger the contractions, the greater will be the amount of electrical activity recorded at the surface. Most studies relating EMG to human performance deal with the activity occurring in large-muscle groups. A nice overview can be found at (Andreassi 2007).

Although being an electrophysiological measure, the muscular activity is directly related with the behavior of the body, and in particular any activity that involves contraction of the muscles, such as gait, key striking and so on.

As far as we know, there is no published work regarding a biometric system based on muscular activity. Therefore, this section describes the state of the art in recording EMG and extracting information from it, focusing on its potential application for person identification/authentication.

Electromyography (EMG) is the recording of muscular activity. It reflects not only how the muscle is, but also the work it is developing. Typical parameters that can be obtained are amplitude, mean frequency, and propagation. This last is related with the anatomy of the muscle, but also with the propagation speed of the signal.

Multi-electrode techniques offer ways to estimate better these aspects. For example, for amplitude, multisite electrode recordings have been proved to improve the quality of the amplitude analyzed. Therefore, site-specific multi-electrode arrays might provide a good way to estimate some underlying anatomical parameters of humans, which might be person-specific, or at least change slowly through time, in the same way muscular anatomy does. Therefore, there is some potential for the development of EMG recording techniques for biometry according to anatomical aspects.

A second aspect to analyze is the potential of these signals for identifying people according to behavioral aspects. This second case has already been reported for facial EMG (Cohn et al. 2002), contrasting it to typical camera-based systems for expression recognition, but a systematic study for different site of EMG signals is -as far as we know- lacking.

Low intrusiveness asks to process EMG signals placing electrodes in a low intrusive place. This excludes directly facial EMG because of the social importance of faces in everyday social interaction. However, there is a large amount of studies of multisite EMG for motor coordination existing (Kleissen 1998) that could be adapted. Therefore, in the same way social behaviors such as facial expressions have subject specific components that can be used for person identification, other EMG information also involving motor coordination might be. These reasons are related with the complexity of the behavior and the correlation between different parts. An example of a particularly frequent behavior that would not interfere with social interaction in an everyday environment is grasping, or tool manipulation. There are quite good techniques for analysis of these activities (Winges and Santello 2005). It would be enough to try to detect the independent components across subjects, instead of the common ones, to do person identification, instead of explaining common patterns in motor coordination.

We did a preliminary test at Starlab which is going to be explained in detail in the next part of this chapter.

Six subjects (3 males and 3 females) participated in this study. The recording device was BIOSEMI ActiveTwo. The sampling rate was set to 2,048 Hz. We used 4 active electrodes placed in the forearms of the subjects (2 electrodes per forearm, see Fig. 7.8). The reference was placed on the right wrist. The electrodes were placed with the help of stickers and tape to make the contact more stable. We use conductive gel in all the electrodes (Fig. 7.9).

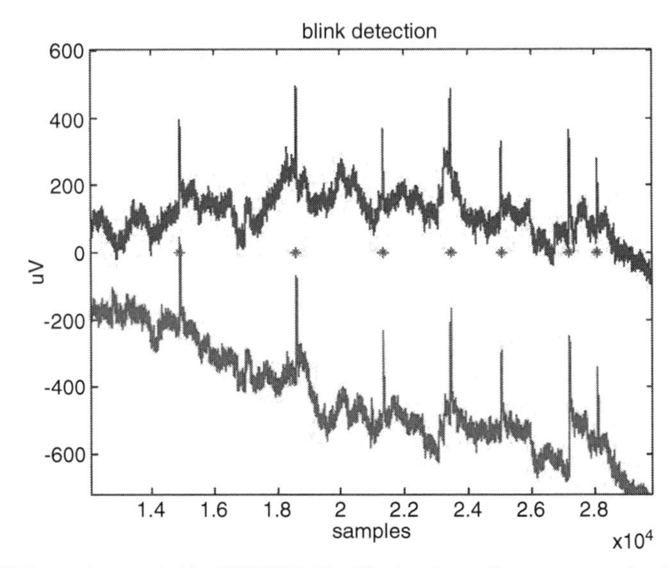

Fig. 7.8 EOG sample recorded by ENOBIO. The *black* and *grey lines* correspond to FP1 and FP2 respectively. The crosses mark the position in the signal where a blink is automatically detected by our algorithm

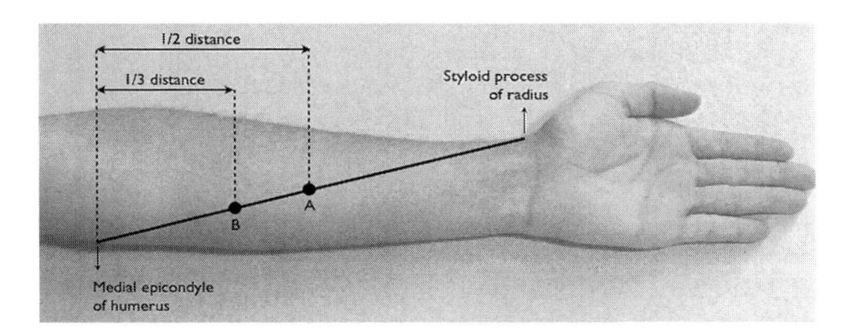

Fig. 7.9 This is a standard configuration to record forearm flexor (flexor carpi radialis and flexor digitorum sublimis) (Andreassi 2007). We perform the subtraction A-B in order to minimize the ECG artefacts. At the end we have two signals, one for each forearm

The subjects were asked to type a random text on a keyboard during 2 min. This task was done two times in two different sessions. By different sessions we mean that the electrodes were remove and replaced. The interval distance between sessions was 1 day for 4 subjects and some hours for 2 subjects.

First of all we reference the electrode A to the electrode B, in order to minimize ECG artefacts. This setup is called bipolar configuration. This was done for each forearm, so at the end we have two signals, one for each forearm. We then applied a band pass filter between 20 and 200 Hz because, although the EMG produces a wide range of frequencies, some experts agree that the maximal activity occurs at the lower end of the spectrum (Goldstein 1972). We are now ready to extract features for each time series. Three types of features are extracted: energy averaged over number of samples, Higuchi Fractal dimension and the Fourier Transform.

7.3.5.1 Energy

It is computed with this formula:

$$E = \frac{1}{N}\sum_{n=1}^{N} x(n)^2$$

Where N is the total number of samples and x(n) is the value of the sample n. This quantity represents the mean energy per sample. If we do not divide by the number of samples, we would have the total energy of the signal. Just as a reminder, the total number of samples is 120 s* 2,048 Hz = 245,760.

7.3.5.2 Higuchi Fractal Dimension

It is an algorithm used to compute Fractal Dimension in real world situation, where data is sampled and so on. The algorithm we use is the one used in the work (Arjunan and Kumar 2007). For a detailed description see (Higuchi 1988)

7.3.5.3 Fourier Transform

We compute the power spectrum density using Welch method (Welch 1967).

Once we have the feature extracted we are ready to present some preliminary results.

In Fig. 7.10 we can see the scatter plot of the energy of the right forearm versus the energy of the left one. We can see that there is a clustering tendency. In fact we can visually group the four takes of the six classes in all cases except in one (the blue round would be considered as belonging to the green class). The classification rate is thus equal to 23/24 = 0.9583.

Regarding the spectrum, we can see in Fig. 7.10 which corresponds to the left forearm, that it is easy to visually discriminate between subjects, except maybe for the blue, which is similar to the yellow and to the red. In fact for the blue one, the inter subject variability is big compared to the intra subject variability. The spectrum of the right forearm is more 'mixed' than the one for the left (see Figs. 7.10, 7.11 and 7.12).

Even thought we do not have a significant number of subjects nor takes, the results show that there is a discriminative potential in EMG based features. Probably by fusing the results from the different features we are able to correctly classify all the takes for the same subject. This work is still to be done in a more systematic way, and we might in the future record more EMG data from a larger data set.

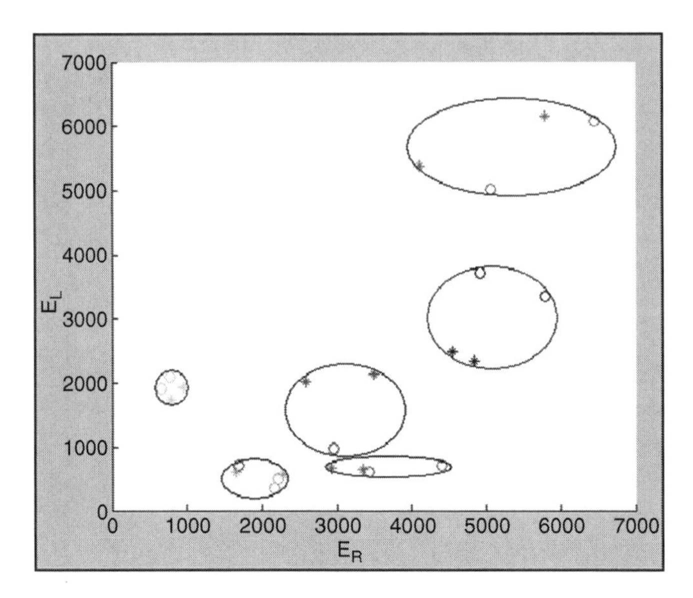

Fig. 7.10 Energy of right forearm versus Energy of the left one. The different colors represent different subjects and the *o* represents the two takes of the first session and the * represent the two takes of the second session

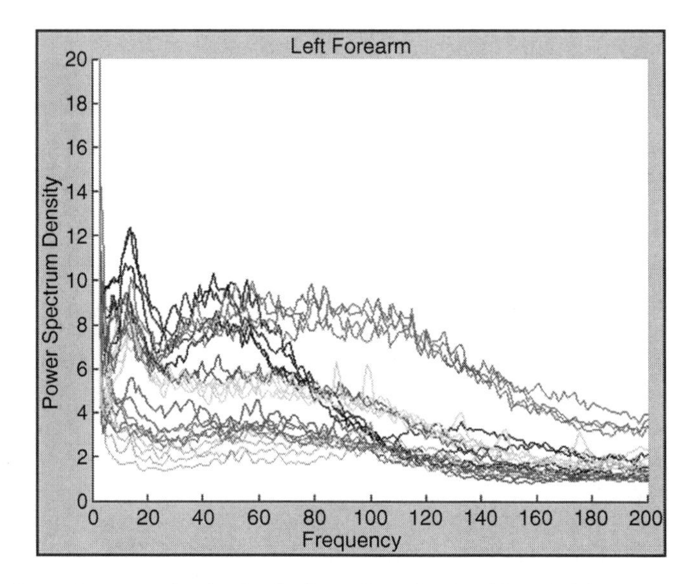

Fig. 7.11 Power Spectrum density for the four takes of each subject (*left forearm*). Each subject is represented by a color

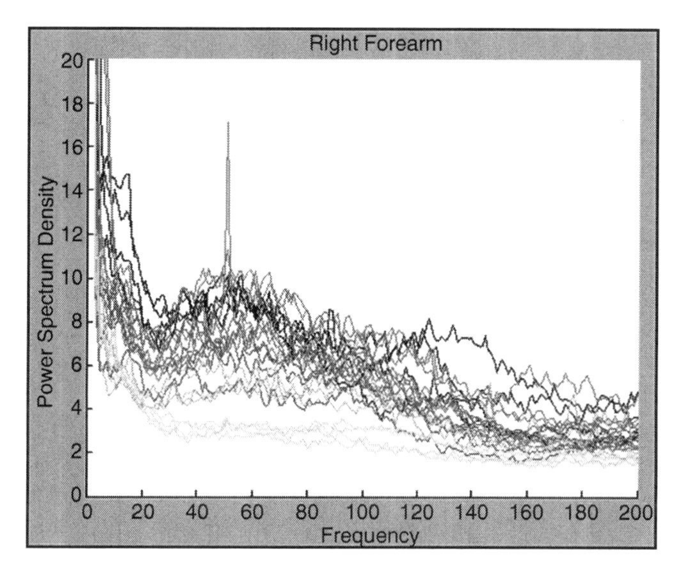

Fig. 7.12 Power spectrum density for the four takes of each subject (*right forearm*). Each subject is represented by a color

7.4 Technology Trends and Opportunities

Much of the current research in electrophysiological biometrics is naturally focused on improving performance, robustness and comfort for the user. It is often pitched as an alternative to existing technologies such as finger-print recognition, iris recognition and voice recognition. This assumes that the application space is the same and that what we offer is "more" security, for any given application, such as secure access.

This, however, is not the case. Electrophysiological biometry is fundamentally different in that it brings added value to existing security applications and is, perhaps more importantly, closely linked to emerging technologies that will generate new opportunities and requirements.

To quickly touch on the first point, what is the added value?

Essentially it is the ability to provide information on the user's physiological status while authenticating that user. This information can then be used to determine affective or emotional state which may have implications for the security of the system. For example, if a user is unusually stressed when accessing the system it may notify security and request a face-to-face follow up to ensure all is well. There are evidences that electrophysiological signals are related with emotions. For instance, "exotic" physiological activities such as gastric myoelectrical activity have been reliably assessed to be related with certain emotions (Vianna and Tranel 2006). Regarding ECG, the classification of 4 basic emotions has already been achieved based on detecting the physiological correlates of them on ECG and respiration data (Rainville et al. 2006). This article uses respiration and ECG signals to discriminate 4 basic emotions, according to the subjective report of them. Since it is possible to extract respiration form ECG, it might be possible to detect these 4 emotions using only an ECG lead. There are also some indications showing that indirect measures of emotions can be obtained from EEG data. For example, phase synchronisation between different cortical areas (Costa et al. 2006) seems to relate closely to emotional intensity, and to be different for different emotional reactions. Non-linear indexes as Kolmogorov complexity might as well reflex emotionally significant activity (Ljubomir et al. 1996).

Looking at emerging technologies we can see that not only is there added value for existing security applications but there is a growing market in applications that rely on electrophysiological signals to carry out their primary task. What this means is that these users are already wearing a sensor system capable of recording electrophysiological signals. This is a game changing shift in the field as it addresses one of the most serious shortcomings of electrophysiological biometry; the need to wear or touch electrodes capable of recording the signals needed to extract the relevant features.

7.4.1 Brain Computer Interfaces

BCIs are beginning to make the jump from the research lab to commercial applications and will soon arrive in your living room.

There are many examples of BCIs being used to control smart homes (Guger et al. 2009) and mobility devices for the disabled such as the Toyota wheelchair controlled by a BCI.[1] Regardless of the type of BCI, be it Synchronous Motor Cortex activation, Steady State Evoked Potentials or any other, we will have access to the raw EEG signal while the device is being used. Not only can we authenticate the user at the beginning of a session but we can continuously authenticate them. This can be transparent to the user and extremely difficult to bypass. As BCIs become more user friendly and comfortable their use will only increase (Riera et al. 2008b). In recent months we have also seen the emergence of cheap commercial BCIs targeted at the game and toy markets. These devices may soon be standard equipment providing access to your brain waves and their biometric potential.

7.4.1.1 Brain Computer Interfaces for Authentication

As far as we know, the approach we describe here has never been used for authentication purposes. In the next lines we will describe a scenario in which a subject, by loading and controlling a Brain Computer Interface (BCI), will be authenticated and will successfully unlock a computer in order to access its information. First of all,

[1] Real-time control of wheelchairs with brain waves http://www.riken.jp/engn/r-world/info/release/press/2009/090629/index.html. Accessed October 26th 2009.

we provide a definition of a BCI: A Brain Computer Interface is a system that translates brain activity into control actions to be performed by artificial effectors, in our case, a computer. There are many types of BCI's and they can be classified depending on many factors, such as level of intrusiveness, based on EEG (endogenous) or based on Event Related Potentials ERP (exogenous), based on motor imagery or based in more complex cognitive tasks, etc...One of the problems of BCI systems is the need for personalization. That is, the system needs to be tuned for each person in order to improve performance. While this is a disadvantage for BCI applications, we can use it as an asset for biometry.

In our model, we will introduce a simple motor cortex based BCI applied to biometric authentication. Most of the BCI's need a training phase, in which the subject is asked to perform certain mental imagery tasks according to some cues provided by a screen. For instance, in our system the subject will first of all decide which 2 motor imagery tasks he/she wants to perform in order to code 2 actions: left and right (the cues would be an arrow pointing right and an arrow pointing left). The subject could choose for instance left hand for left action and right hand for right action but many more combinations could be used choosing imagery tasks from the next list:

- Right hand
- Left Hand
- Right foot
- Left foot
- Both hands
- Both feet
- Tongue

In total, we have 42 possible combinations. Each subject will choose his own imagery tasks, and he/she would keep it for himself, in the sense that only he/she should know his/her imagery tasks. Once the subject has performed the training which can be considered as the enrolment into the biometric system, all the EEG recorded data will be used to select the best combination of features, classifiers and electrodes for each particular subject in order to maximize the classification rate of his/her training set.

Now the subject is ready to use the BCI in authentication mode. In order to unlock the computer, the subject will first of all claim his identity in order for the system to load his personal BCI (with the optimal combination of features, classifiers and electrodes for that particular subject). Then he/she will have to control a virtual locker in order to provide a password only known by him/her. In order to input the digits of the password, the subject will need to move the cursor to the corresponding number by performing motor imagery tasks (just as during the enrolment phase). If he/she accomplishes to do so in a limited amount of time, then the computer will be unlocked, and the subject authenticated.

There are two levels of security: the password and the control of the BCI. The biometry involves two aspects: on the one hand, the features, classifiers and electrodes locations are personalized, on the other hand, only the subject will know what imaginary movements he/she needs to do (hands, feet or tongue as explained before).

7.4.2 Telepresence Systems

As technology advances and environmental and political pressure increase we are beginning to see the first viable telepresence systems such as the one offered by Cisco.[2] These systems show huge potential for the reduction of business travel and the corresponding environmental impact.

However these systems are not just about video conferencing, research on Presence, VR and advanced interfaces is paving the way for fully immersive systems where not only will your image and voice be transmitted but also your physiological and emotional state. This augmented representation of the user can compensate for the lack of the personal multi-modal interaction that we are used to in daily life.

It can also address another issue associated with digitally reconstructed avatars as representatives; that of trust. If the person that claims to be speaking to you is authenticated by their own heart or brain signals we can reintroduce some sense of a flesh and blood person.

7.5 Vision for the Future and Risks

In previous chapters we have discussed emerging technologies and trends that will have a profound influence on electrophysiological biometrics. One of the more important is BCI and from our point of view, the future regarding electrophysiological biometrics is very much related with the future of the BCI technologies. There are many different ways to classify the various approaches to BCI but level of invasiveness is one of the most fundamental.

There are essentially three levels of invasiveness:

- Non-invasive BCI typically using external EEG electrodes.
- BCI based on electrocorticogram (ECoG) which is obviously more invasive, since now the electrodes are placed directly on the surface of the brain (cortex), having previously removed a piece of the skull.
- And the most invasive BCI approach of all where the electrodes are placed inside the brain (deep brain electrodes), by means of long needles with the recording part in the tip of the needle.

In any BCI the key to success is being able to distinguish your signal or feature of interest from the background noise (everything else that is going on). The versatility of the BCI depends on the number of separate states that can be classified from that signal. In many cases, such as motor cortex BCIs (Wang et al. 2006), this depends on being able to record a clean, well localized signals corresponding to

---

[2]Cisco Telepresence Solution http://www.cisco.com/en/US/netsol/ns669/networking_solutions_solution_segment_home.html. Accessed October 26th 2009.

well defined activation patterns. This is obviously easier to do on the cortical surface itself (ECoG) than on the surface of the scalp (EEG).

So here we have two contrary requirement drivers, on the one hand performance improvement pushes us towards invasive techniques while on the other, user acceptance and comfort push us towards non-invasive techniques. Obviously opening your skull in order to install a BCI is unacceptably aggressive and so we try to develop non-invasive EEG systems that apply ever more advanced signal processing and machine learning based classification techniques.

Looking to the future however we can ask, will this always be the case?

Recent work (Song et al. 2009) has shown significant improvement in invasive ECoG devices. These systems show great promise and may soon be seen as an appropriate response to certain circumstances. Pace makers and cochlear implants have become standard and the procedure for implanting them routine. There is every reason to expect that the same will be true of cortical implants if the benefits justify the risk. A reliable and versatile BCI that allows a paralyzed patient to interact more fully with their environment may be sufficient motivation.

In the future, both the electrode size and the procedure to implant them will surely improve and thus probably making the use of deep brain electrodes and implanted ones more common than nowadays. The electrodes of the future will be very small, will consume very few energy (they could even be powered by the subject's body movements or temperature changes) and of course wireless.

So we see that even in the most invasive case there is reason to expect an increase in uptake of these technologies and therefore an increase in the number of users that may take advantage of electrophysiological biometrics.

While there are many advantages to "always on" authentication and many advantages to having access to background physiological information we should not forget that for most normal use scenarios, where privacy and anonymity are valued, this may not be appropriate.

These issues must be very carefully studied and their implications clearly understood before any such system becomes widely implemented. Ensuring the privacy of Personal Health Records is a major concern for the main players in this arena as has been seen with the launch of Google Health and MS Health-Vault. These concerns will surely be raised for any system that has direct access to the physiological signals of a user.

7.6 Conclusion

Although the electrophysiological biometry is still in his youth and any commercially available device does not exist yet, this approach to biometric is very interesting for several reasons.

First of all, the possibility to perform a continuous authentication is very attractive because that way the subject does not need to follow any particular protocol in order to get authenticated, making the system transparent for the user. Of course in order to minimize the obtrusiveness, the recording device should be wearable and wireless. Moreover, such a device should be as small and comfortable to wear as possible. There are devices nowadays that fulfill these requirements, but certainly in the future, the miniaturization of both the electrodes and the electronic compounds will make those systems much less obtrusive. The possibility of implanted electrodes is also very attractive but in this case a lot of ethical issues arise: do we want always on systems? How will the privacy and anonymity of the data be handled? Will the society be going towards a 'big brother' type of society?

A second interesting feature of the electrophysiological biometry systems is their universality, at least for ECG and EEG. We can affirm that every living person has a beating heart and a brain that produces electrical signals. Some other standard biometric features are not universal or hard to collect in certain conditions. For instance for voice recognition, the subject needs to be able to speak, and even if he/she can speak but suffers from aphonia the biometric test would probably fail. The same applies to fingerprint recognition in which the fingerprint collection might fail if the finger skin of the subject is dry, or even if a small wound is present in his/her finger tip.

The possibility to extract more information besides the identity of the subject is a very interesting added value to the electrophysiological biometry. There are evidences that EEG, ECG and EMG can provide information related to the emotional state of the subject. This can be very useful in order to decide if a subject is in the proper conditions to fulfill his/her work. That becomes obviously interesting if the task of the subject requires a high concentration level and is potentially dangerous such as truck driving, nuclear plan controller or traffic air controller. Also in the future, the telepresence and virtual reality environments would become more and more used. In this field, the electrophysiological biometry can authenticate the avatars continuously, so we could be sure we are speaking with the right person in the virtual world and thus being more secure if we are sharing private and/or confidential information. We could also get access to the emotional information of the avatars, something related with the field of augmented reality, where humans can access information that are hidden to the 'common' senses. This could have many applications in the field of virtual reality storytelling (reference?) and virtual reality entertainment.

Many studies have been done with EEG and ECG (Llobera 2007). The challenge now is to take these systems outside the laboratories and study if the performance is maintained in real world applications where more artifacts are likely to appear. At least at this stage we can claim that the results are promising for these two modalities.

Regarding EMG, the results are promising as well, but very preliminary since a study with much more subjects should be done in order to extract deeper conclusions. It is interesting to note that this study is the first in which EMG is used a biometric feature, as far as the authors know.

Finally EOG has also been tested as a biometric feature, but in this case the results are poor. There is some biometric potential in EOG, but the performance does not match other biometric modalities. Probably the use of improved classifiers techniques, the search for new features in EOG, and the use of electrodes below the eyes (and in the sides of the eyes) to have access to vertical and horizontal movement separately and also to blinks would improve the performance, but this study is still to be done.

Acknowledgments The authors wish to acknowledge the ACTIBIO project, a STREP collaborative project supported under the 7th Framework Program (Grant agreement number: FP7-ICT-2007-1-215372) in which Starlab is actively involved. ACTIBIO aims at authenticating subjects in a transparent way by monitoring their activities by means of novel biometric modalities.

References

Andreassi, J.L. 2007. *Psychophysiology: Human behavior and physiological response*, 5th ed. Mahwah/London: Lawrence Erlbaum Associates Publishers.

Arjunan, S.P., and D.K. Kumar. 2007. Fractal based modelling and analysis of electromyography (EMG) to identify subtle actions. *Conference Proceeding of IEEE Engineering in Medicine and Biology Society* 2007: 1961–1964.

Biel, L., et al. 2001. ECG analysis: A new approach in human identification. *IEEE Transactions on Instrumentation and Measurement* 50: 808–812.

Chang, C. 2005. Human identification using one lead ECG. Master thesis. Dep Comput Sci Inf Eng. Chaoyang Univ Technol (Taiwan).

Cohn, J., et al. 2002. Individual differences in facial expression: Stability over time, relation to self-reported emotion, and ability to inform person identification. In *Proceedings of the International Conference on Multimodal User Interfaces*. Washington, DC: IEEE Computer Society.

Costa, T., et al. 2006. EEG phase synchronization during emotional response to positive and negative film stimuli. *Neuroscience Letters*. doi:10.1016/j.neulet.2006.06.039.

Galvani, L. 1791. De viribus electricitatis in motu musculari: Commentarius. Bologna: Tip. Istituto delle Scienze, p 58.4 tavv. f. t.; in 4.; DCC.f.70.

Goldstein, I.B. 1972. Electromyography: A measure of skeletal muscle response. In *Handbook of psychophysiology*, ed. N.S. Greenfield and R.A. Sternbach, 329–365. New York: Holt, Rinehart & Winston.

Graff, C., et al. 2007. Physiological signals as potential measures of individual biometric characteristics and recommendations for system development. Deliv D2.1, EU IST HUMABIO Project (IST-2004-026990).

Guger, C., et al. 2009. Brain-computer interface for virtual reality control. Proc ESANN: 443–448.

Higuchi, T. 1988. Approach to irregular time series on the basis of the fractal theory. *Pfysica D* 31: 277–283.

Israel, S., et al. 2005. ECG to identify individuals. *Pattern Recognition*. doi:10.1016/j.patcog.2004.05.014.

Kandel, E. 1981. *Principles of neural science*. New York: Elsevier.

Kleissen, R. 1998. Electromyography in the biomechanical analysis of human movement and its clinical application. *Gait & Posture* 8: 143–158.

Kyoso, M. 2001. Development of an ECG identification system. In *Proceedings of 23rd Annual International IEEE Conference of Engineering in Medicine and Biology Society*, 4: 3721–3723.

Ljubomir, A., et al. 1996. Non-linear analysis of emotion EEG: Calculation of Kolmogorov entropy and the principal Lyapunov exponent. *Neuroscience Letters*. doi:10.1016/S0304-3940(97)00232-2 DOI:dx.doi.org.

Llobera, J. 2007. Narratives within Immersive Technologies. arXiv:0704.2542.

Marcel, S., and J. Millán. 2007. Person authentication using brainwaves (EEG) and maximum a posteriori model adaptation. *IEEE Transactions on Pattern Analysis and Machine Intelligence*. doi:10.1109/TPAMI.2007.1012.

Mohammadi, G., et al. 2006. Person identification by using AR model for EEG signals. In *Proceedings of 9th International Conference on Bioengineering Technology*. Czech Republic.

Moukabary, T. 2007. Willem Einthoven (1860–1927): Father of electrocardiography. *Cardiology Journal* 14: 316–317.

Neher, E., and B. Sakmann. 1992. The patch clamp technique. *Scientific American* 266(3): 44–51.

Palaniappan, R., and S.M. Krishnan. 2004. Identifying individuals using ECG beats. In *Proceedings of the International Conference Signal Processing Communication*, 569–572. Bangalore. ISBN: 0-7803-8674-4.

Paranjape, R., et al. 2001. The electroencephalogram as a biometric. *Proceeding of the Canadian Conference on Electrical and Computer Engineering*. doi:10.1109/CCECE.2001.933649.

Poulos, M., et al. 1998. Person identification via the EEG using computational geometry algorithms. In *Proceedings of the Ninth European Signal Processing*, 2125–2128. Rhodes. ISBN 960-7620-05-4.

Poulos, M., et al. 1999. Parametric person identification from EEG using computational geometry. In *Proceedings of the 6th International Conference on Electron, Circuits and System (ICECS'99)*. doi: 10.1109/ICECS.1999.813403.

Poulos, M., et al. 2001. On the use of EEG features towards person identification via neural networks. *Médical Informatics & the Internet in Medicine*. doi:10.1080/14639230118937.

Poulos, M., et al. 2002. Person identification from the EEG using nonlinear signal classification. *Methods of Information in Medicine* 41: 64–75.

Rainville, P., et al. 2006. Basic emotions are associated with distinct patterns of cardiorespiratory activity. *International Journal of Psychophysiology*. doi:10.1016/j.ijpsycho.2005.10.024.

Riera, A., et al. 2008a. Unobtrusive biometric system based on electroencephalogram analysis. *Hindawi Journal of Advances in Signal Processing*. doi:10.1155/2008/143728.

Riera, A., et al. 2008b. STARFAST: A wireless wearable EEG/ECG biometric system based on the ENOBIO sensor. Phealth *Proceedings of the 5th International Workshop on Wearable Micro and Nanosystems for Personalised Health*.

Ruffini, G., et al. 2006. A dry electrophysiology electrode using CNT arrays. *Sensors and Actuators*. doi:10.1016/j.sna.2006.06.013.

Ruffini, G., et al. 2007. ENOBIO dry electrophysiology electrode; first human trial plus wireless electrode system. In *Proceedings of the 29th IEEE EMBS Annual International Conference*. Lyon. 10.1109/IEMBS.2007.4353895.

Song, Y., et al. 2009. Active microelectronic neurosensor arrays for implantable brain communication interfaces. *IEEE Transactions on Neural Systems and Rehabilitation Engineering* 17(4): 339–345.

Swartz, B.E. 1998. Timeline of the history of EEG and associated fields. *Electroencephalography and Clinical Neurophysiology* 106: 173–176.

Vianna, E., and D. Tranel. 2006. Gastric myoelectrical activity as an index of emotion arousal. *Psychophysiology*. doi:10.1016/j.ijpsycho.2005.10.019.

Waller, A.D. 1887. A demonstration on man of electromotive changes accompanying the hearts beat. *The Journal of Physiology* 8: 229–234.

Wang, Y., et al. 2006. Phase synchrony measurement in motor cortex for classifying. Single-trial EEG during motor imagery. In *Proceedings of the 28th IEEE EMBS Annual International Conference*. New York. doi:10.1109/IEMBS.2006.259673.

Welch, P.D. 1967. The use of fast fourier transform for the estimation of power spectra: A method based on time averaging over short, modified periodograms. *IEEE Transactions on Audio and Electroacoustics* AU-15(2): 70–73.

Winges, S., and M. Santello. 2005. From single motor unit activity to multiple grip forces: Mini-review of multi-digit grasping. *Integrative and Comparative Biology*. doi:10.1093/icb/45.4.679.

Zhang, D.D. 2000. *Automated biometrics: Technologies and systems*. Heidelberg: Springer.

Chapter 8
Intelligent Biometrics

Farzin Deravi

8.1 The Promise of Biometrics

Imagine approaching an automated cash machine and simply asking for money from your account which is then duly delivered to you – perhaps with a kind personal greeting. No cards, no PINs. Or just going through the public transport system, imagine never needing to buy a ticket or re-charge a token. "The system" knows you and knows which account to debit every time you pass in and out of the trains and buses. And when you get home, you "plug into" your home virtual reality system which takes you into another "world" where you can socialise with friends who may be geographically remote yet represented and uniquely recognised and where you may conduct further social and commercial transactions.

Whether such a future may appeal or not, the key to its realization is the availability of reliable and trusted recognition of personal identity. Furthermore, despite the complexities of system design, such recognitions must be virtually effortless from the perspective of the users. Biometric technologies, allowing the automatic recognition of identity through the measurement of physiological and/or behavioural traits, are therefore key components in realising such futures where our identity is easily and reliably asserted as required and dependably protected when necessary.

This chapter will explore the challenges facing such visions of a biometrics-enabled future and suggests possible solutions to overcome them – where Ambient Intelligence (AmI) is pervasive and is used to create environments that are sensitive and responsive to the presence and activities of people (Aarts et al. 2001).

F. Deravi (✉)
School of Engineering and Digital Arts, The University of Kent,
Jennison Building, Canterbury, Kent CT2 7NT, UK
e-mail: F.Deravi@kent.ac.uk

E. Mordini and D. Tzovaras (eds.), *Second Generation Biometrics: The Ethical, Legal and Social Context*, The International Library of Ethics, Law and Technology 11,
DOI 10.1007/978-94-007-3892-8_8,

8.2 The Challenge of Biometrics

The promise of reliable and efficient human recognition has been the driver for the development of biometric technologies in recent years. A further impetus for such development has been the growing threat from terrorism and crime as well as the increasing reliance on information networks for communication and commerce. In these expanding fields of activity and concern it is essential to be able to establish and monitor the identity of individuals.

However, despite the rapid development and increasing deployment of biometric technologies, there are serious concerns regarding their ability to effectively rise up and meet the challenge of current, let-alone future application requirements. These include the issue of recognition accuracy, especially given the large populations involved in case of national and international deployments, as well as issues related to security and privacy vulnerabilities that these technologies bring with them as side-effects (Jain et al. 2004).

In particular, in the area of security, it has become apparent that the promise of "uniqueness" that biometrics has often been associated with, can be undermined by enterprising fraudsters who could spoof such systems by using artefacts that can mimic the human physiological or behavioural characteristics, often with astonishing ease (Matsumoto et al. 2002; Matsumoto 2004; Thalheim et al. 2002; Schuckers 2002).

These issues and challenges, have to be countered with the development of the next wave of biometric solutions if the current applications are to be adequately supported. To address them, technological solutions may have to be co-developed with societal and legal remedies as an integrated whole. In the next section we will review some of the possible technological solutions to address the current challenges facing biometric technologies. The paper will then go on to envisage some future application scenarios and suggest how these solutions may be adapted to cope with the evolving user requirements.

8.3 Technological Remedies

Focusing on the twin challenges of accuracy and security, a set of key technological approaches have been proposed that may provide a basis for solutions for a range of applications. In this section we will present an overview of some of these technological approaches and examine the new challenges they bring in their turn.

8.3.1 Multibiometrics

One important technological approach that has been gaining ground is the integration or fusion of information from multiple sources of identity information. There is a wide, extensive and varied literature on such multibiometric identification systems and different types of such systems have been categorised (Ross et al. 2006). For example, information may be gathered from multiple biometric modalities – parts of the physiology (e.g. face, finger) or behaviour (e.g. gait, handwriting) and then combined at different levels (e.g. sample, feature, score or decision) to produce multimodal systems with enhanced performance and greater accuracy.

Alternatively systems may make use of additional biometric information, perhaps from the same biometric modality, but from multiple sensors to capture biometric information (multisensorial) or capture multiple samples captured from the same modality and with the same sensor (multiinstance). Combining identity information from such multiple sources has been shown to be helpful in improving the recognition performance and robustness of biometric systems in a number applications.

8.3.2 Adaptive Biometrics

A range of inter-related sources of variability are often present in most realistic scenarios where biometric systems are involved and are likely to affect the performance and overall effectiveness of such systems. These sources include environmental conditions (e.g. background lighting and noise), users' physiological/behavioural characteristics, users' preferences, variability of the communication channels in remote applications, and so on. If we focus on the users' biometric characteristics alone, it is clear that with a widening user-base it is important to consider the impact of "outliers" – those users who find it difficult or impossible to use the system. Failure to enrol on biometric systems or to consistently provide useable samples for biometric comparison may be due to a range of factors including physical or mental disability, age, and lack of familiarity or training in the use of particular biometric systems. Such "outliers" may compose a significant proportion of the population and their biometric exclusion may simply render the whole project untenable. In some key applications (e.g. a biometric system that may have to be deployed for authentication of a large population, such as in a national identity card scheme) it is essential that no one is excluded due to their inability to provide biometric samples because of age or disability, and therefore measures must be introduced to handle such outliers in a way that does not reduce the security or usability of the system.

While in most reported work, attention is generally focused on a multibiometric recognition setup based on a fixed set of information sources, it is clearly possible to adopt a more flexible approach in choosing which sources to integrate depending on individual user needs and application constraints. Such an approach removes, or at least reduces, the barrier to use by 'outlier' individuals and difficult environmental conditions, thus facilitating universal access through biometrics. Multibiometric systems offer the possibility for adapting the biometric information acquisition stage to suit the needs of particular individuals, for example by offering a selection and choice of biometric modalities to best match individual preferences and constraints, and thus reduce and perhaps even eliminate the size of the outlier population. Of course, in any implementation care must be taken so that the information regarding the selection of the modalities by outlier individuals does not compromise their privacy.

An adaptive multibiometric approach to identity recognition offers the system designers and users a measure of flexibility in the face of uncertain environmental conditions and user needs and capabilities. There are also clearly performance advantages with a more flexible multibiometric structure allowing an element of re-evaluation and adaptation in the information fusion process (Chibelushi et al. 1999). Incorporating an adaptive fusion strategy allows the system to cope better when there is a mismatch between the conditions at the time of system development and training and when it is used in the field thus avoiding significant drops in accuracy and generally producing more robust solutions.

All such multibiometric systems, if properly designed and adequately trained to cope with the type of data they may face in their target application, hold the promise of significantly increased recognition performance. One can also consider combining within the same fusion framework information of a non-biometric nature. Such "soft biometric" information can also contribute to improved recognition performance (Jain et al. 2004).

However, it must be admitted that there is an, often unspoken, tradeoff between the above two advantages of multibiometrics. The increase in accuracy promised by the combination of several modalities cannot be simultaneously enjoyed together with the flexibility that can result from a selection from an available menu of biometric modalities. Nevertheless, the multibiometric approach provides the facility for fine-tuning such tradeoffs and reaching the right bargain between flexibility and security to best match the application.

8.3.3 Liveness Detection

An important feature of most biometric modalities is that they are openly accessible by suitable sensors within range. This makes it very easy to capture biometric samples from people even when they are unwilling or unaware of such capture taking place. Coupling this with recent work on creating artificial biometric artefacts and samples to gain unauthorised access to biometric systems it is clear that there is an "existential" threat to all biometric systems if the problem of sensor-level spoofing is not effectively countered. In an ambient biometric future, where sensors are ubiquitous, the nature of this challenge will be substantially magnified (Thalheim et al. 2002; Matsumoto et al. 2002).

The technological response to this threat has been to develop techniques to detect if the biometric modalities being measured by a sensor are coming from a live subject and not a fabricated artefact. Such liveness detection techniques have been developed for individual modalities but their effectiveness has yet to be fully established. This is in part because the manufacturers are naturally reluctant to reveal the techniques used and also because evaluating liveness detection systems is still at its infancy (Matsumoto 2004).

A multibiometric approach can be helpful in this important arena of liveness detection as it requires samples from multiple sensors, thus creating additional barriers to sensor-level spoofing attacks. The very fact that information may be required from different modalities and sensors will raise the bar for any fraudster who will now need to devise additional artefacts and strategies. The use of multiple samples and sensors can give the biometric system designers increased flexibility for developing sophisticated liveness detection algorithms and strategies that may not be possible with a mono-biometric system.

8.3.4 Biometric Sample Quality

As one considers unattended and embedded biometric systems, it may be difficult to control the environmental conditions that may be critical to the performance of particular modalities. Here it becomes useful to not only capture a range of identity information but also to make an assessment of their "quality". For example, the lighting may not be correct for a high quality facial image capture, but the ambient sound levels are acceptable for a capturing a good voice sample. Biometric sample quality information can be utilised to further enhance the fusion process. By making an estimate of the quality of the live biometric sample and using this additional information to adapt the operation of the fusion module (which may have been trained earlier incorporating knowledge from both biometric samples and their associated quality) better recognition accuracy can be obtained. Having access to multiple modalities and a measure of their sample quality will allow the appropriate weighting of the evidence to achieve a quality weighted fusion of information leading to more robust decisions in the face of variabilities that often beset biometric systems. The attractiveness of such systems has lead to increasing interest and research in biometric sample quality measures and quality-weighted fusion algorithms (Nandakumar et al. 2006).

8.3.5 Biometric Encryption

Finally, another important technological component that is likely to play a critical role in future biometrics-enabled systems is that of encryption. Encryption technologies have been used for a long time in a range of security applications. In the field of biometrics they have emerged as an important guarantor of data privacy. An important danger in biometric systems is for a database of biometric references (templates) to be compromised and become accessible to unauthorised parties. Such templates may be traced back to individuals and their biometric samples and therefore raise the danger of spoofing attacks and/or function creep where personal data, meant for a particular application, becomes accessed for illegal or unintended applications (Jain et al. 2004).

To avoid such outcomes there has been an increasing interest in template-protected biometric systems (Ratha et al. 2002), where encryption technologies are used to protect the stored data. Encrypting the biometric samples, or building secondary secret keys that uniquely link to them, make it impossible to use the stored identity data to recover the original biometric samples. Such techniques are essential as we move towards ambient and pervasive biometrics-enabled applications to ensure the trust and protection of users and the continued usability of their personal biometrics.

So, in summary, technological components such as multibiometrics, liveness detection, biometric sample quality assessment and biometric encryption are likely to play an important role in the transition of biometrics from controlled applications to more unrestricted and ambient scenarios. They bring with them a range of additional challenges that need management through the design of systems with a greater degree of intelligence.

8.4 Intelligent Biometrics

While multibiometrics has the potential for overcoming many of the challenges facing identity recognition in current and future applications it presents two sets of complexities that need to be addressed. The first of these are related to the interaction of biometric users with the system. Given that instead of a single sensor there may now be several sensors for the acquisition of biometric data, and possibly more than one instance of capture needs to be accomplished, it is clear that there could be a greater burden on the user for interacting with the now more complex biometric acquisition system.

Another source of complexity is the interaction of the biometrics authentication subsystem with the rest of the application which could now become richer in possibilities. This will be explored later in the context applications requiring remote authentication, where the operating conditions and application requirements can vary widely and would have to be considered alongside user requirements and preferences.

8.4.1 Software Agents

Now to cope with this increase in complexity it may be possible to deploy the paradigm of *intelligent software agents* to design and implement biometric systems that can dynamically respond to application and user needs. Intelligent autonomous software agents and multi-agent systems have emerged as a rapidly expanding research field (Wooldridge and Jennings 1995). "Agents" can then be defined as software (sub-)systems that monitor and act upon some environment and are capable of autonomous action while pursuing the interests of some user or users and directing their activity towards accomplishing some goals. Such agents typically incorporate knowledge about their users' aims and objectives. This may be achieved using a pre-supplied knowledge-base as well as through a learning system that accumulates knowledge gained by the experience of the agent. This knowledge can then be used by the agent to accomplish the goals set by their users. In pursuit of their "missions", agents may be designed to be pro-active by exploring and exploiting any opportunities that may be available. They may also cooperate as well as compete with other agents and may have other interesting properties such as mobility.

8.4.2 Multi-agent Systems

Multi-agent systems (MAS) may be defined as a group of interacting agents (Jennings et al. 1998). Such systems can be used to tackle applications where multiple, and possibly contending, goals can be identified and where multiple perspectives of a problem-solving situation may be exploited. It may not be possible to handle such applications effectively with just one agent and subdivision of the overall task between a set of independent agents may result in the best solutions. Interactions in a MAS may include co-operation, co-ordination and negotiation between agents.

The proliferation of Internet-based applications has been a driving force for research and development of multi-agent systems (Fatima et al. 2006) where negotiating agents are of particular importance in electronic commerce. Negotiating agent systems, when applied to user authentication applications, can help strike a bargain between the needs of the information provider for establishing sufficient trust in the user on the one hand, and the confidentiality of the user's personal information and the ease of use of the system on the other. A different balance between trust and privacy may need to be achieved for each different service, transaction or session and may even need to be dynamically modified during each interactive session. Multi-agent systems can provide an effective framework for the design and implementation of such authentication bargaining systems.

Additional areas of active research and development in the field of intelligent agents include specialist programming and agent communication languages and software development environments. The design of the overall architecture for agent-based systems, with layered or hybrid architectures, involving reactive, deliberative and practical reasoning architectures, is also of considerable interest (Weiß 2001).

8.4.3 Biometric Agents

The development of multibiometric technologies raises the level of system complexity and the burden on the biometric system users to conveniently interact with such systems. In multimodal systems, the interaction effort of providing a biometric sample for only one sensor is now increased as there is a set of sensors for the user to interact with. Not only is there more effort required from the user but also there are more choices to be made in the form and sequence of the interaction with the system. It is this increase in the choices available to the users and system designers that suggest intelligent agents as an effective way forward for designing and managing intelligent and adaptable user interfaces with multi-agent architectures to facilitate the negotiation of trust, security and privacy requirements of system users.

With an agent-based biometric system, the dynamic selection of a variable set of biometric modalities can be accommodated to match the demands of a particular task or the availability of particular sensors. For example, a multi-modal system should be able to deal with situations where a user may be unwilling or simply unable to provide a certain biometric sample, or where a preferred biometric modality cannot support a required degree of accuracy. The deployment of a multimodal approach, where it is possible to choose from a menu of available modalities and modes of interaction, can therefore help to overcome barriers to access. In the case of users with disabilities, where the use of a particular modality (e.g., speech) may be difficult or impossible, identity information is captured through alternative sensors to suit the user constraints. Similarly, when environmental conditions, such as acoustic noise or background illumination, may be challenging to the use of specific modalities such as voice or phase, an adaptive agent-based approach may change the mode of interaction with the users as well as the algorithms used for fusing the available information to best effect.

In applications where remote and unsupervised biometrics-enabled access is required, it is essential to build in protection against fraudulent attacks. In particular, it is essential to establish "liveness" of the biometric sample to protect against spoofing and replay attacks. As we have argued earlier, this is an important consideration due to the ease with which for many modalities biometric samples can be recorded and reproduced, even without the subjects' active cooperation, to gain unauthorized access. Despite the progress that has been reported in integrating liveness detection for individual modalities it is likely that an agent-managed multibiometric framework provides a platform with additional flexibility to support more advanced counter-spoofing measures. For example, the agent interface can be deployed to provide a sophisticated challenge/response mechanism making it much more difficult to use sensor-level replay attacks and much easier to maintain the appropriate level of confidence in the liveness of samples.

In remote and networked applications of biometrics technologies an important consideration is to ensure that the legitimate rights and requirements of the users for the privacy of their personal information are respected. At the client side, the users will therefore wish to limit what they reveal by way of personal biometric information to only the minimum of what is necessary for establishing their access rights. Simultaneously, there is the need to establish the identity of user on the server side with as much confidence as may be required for a particular type of transaction or information access. In general, these goals at the server and client sides are in contention and a negotiating multi-agent architecture may be effectively utilized to engage in such bargaining on behalf of the users at the server and client sides.

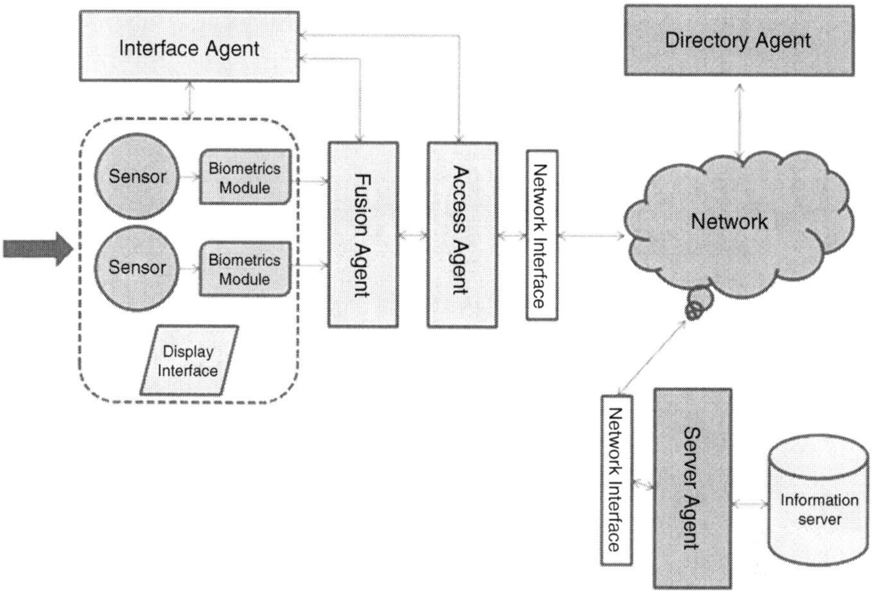

Fig. 8.1 A multi-agent networked architecture for biometric authentication

8.4.4 *Agent Architectures*

To enhance the performance of biometrics-enabled systems and services the agent paradigm may be employed in a number of ways. Its effectiveness may be best illustrated in a multi-modal biometric system for networked authentication and information access. Here a set of autonomous agents are deployed to handle the management of the user interface, the information fusion process and the negotiation between the information user and information server across a network. An example of such a multi-modal biometric system for a health-care application has been the IAMBIC project (Deravi et al. 2003) which is outlined below to illustrate possible applications of intelligent agent technologies in a biometric authentication setting (Fig. 8.1).

A set of agents will be cooperating on the client side of such a client-server architecture to address the user's specific requirements and constraints and to manage the user interface. A User Interface Agent coordinates the direct interaction with the user. Its goal is to establish, according to the requirements of the current transaction as well as the past user choices and behaviour (based on their claimed identity), the set of biometric measurements that must be obtained from the user. Additionally, it may be tasked to assess the quality and reliability of the measurements from each of the available biometric recognition modules. This agent can also adapt the mode of interaction with the user according to the user constraints and characteristics such as computer literacy, familiarity with the system being used, and so on.

The Interface Agent may also be used for the capture of other important non-biometric information. Additional environmental data may be captured by the available sensors (e.g., for the voice modality a sample of background noise may be captured) the analysis of which may determine the quality of any acquired data. Such acquisition conditions information can be used to help the agent to analyze any possible systematic enrolment and/or verification failures. The results can then be fed back to the user or to system operators to improve future performance. The interface may offer immediate suggestions to a user who is finding it difficult to provide useable samples on how to improve their interaction with the system. A user whose performance has been declining over a period of time may be automatically invited by the interface agent to re-enroll on to the system thus ensuring that the biometric template ageing effects are minimized.

Additionally, the acquired samples may be assessed for quality and associated with appropriate quality scores (cf. Sect. 9.3.4) and this information can be passed on to the fusion stage. The Interface Agent can directly interact with the individual biometric modules that produce features, matching scores and/or decisions and can change the acquisition parameters of these modules as required. Depending on the level at which the fusion takes places (sample, feature, score or decision) (ISO/IEC 2007) the appropriate information is then transmitted from to the Fusion Agent to manage the fusion process.

The integration of the biometric and other information obtained through the interface is handled with a Fusion Agent. Its main goal is to combine the available information using the best available technique and its optimal parameters to ensure the best possible performance. The design of the Fusion Agent requires knowledge of the types of biometric modalities available at the interface, as well as of their corresponding characteristics and of the levels of confidence in claimed identity that they can typically generate. This agent may have a set of different trained fusion algorithms to choose from. Biometric data as well as sample quality and environmental information obtained from the user by the Interface Agent are passed on to the Fusion Agent which in turn can produce an overall confidence score. This and other biometric data may then be passed to the Access Agent to facilitate its negotiation with the Server Agent.

Negotiating the access to the required data (e.g., medical records or other sensitive data) on behalf of the user is the responsibility of the Access Agent. Once this agent receives a request to access information from the Interface Agent, it determines the best remote data location (in the event that the data can be found in different places), contacts the sources of the desired information and negotiates its release with the appropriate Server Agent(s). The Access Agent has as its goal the release of the requested resources or information while providing the minimum amount of sensitive client-side user information. The goal of the Server Agent is to ensure that the requested information is only released to authorized users. Its goal is to ensure that sufficient confidence is reached in the identity of the claimed user. What may be considered as sufficient confidence is part of the Server Agent's knowledge base and may depend on the sensitivity of the data requested and the class of user who is accessing the information. If the information provided by the Interface and Fusion

Agents is not sufficient to satisfy the Server Agent it may enter into negotiation with them through the Access Agent. As part of this negotiation, a re-acquisition of biometric samples, perhaps using different modalities, may be requested. In some applications it may be essential that no raw biometric information is exchanged across the network. In these applications template protection and encryption strategies (c.f. Sect. 9.3.4) may be deployed by the agents to ensure user privacy is maintained.

Optionally, a Directory Agent may be used for cataloguing all relevant information about the location of resources and services within the network. In a healthcare application, for instance, this agent may store information on which databases contain particular information about the patients, practitioners and medical tests. Additionally, information may be stored with regards to databases of biometric identity information for comparison and authentication as well as information regarding where suitable and trusted algorithms for matching, fusion and sample quality assessment may be obtained to facilitate the agents' tasks. In providing pointers to such information, this agent may also suggest the best way of accessing the required information (for example, in the situation where several databases contain the information specified), based on considerations such as network traffic, bandwidth and cost.

A community of interacting agents such as the one described in the IAMBIC project can be implemented using a number of different approaches. These include methodologies for agent and interaction modelling, knowledge representation schemes and languages for inter-agent communication (Weiß 2001; Zambonelli et al. 2003; Chaib-draa and Dignum 2002). An important aspect of agent communication, especially in the contexts where biometrics and other personal information may be involved, is to ensure the security and privacy of the information exchanged between agents. The use of encryption and secure communication techniques is therefore an important consideration in the application of agent technologies to biometrics applications.

8.5 Biometrics in Virtual Worlds

As we envisage biometrics-enabled environments that seamlessly integrate knowledge of personal identity it is not hard to imagine a transition of such identity-aware environments and interactions from the real world into the virtual environments of multi-user games and virtual worlds. Millions of people currently inhabit multiplayer online games and virtual worlds. These are effectively large-scale virtual-reality environments and their population has been consistently and rapidly growing since their inception in the 1990s (Castronova 2002).

Three-dimensional (3D) virtual worlds such as Second Life (2009); There (2009) and Active Worlds (2009), go beyond the category of games and can be considered as extensions to reality and not only an escape from it through games (Balkin 2004).

These computer-based simulated environments are populated by multiple users who interact with each other and the environment via avatars. An avatar is a human representation in the form of a three-dimensional model. Additionally, the environment may contain Embodied Conversational Agents (ECAs) that could be designed and placed by the creators and users of the virtual world. Persons controlling one or more avatars can interact with the virtual world using an audio-visual interface.

An important example is Second Life (SL), a virtual world accessible via the Internet and developed by Linden Lab. SL, which was launched in 2003, is a persistent 3D world which enables its users, who are called residents to interact with other users through avatars. Its key features include construction tools to enable residents to construct 3D objects and scripting tools for developing interactive content. It also has its own currency, Linden Dollar, which has given rise to an effective economy that spans both real and virtual worlds as the Linden dollars can be converted to US dollars.

With its increasing popularity several companies and organisation including universities and government embassies have bought "land" in SL and opened up premises there. With the development of such a hybrid economy and rich social interaction it is not surprising to note the emergence of crime and the need for establishing personal identities of the people, avatars and agents in these virtual worlds. For example, disruptive residents, known as 'griefers', have been known to negatively impact users experience in virtual worlds (Kemp and Livingstone 2006).

There have also been reports of virtual terrorist attacks against buildings and avatars and also there have been reports of extremists using SL for "community building" (Gourlay and Taher 2007). The virtual currency in SL also opens the door for possible money laundering and funds transfer that may be difficult to monitor by the authorities.

8.5.1 Biometric Links to Reality

These developments highlight the need for better tools for personal identification geared to such hybrid worlds. First it is essential to be able to link real identities with virtual identities. In such online multi-user environments, biometrics can play an important role in establishing and maintaining a tight binding between real and virtual identities where such a binding is required.

Given the remote access scenarios that are invoked in such applications, an intelligent agents framework such as the one described in this chapter may be an effective basis for linking virtual and real identities, in a robust yet adaptable way.

8.5.2 Virtual Biometrics for Virtual Worlds

Also it can be envisaged that elements within the virtual environment, such as other avatars or their agents, the ECAs, may wish to perform a biometric verification or identification of other avatars that may be present in the environment. Here, one can envisage the assignment of a synthetic and virtual biometric characteristic to an avatar.

Such an assignment of tightly controlled by the authorised creators and maintainers of heterogeneous interacting virtual worlds (e.g. Linden Lab) can provide another means for the recognition of individuals within and across these worlds.

The creation of synthetic biometric characteristics is readily feasible with examples of fingerprint, voice and face synthesis having achieved prominence in recent years (Cappelli 2009; Bailly et al. 2003).

One can then also envisage a biometrics industry in such virtual settings with some companies offering synthetic identities and others offering multibiometric agent-based verification systems to facilitate transactions.

8.6 "Biometrics" Beyond the Living

Beyond biological life one may consider non-biological entities as having a historical life. A porcelain vase, for example, may be considered to have been "born" on a particular date, it may "live" for a while – as it "ages" – and finally it may "die". With very similar techniques to that used for human biometrics technologies, one could establish and monitor the identity of such non-biological entities. Such an assemblage of a large number of atoms may be considered to be distinguishable, even unique, from most if not all other entities.

Evidence of such an approach for uniquely identifying non-biological entities are emerging. For example, technologies for uniquely identifying paper, even before any content is written on it, have been presented. In (Clarkson et al. 2009) a novel technique is presented "for authenticating physical documents based on random, naturally occurring imperfections in paper texture". Using an off-the-shelf scanner, the authors were able to generate a "concise fingerprint" that "uniquely identifies the document". The ability to uniquely identify such objects has a wide range of applications, including detection of forgeries and counterfeits and applications in health-care and hygiene.

This work shows that ordinary objects, such as pieces of paper, can be considered to possess something akin to biometric identifiers allowing individual objects to be uniquely identified. It is easy to imagine extending this work to cover objects other than paper, given appropriate sensors and scanners and the design of corresponding feature extractors. With the increase in the resolution of sensors it is likely that more detail can be utilised in the object recognition systems of the future providing richer sources of information to identify individual objects.

8.7 Conclusions

In this chapter we have taken an overview of the challenges facing a more pervasive and ambient use of biometric person recognition technologies. The challenges currently facing the deployment of biometric systems are only likely to be further amplified when one considers a future where biometrics-enabled embedded systems are used to seamlessly monitor and mediate human activities with ever large user communities and increased requirements for accuracy, security, privacy and usability. We have also considered how the notion of biometric identity may be extended to virtual worlds and inanimate objects, blurring the boundaries that limit the application of identity recognition techniques.

Some key technological ingredients that are essential to addressing the challenges facing ambient biometrics are highlighted and the paradigm of intelligent software agents has been suggested as the glue that can bind these technological components together providing the support needed for the management of multimodal biometrics within an overall framework for trusted and privacy-preserving information exchange.

However, to reach the vision presented at the start of this chapter and prevent its opposite from coming true, it is likely that technological remedies alone will not be enough. Considerations of societal and legal remedies are likely to play an equally important role if we are to avoid a future where biometric technologies are used against the interests of the citizen or are abandoned as being unworkable. With the advent of new and ever more demanding application scenarios, the need for technological innovation will be sustained and must be faced in a multidisciplinary and holistic way to ensure that future services and applications can guarantee an acceptable level of user experience and approval.

References

Aarts, E., R. Harwig, and M. Schuurmans. 2001. Ambient intelligence. In *The invisible future: The seamless integration of technology into everyday life*, ed. P.J. Denning, 235–250. New York: McGraw-Hill.

Active Worlds. 2009. http://activeworlds.com/ Accessed 3 Sept 2009. (Archived by WebCite® at http://www.webcitation.org/5jVolymNk).

Bailly, G., M. Bérar, F. Elisei, and M. Odisio. 2003. Audiovisual speech synthesis. *International Journal of Speech Technology* 6: 331–346.

Balkin, J.M. 2004. Virtual liberty: Freedom to design and freedom to play in virtual worlds. *Virginia Law Review* 90(8): 2043–2098.

Cappelli, R. 2009. Synthetic fingerprint generation. In *Handbook of fingerprint recognition*, 2nd ed, ed. D. Maltoni, D. Maio, A.K. Jain, and S. Prabhakar, 271–302. London: Springer.

Castronova, E. 2002. On virtual economies. *CESifo Working Paper Series* No. 752.

Chaib-draa, B., and F. Dignum. 2002. Trends in agent communication language. *Computational Intelligence* 18(2): 89–101.

Chibelushi, C.C., F. Deravi, and J.S.D. Mason. 1999. Adaptive classifier integration for robust pattern recognition. *IEEE Transactions on Systems, Man, and Cybernetics - Part B: Cybernetics* 29(6): 902–907.

Clarkson, W., T. Weyrich, A. Finkelstein, N. Heninger, J.A. Halderman, and E.W. Felten. 2009. Fingerprinting blank paper using commodity scanners. In *Proceedings of IEEE Symposium on Security and Privacy*, 301–314. Oakland: IEEE Computer Society Press.

Deravi, F., M.C. Fairhurst, R.M. Guest, N. Mavity, and A.D.M. Canuto. 2003. Intelligent agents for the management of complexity in multimodal biometrics. *International Journal of Universal Access in the Information Society* 2(4): 293–304.

Fatima, S.S., M. Wooldridge, and N.R. Jennings. 2006. Multi-issue negotiation with deadlines. *Journal of Artificial Intelligence Research* 27: 381–417.

Gourlay, C., and Abul, Taher. 2007. Virtual jihad hits second life website. *The Sunday Times*, 5 August 2007.

ISO/IEC TR 24722:2007. 2007 Information technology — Biometrics — Multimodal and other multibiometric fusion, ISO.

Jain, A.K., S. Pankanti, S. Prabhakar, J. Hong, and A. Ross. 2004. Biometrics: A grand challenge. In *Proceedings of the 17th International Conference on Pattern Recognition (ICPR'04) –* Volume 2, 935–942. Cambridge.

Jennings, N.R., K. Sycara, and M. Wooldridge. 1998. A roadmap of agent research and development. *Int. Journal of Autonomous agents and multi-agent systems*, 1(1): 7–38. http://eprints.ecs.soton.ac.uk/2112/.

Kemp, J., and D. Livingstone. 2006. Putting a second life "Metaverse" skin on learning management systems. In *Proceedings of the Second Life Education Workshop at the Second Life Community Convention*, 13–18. San Francisco: The University of Paisley.

Matsumoto, T. 2004. Artificial fingers and irises: Importance of vulnerability analysis. In *Proceedings of the 7th International Biometrics Conference,* London.

Matsumoto, T., H. Matsumoto, K. Yamada, and S. Hoshino. 2002. Impact of artificial gummy fingers on fingerprint systems. In *Proceedings of SPIE Vol. #4677, Optical Security and Counterfeit Deterrence Techniques IV*, 275–289. San Jose.

Nandakumar, K., Y. Chen, A.K. Jain, and S.C. Dass. 2006. Quality-based score level fusion in multibiometric systems. In *Proceedings of the 18th International Conference on Pattern Recognition (ICPR'06)* Volume 4, 473–476. Hong Kong.

Ratha, N., J. Connell, and R. Bolle. 2002. Enhancing security and privacy of biometric-based authentication systems. *IBM Systems Journal* 40(3): 614–634.

Ross, A.A., K. Nandakumar, and A.A. Jain. 2006. *Handbook of multibiometrics*. New York: Springer.

Schuckers, S.A.C. 2002. Spoofing and anti-spoofing measures. *Information Security Technical Report* 7(4): 56–62.

Second Life. 2009. http://secondlife.com/. Accessed 3 Sept 2009. (Archived by WebCite® at http://www.webcitation.org/5jVoZ8tCC).

Thalheim, L., J. Krissler, and P.M. Ziegler. 2002. Body check: Biometric access protection devices and their programs put to the test. *c't magazine*.

There. 2009. http://www.there.com/. Accessed 3 Sept 2009. (Archived by WebCite® at http://www.webcitation.org/5jVoiiU2j).

Weiß, G. 2001. Agent orientation in software engineering. *Knowledge Engineering Review* 16(4): 349–373.

Wooldridge, M., and N.R. Jennings. 1995. Intelligent agents: Theory and practice. *The Knowledge Engineering Review* 10(2): 115–152.

Zambonelli, F., N.R. Jennings, and M. Wooldridge. 2003. Developing multiagent systems: The Gaia methodology. *ACM Transactions on Software Engineering and Methodology* 12(3): 317 370.

Part III
Identity, Intentions and Emotions

Chapter 9
Behavioural Biometrics and Human Identity

Ben A.M. Schouten, Albert Ali Salah, and Rob van Kranenburg

Abbreviations

ADABTS	Automatic detection of abnormal behaviour and threats in crowded spaces
GSR	Galvanic skin response
HUMABIO	Human monitoring and authentication using biodynamic indicators and behavioural analysis
ICT	Information and communication technologies
OECD	Organization for economic cooperation and development
PIR	Passive infrared
RFID	Radio frequency identification

9.1 Prelude

Biometrics is a key fundamental security mechanism, which links the identity of an individual to a physical characteristic or action of that individual, using methods that focus upon the individual variations between members of a given population. Currently mainstream biometrics that are being exploited in commercial systems include fingerprint and face recognition, speech verification, dynamic signature

B.A.M. Schouten (✉) • R. van Kranenburg
Fontys University of Applied Sciences, Rachelsmolen 1, R1 5612 MA Eindhoven, The Netherlands
e-mail: ben.schouten@fontys.nl; r.vankranenburg@fontys.nl

A.A. Salah
Computer Engineering Department, Bogazici University, 34342 Bebek, Istanbul, Turkey
e-mail: salah@boun.edu.tr

E. Mordini and D. Tzovaras (eds.), *Second Generation Biometrics: The Ethical, Legal and Social Context*, The International Library of Ethics, Law and Technology 11,
DOI 10.1007/978-94-007-3892-8_9, © Springer Science+Business Media B.V. 2012

Fig. 9.1 Some commercially available sensors to detect vibration, rotation and humidity (from *left to right*); see www.phidgets.com

recognition, iris and retinal scanning, hand geometry and keystroke dynamics. In general, there are two types of biometrics: behavioural and physical. Behavioural biometrics focuses on how a human characteristics evolves over time (handwriting, gait, etc.), while physical and more traditional biometrics can be seen as an imprint of a certain physical property (face, iris, etc). Combinations are also possible when different biometrics are fused (so called *multi biometrics*).

The interest in behavioural biometrics is rapidly growing. New advanced sensor technologies enable different bodily behavioural characteristics (heart beat, electrical skin conductance, etc.) to be analyzed for authentication, and the robustness of these techniques is rapidly improving. Moreover, new and networked sensors have been introduced in smart environments, capable to detect physical properties (like pressure, temperature, etc.), motion and motion-based properties, contact properties, and presence (e.g. radio frequency identification (RFID), passive infrared (PIR) sensors etc.), and come commercially available, see Fig. 9.1. An overview of these is given in (Cook and Das 2005). The fusion of these characteristics over time and place are very promising for biometrical authentication (Li et al. 2009).

With the availability and advances of this new sensor technology and the improved network capabilities, there is a growing interest in intelligent distributed sensor networks. Such proliferation of technology has immediate implications on biometrics technology, which requires this kind of infrastructure to extend the capabilities offered by the biometric system, particularly in terms of increased accuracy and decreased intrusiveness (Tistarelli et al. 2009). Taken together with the multiplicity of digital identities most people in modern societies maintain, the future lifestyle in a digitally enhanced environment will obviously require more and more biometric technology to protect information and to ease access to personal resources. The ISTAG report published in July 2009 stresses this point aptly: "*Citizens need to be assured of the security of the complex systems that they do not control and on which they depend. The information society is becoming more fragile with respect to the threats of a totally networked world*" (ISTAG). In this chapter we discuss the implications of these trends.

9.2 Introduction: Identity and Body

In our networked society one of the most crucial questions in many transactions or engagements is the identity of the entity (person) with whom the transaction is being conducted. Historically our acquaintances are very much local: personal relationships, face-to-face contract signings, notaries, and third party counsels are used to help establish trust in our communications. There are currently two mainstream trends in identity management: a technology-driven approach and a sociology-driven approach, respectively. In the definition of Goffman (1959) identity is based on interaction; a fluid, active process, depending on context of actions (gender, class, ethnicity etc.). It consists of independent and partial sub-identities, which are to be constructed anew in everyday life. In this way information and communication technologies (ICT) can be seen as tools to support these actions. Facebook and Second Life are examples of this. Lamb and Davidson (2002) state that individuals build and maintain social networks through which they "negotiate" their identities.

In a second definition of identity, Hayles posits information over the material itself, and erases the traditional boundaries of body and personality (Hayles 1999). As the breathing medium of information, communication becomes the defining characteristic of the human. This definition of identity is perhaps inevitable, as (historically) the information processing paradigm dominated cognitive sciences and reduced the human mind into a black box that processes data and produces information for a while. The metaphor prevailed in shaping the notion of identity, and the actual body became almost an afterthought. However, in the light of accumulating research evidence, the body had to be re-introduced through theories of embodied cognition, thereby reasserting the dynamic nature of the human organism that creates itself historically in constant feedback loops within its physical and social setting (Varela et al. 1992).

In the more biometrical or technology-driven practice (or equivalently, in more conventional practice), identity is seen as a relatively stable set of personal data, occasionally divided into subsets (partial identities), but mostly constituted of sensitive personal data, which, as such, needs protection, privacy and control. This approach treats the body as a source of multimodal patterns that can be predicted and verified. In all cases, given the direction of development, two types of digital data are being communicated: bodily data and non-bodily data, respectively. Both can be used for authentication; however, biometrics focuses on the first category.

With the advance of biometrics, medical science and other disciplines, the cardinality of data related to the body is ever growing, allowing us not only to measure the health and functionality of our body, but also its appearance. Moreover, with the influence of the advertisement sector and TV commercials, the bodily appearance becomes a playground, and subject to constant change with programs like Adobe Photoshop or other visual manipulation tools. The status of body in terms of its information content and the implications of its digitalization with respect to biometrics are the two topics under study in this chapter.

The remaining part of the chapter is structured as follows. In Sect. 9.3, we discuss the new trends in biometrical research, focusing on behavioural biometrics, remote biometrics and multibiometrics. Section 9.4 elaborates on disembodied, ephemeral scenarios and use-cases and the impact of these new technologies for engineering the body, as well as identity management and bodily aspects. In Sect. 9.5 we discuss the use of biometrics for an ambient lifestyle and we conclude in Sect. 9.6.

9.3 Second Generation Biometric Modalities

9.3.1 Behavioural Biometrics and Applications

With increased availability of cheap and innovative sensors, is has become possible to derive correlations from many sensors and construct prototypical patterns of behaviour, which can be employed to authenticate a person, as well as to derive a host of associations and inferences about a person. We will call this *behavioural biometrics*. What is learned from such behavioural patterns usually pertains specifically to a particular sensor setup, and thus it is difficult to generalize or 'hijack' this kind of information, although analysis can be carried out to learn many more things than ordinarily indicated by the sensor readings. The type of personal and interaction information collected this way is a rich source for mining all kinds of social signals, and opens new vistas in marketing and business intelligence (Pentland 2008). Pattern recognition methods are adapted to find spatio-temporal patterns in multiple streams of sensor data for automatic analysis of human behaviours and habits in these settings. These methods include search for recurrent event patterns (Magnusson 2000; Tavenard et al. 2007), clustering time series generated by low-resolution sensors using Markov models (Wren et al. 2006), using compression algorithms for extracting patterns (Cook 2005), and eigen-analysis of behaviours (Eagle and Pentland 2006).

The modern mobile phone is already equipped with many such sensors. In a revealing study, Eagle and Pentland have equipped a large number of students with smart phones, and collected simple behaviour data for over a year (Eagle and Pentland 2006). The data included information about when the phone is turned on, or whether a conversation is carried out, location information, and other simple sensor reading. A correlation analysis of behaviour patterns proved to be sufficient to determine for instance with good accuracy, to which group (e.g. management vs. engineering students, junior vs. senior) a particular student belonged. It is thus possible to create a behaviour template of the user of a system, and authenticate the user with this, or at least reject a large number of attempts to use the system based on deviations from the user's normal behaviour.

Another good platform is the sensor network setting, for instance an ambient intelligence environment like a smart home or a smart car (Cook 2005). It is possible to perform biometric authentication by correlating many simpler sensors rather than collecting data directly revealing the identity. The benefits of this setup are multiple; it becomes possible to authenticate groups of users (for instance to prevent access of children to potentially dangerous areas) and the perceived intrusiveness of simple sensors is much lower than for instance cameras observing the environment. Comparing data streams emanating from a sensor network will uncover meaningful associations, which might have a significant practical value in contributing to the robustness of identification process via traditional biometric modalities.

The recent FP7 research project ACTIBIO explores the possibility of continuously determining and verifying the identity of a user in typical and non-obtrusive scenarios, for instance during activities observed in a working environment (Ananthakrishnan et al. 2008). Possible novel biometric modalities include grasping patterns, facial actions, hand and body movement patterns, and keyboard typing behaviour. Its precursor project HUMABIO (Human Monitoring and Authentication using Biodynamic Indicators and Behavioural Analysis) has proposed a posture analysis authentication mechanism for preventing the hijacking of heavy goods vehicles (Damousis et al. 2008). For digital environments, it is possible to define biometrics that do not require additional sensors. For instance mouse movements (Ahmed and Traore 2007) and keystroke dynamics (Monrose and Rubin 2000) are behavioural cues that can lead to identification. However, these modalities contain a high variance, and thus are rarely usable as stand-alone modalities.

9.3.2 *Patterns of the Body and New Modalities*

Traditional biometric modalities are the ones that people use for identifying other people. Computers have access to sensors that go beyond these modalities, making novel biometric applications a possibility. For instance brain patterns, which are distinctive to individuals, can be a potential modality for authentication (Marcel and Millan 2007). The American company Emotive Communications, Inc sells headsets that can be used to navigate through a game by simple imagination (www.Emotive.com).

A new biometric modality that has come into consideration is the gait of a person (Boyd and Little 2005; Sarkar et al. 2005). The gait has been analyzed before to determine the activity type of a person (running, walking, etc.), but its use for biometric purposes requires more advanced techniques, resistant to variations due to shoe type, clothes, walking surface type, and view point. According to the extensive HumanID evaluation, shoe type has a small but statistically significant effect, followed by camera view point changes and carrying a briefcase. Matching over different time periods and surface type have also been shown to affect the authentication rates greatly (Sarkar et al. 2005).

Some novel biometric modalities are derived from research originally started for other purposes. For instance tongue diagnosis is an important method in Traditional Chinese Medicine, which makes automatic tongue image analysis an interesting application (Zuo et al. 2004). Once the analysis techniques are developed, it becomes possible to ask the question of whether or not it is possible to authenticate a person by his or her tongue image.

In a recent and excellent review of novel biometric modalities (Goudelis 2008) enlists over 20 different non-traditional approaches to person authentication. Most of these new modalities do not enjoy the extensive testing traditional modalities like face recognition received, but they certainly point out to different possibilities for different application requirements, as each of them has distinct advantages and disadvantages. For instance thermal images of the face are robust to surface modifications like make-up and possibly aging, and it can be operated in darkness (Socolinsky et al. 2003). Near-infrared imaging, on the other hand, is illumination resistant, and ideal for controlled indoor scenarios (Buddharaju et al. 2007).

Usability is of great importance for a biometric modality. Subsequently, many systems rely on biometrics that can be easily acquired or natural for a person to present. Biometrics based on palm print, finger vein patterns, nail texture, skin spectroscopy, hand or finger texture all rest on the idea that presenting the hand is natural and fast. Acquisition convenience and accuracy needs to be balanced for a given scenario. For instance dental images may be highly accurate in identifying persons (Chen and Jain 2005), but the acquisition of the image is difficult. X-ray imaging is not an option for everyday usage, because of the radiation exposure. On the other hand, ear images are easier to acquire, but the discriminativeness of the ear is not as high, and their uniqueness is contested (Chang et al. 2003). Biometrics that rely on patterns of DNA, ECG and EEG do require special equipment, and their use remains restricted to specific scenarios.

There are already a plethora of biometric possibilities, and we can very well expect new modalities to be considered in the future. The choices are further tailored to application scenarios by taking combinations of modalities.

9.3.3 Multi Biometrics and Soft Biometrics

The future of biometrics involves adapting biometrics to ever more challenging situations. For this purpose, several extensions to conventional biometric systems are relevant. In this section we look briefly at recent research in multi-biometric fusion and soft biometrics.

Multi-biometrics, i.e. consolidating the evidence by multiple biometric sources, is a primary way of adjusting the security-convenience trade-off in a biometrics system. It can be implemented by authenticating a user on a number of multiple modalities at the same time (parallel scheme), or in a cascade (serial scheme), where a user only has to submit a second (and subsequent) biometric signal in the case of doubt (Ross et al. 2006). In parallel architectures the security is increased by reducing the false accept. Disadvantages are higher financial costs and larger user involvement, as the evidence acquired from multiple sources is simultaneously processed in order to authenticate an identity (Maltoni et al. 2003).

Information fusion in biometrics is useful for two main purposes. Firstly, it may be the case that the design specification of the biometric system requires a range of operational beyond the technological provisions of a single biometric modality, either in terms of security, or user convenience. Through multiple biometrics, it becomes possible to design systems that fit more demanding requirements. In terms of user-convenience, we should also mention that multiple biometrics may be essential to prevent discrimination of users. Some biometric modalities (like fingerprints) are not applicable to for a small percentage of the population (Newham 1995), and consequently, the introduction of these modalities will discriminate these users. It is easily conceivable that a company, instead of implementing a costly backup strategy for these cases, just replaces the employees that do not conform to the requirements of the biometrics system used in the company. Multi-biometric fusion offers a way out by providing alternative authentication paths, at the cost of making the security of the system equal to the security of the weakest modality. As a compromise, it is possible to allow a small number of known users through a single modality, whereas 'normal' users will be authenticated through multiple biometrics in parallel.

Biometric information can be fused at different levels, including fusing the raw data, features, match scores, or decisions of individual matchers. Dynamic Bayesian networks are popular for probabilistic fusion of biometric evidence, allowing to cope with uncertainties in the input (Maurer and Baker 2007). An important issue is the evaluation of the statistical correlation of the input to such fusion systems (COGNIRON; Salah et al. 2008). Another dimension of fusion is the architecture, which can be serial or parallel. In (Gökberk et al. 2005), a serial (hierarchical) fusion scheme is considered for 3D face recognition in which the large number of possible classes is first reduced by a preliminary classifier that ranks the most plausible classes, followed by a second and more specialized tier of classifiers. This scheme is contrasted with a parallel fusion scheme in which all classifiers outputs are evaluated and fused at decision level. The parallel approach has increased real-time operation cost, but its accuracy is superior to that of the serial, and both fusion approaches excel in comparison to individual classifiers.

The quality of biometric samples used by multi-modal biometric experts to produce matching scores has a significant impact on their fusion. The quality depends on many factors like noise, lighting conditions, background, and distance to sensor (Tabassi et al. 2004; MIKR 2005). In Poh et al. (2009), 22 multi-biometric systems are assessed for the inclusion of quality information, as well as the cost of using additional modalities. The comparative evaluation suggests that using all the available biometric sensors will definitely increase the performance. The consequences however are increases costs in terms of acquisition time, computation time, the physical cost of hardware and its maintenance cost. These costs are alleviated to a certain extent in serial fusion schemes, where a fusion algorithm sequentially uses match scores until a desired confidence is reached, or until all the match scores are exhausted, before outputting the final combined score. In practice, the scenario may correspond to two settings; one in which multiple biometrics are acquired at the same time (at no additional cost in terms of user convenience) and the benefit is in processing time, and one in which the user is repeatedly queried until authentication occurs (or fails).

A second enhancement to ordinary biometrics is the inclusion of *soft biometrics* (Jain et al. 2004), which are easily measurable personal characteristics, such as weight and fat percentage, which can improve the performance of biometrics in verification type applications. Studies show that such simple physiological measurements can be used to support biometric recognition. Furthermore, most soft biometric traits are unobtrusive, posing no risk of identity theft, and they can be obtained via cheap sensors and simple methods. Their simplicity and weak link to identity are positive aspects with respect to usual negative connotations of biometrics, which make them especially adequate in applications where convenience is more important than security. A typical example is a weight-sensing car seat that can differentiate between two typical users of a car and allows customization based on this simple information. This seat may also prevent the child of the family from starting the car. In certain environments with a small number of subjects, like a smart car or a smart home environment, these features are robust enough to perform authentication or at least capable of determining partial identity classifications like gender, age group and such, thereby enhancing forms of anonymous identity management.

9.3.4 Biometrics from a Distance and Transparent Biometrics

The vast progress in sensor technology and computer vision enables the capture of biometrical traits from a distance for certain modalities like face recognition and iris recognition. The use of CCTV camera networks for security is an example of advanced face recognition at a distance using distributed sensor networks. Such sensor systems can be used to explore the correlations of biometric traits over time and place (spatio-temporal correlations). The processing methods depend on the configuration of sensors; for instance methods for tracking and authentication people from camera input are different for the case of a single-camera and the case of multi-camera (Fleuret et al. 2008). With present technology, it is possible to track people by using simple background-foreground separation in combination with colour features when the scene is not crowded (Kang et al. 2004; Mittal and Davis 2003). In more crowded environments, several methods are developed for segmentation and movement filtering to deal with occlusions (Haritaoğlu et al. 2000; Black et al. 2002). Tracked subjects can be authenticated remotely with face and gait recognition approaches.

What these techniques have in common is the fact that no explicit action is required from the user during the authentication (in contrast to for instance presenting a finger during the crossing of a border). We will call this *transparent biometrics* (Tangelder and Schouten 2006). In general, biometric recognition techniques at a distance are less robust as a consequence of uncontrolled occlusions, movements of objects and subjects, lighting variations, acquisition noise, or simply because smaller templates due to distance. To enable authentication in these more difficult conditions, it becomes necessary to consult more modalities, through fusion and cross correlations in time and space to improve the performance of such systems, in addition to using more powerful (and expensive) sensor sets. Many machine learning methods are used to improve offline template construction, to adapt the algorithms to operation conditions, and to integrate spatio-temporal information probabilistically see (Salah 2009) for a review of machine learning methods as applied in biometrics research).

9.4 The Impact of Behavioural Biometrics on Society

As should be obvious from our survey in the previous section, new technological developments, especially those focusing on behavioural biometrics, soft biometrics and transparent biometrics, open up many possibilities of gathering and storing physical information about individuals. In this section we will elaborate on the impact of new behavioural biometrics on our society.

The first and foremost implication of this proliferation of biometric modalities is the possibility of using gathered information for classifying a subject into arbitrary subclasses. These classes can include gender, age, economic or sociological status (Eagle and Pentland 2006), as well as spontaneous evaluation of behaviours, including the analysis of (potential dangerous) behaviour. These classifications must follow pre-defined models, or pre-selected sets of samples that are manually classified into subgroups. The central question is who defines these models and on which assumptions they are based. Such models are employed for instance in a recent European research project (FP7) *Automatic Detection of Abnormal Behaviour and Threats in crowded Spaces* (ADABTS). ADABTS aims to facilitate the protection of EU citizens, property and infrastructure against threats of terrorism, crime, and riots. In another project called *Samurai* (short for "suspicious and abnormal behaviour monitoring using a network of cameras and sensors for situation awareness enhancement"), a surveillance system is developed for monitoring people and traffic at critical infrastructures (Samurai Project).

It is important to state that behavioural biometrics in this context is fundamentally different from traditional biometrics, which are based on the actual imprint of physical characteristics. If dissociated from the identity, the imprint cannot be used for authenticating the subject. Conversely, one behaviour alone is rarely enough for establishing identity. This is especially important when we take into account the possibility of adjusting and changing of behaviours with the purpose of conforming to a biometric setting. This possibility harbours a certain degree of danger to personal freedom of individuals.

9.4.1 Reality-Changing Implications

The extreme empowerment of the control-state, which is a generic argument against any conceivable technology in the use of power holders and policy makers, is often heard in the context of biometrics. These concerns include *big brother* scenarios (the feeling of control and less freedom), privacy and security aspects like the storage of personal data in (central) databases, and a host of other issues.

Privacy concerns are not new to the biometric community. In particular, Marek Rejman-Greene (2005) defined several criteria, which we summarize under three headings: (1) the authentication process should be accurate and data should not be kept longer then necessary, (2) the biometric (and identity) data should not be processed further than for a specific and lawful purpose (Purpose Principle) and (3) the use of biometrics should be proportional, adequate and relevant. Unfortunately, these principles are often only taken into consideration for evaluating existing applications and find limited use in the design of new applications.

In light of the latest technical development in biometrics we can predict a stronger tension between public interest and individual privacy, especially in the case of distance-based biometrics, where the user is non-obtrusively observed and authenticated over a distance (Tistarelli et al. 2009). As a consequence of these technologies, a user can be tracked and traced 24 h a day, 7 days a week. The growing resentment for hundreds of public cameras installed on the streets of London is but a small example of what could be in store. Through the violation of the purpose principle, the existence of the dense surveillance structure creates a situation where citizens are held accountable over their actions, regardless of any private aspects of their activities. Imagine an ordinary city life where each citizen commits little crimes and trespasses every now and then, ranging from littering the streets to crossing an empty street on a red light. When the state is given the power to selectively punish a citizen for all such crimes, the oppressive nature of all-around surveillance is revealed.

Moreover, with the new behavioural biometric technology, function creep becomes more likely in general. As an example, take CCTV cameras which are installed to serve public interest. The same data can be used for abnormal behaviour analyses, as in some of the currently running EU-funded research projects. Unless proper legislation is in place in accordance with the EU directive on Data Protection (or other frameworks like the Use Limitation Principle of the Organization for Economic Cooperation and Development (1980)), these applications will not be accepted by the informed citizens.

9.4.1.1 The Human Aspect

In the modern notion of technology, the end user has a crucial role, especially with regards to the environment and to sustainability. Before going any further, let us make a distinction between the **user** and the **end-user**. In a typical biometric application the user of the system is the one who deploys the system (for instance the airport authority), while the end-user is controlled by the system (for instance the passenger) (Schouten and Salah 2008). This distinction is relevant in elucidating the objectives of systems and for issues of decision control. Much of the concern for technology originates from the lack of meaning associated with applications on the side of end-users. According to Mordini (2007), present technology is developing without a sound cultural framework that could give technology a sense beyond mere utilitarian considerations.

An even more important realization is that the immersive and surrounding technologies we create around us do not remain passive objects of action and manipulation (van Oortmerssen 2009). They become a part of the everyday existence, subtly infusing our reality with their basic assumptions and the logic that dictated their creation and operation. For any technology put into operation, including the biometric technology, it is erroneous to think of an external reality mildly accepting a new concept into its bosom; instead, we conceive of the new technology with its imposed and implied behaviour patterns as establishing a new equilibrium with the existing culture, changing it (hopefully) in relatively small ways. Yet, there is always the possibility that one small change is one too many, and it is well conceivable that a cascade of consequences follows from the small kernel of discomfort introduced through the novel technology.

The extreme empowerment of the control-state is not as scary as the reality-shift scenario, as the latter implies a wholesale and invisible reconstruction of meaning. This, in itself, may be seen as natural, since there is an inevitable momentum and slow but continuous morphing of the culture, as new cultural artefacts are generated and absorbed into the public consciousness. Some of these changes are necessarily detrimental to the existing set of values, a snapshot of which is a static picture of the culture. Yet, once absorbed, these changes are seen from a different and more favourable perspective. The development of camera, for instance, is the primary enabler of most surveillance technologies, although this was not a foreseen result at the time of its conception. Once accepted, it has changed the culture fundamentally.

As the new technology gets more transparent, and vanishes into the backdrop of the existing cultural behavioural codes, the changes it calls for (from its users and end-users) are reduced, perhaps to non-existence. This transparency does not necessarily mean that the reality-shift introduced by this technology is minor. Quite on the contrary, a transparent biometric technology converts all ordinary existence into existence under surveillance. It is not the *actualization* of the technology, but the *implications of its possibility* that are damaging in this case. The data that are generated from surveillance can be discarded immediately, but this is completely irrelevant. For instance, the tools for analyzing surveillance data from thousands of cameras in London are not developed yet, but the presence of the cameras is enough to instil a feeling of paranoia in the collective subconscious of the population. The removal of visible cues that indicate surveillance may even be more damaging, as it leaves open a possibility of their existence at any given location.

Moreover, there is another consequence of making the biometric technology ubiquitous. If understood as a challenge, the surveillance technology can prompt misbehaviour. The cameras that are visibly observing people imply reaction for punishable actions. In the absence of consequent reaction, the camera becomes an empty taunt. This is also true for other biometric scenarios, where the collected biometric is not immediately linked to a clear and unchallenged purpose. Even access control scenarios are not immune to this danger, particularly for situations in which the collected biometric is excessive with respect to the perceived security requirements. This is one risk that the designers of multi-biometric systems need to take into account.

We would like to mention one last aspect of biometrics, particularly pertaining to behavioural biometrics. Irma van der Ploeg speaks of another reality shifting scenario in "*The Machine Readable Body*": the *informatization of the body,* or the digitalization of physical and behavioural attributes of a citizen and the distribution of these attributes across information networks (van der Ploeg 2005). With improved biometrical technologies, the amount of bodily data will grow exponentially, but more importantly, it will become more and more feasible to use these data in other settings. In our technological history we have seen earlier examples, where digital information of natural processes was used to influence these processes and their objects. Genetically manipulated crops or cattle are the first such examples that come to mind, of the many that exist.

Currently videos can be found on YouTube that demonstrate how to "graphically enhance" ordinary females, using Photoshop, to look like a photo model. This practice is initially only limited to the digital domain and/or used in plastic surgery. In the future we might foresee a situation where biometrical data can directly be used to analyze the body and change it according to the desired values of bodily appearance.

9.4.2 Identity: Accountability and Control

Digital technology has changed our notion of identity. Current biometric practice favours governmental applications and reinforces a centrally controlled identity. John Torpey (1999) argues that modern states, and the international state system of which they are a part, have expropriated from individuals and private entities the legitimate means of movement. In contrast and in a more fluid process we currently see within virtual communities (e.g. Facebook) identity being established and negotiated. The striking difference between biometrical identity and this *social* identity is the role of the end-user. Thomas Erickson and Wendy Kellog (2000), in their article "*Social Translucency: An Approach to Designing Systems that Mesh with Social Processes*," introduce three core aspects: Transparency, Awareness and Accountability, respectively, as essential properties of systems for the collaboration and communication of large groups. We would like to propose another quality, namely Control. In this case control is (or should be) given to the end user for applications related to identity management.

Accountability is a concept in ethics with several meanings. It is often used synonymously with such concepts as responsibility, answerability, enforcement, blameworthiness, liability, and other terms associated with the expectation of account-giving. As an aspect of governance, it has been central to discussions related to problems in both the public and private domains.

Accountability is defined as "A is accountable to B when A is obliged to inform B about A's (past or future) actions and decisions, to justify them, and to suffer punishment in the case of eventual misconduct" (Schedler 1999).

In his definition of **control**, the American Psychologist James Averill distinguishes three types of control (Averill 1973): (1) Informational control, to be informed about (the functionality of) a system, (2) Behavioural control, the user has influence on the behaviour of the system and (3) Decision control to have different options and the ability to choose among them.

It is important to see how accountability and control are assigned to the different stakeholders in biometrics and the difference in objectives between them. For end-users, convenience and privacy might be the reasons to use an application, whereas for the user or authority of the system, throughput or security might be the main arguments. Although (identity) data are private information in many countries, the interpretation of these data, and the consequences it has in the process of authentication are defined by the user (authority). The end-user has no other options apart from being held accountable over his or her actions. As there is little communication and/or interaction in the process of authentication, mistakes in both biometric recognition as well as the storage of identity data, are hard to prevent or correct. More importantly, as long as accountability and control are not (partially) assigned to the end-user, there will always be a fundamental difference in the objectives between the different stakeholders, for instance in cases where private data is exposed in the public domain (Fig. 9.2).

The following pictures show an abandoned police station. These are documentary images that are part of the project *Special Attention* by the artist Jimini Hignett. This police station is located in the wider Detroit area, in the United States of America. Fingerprints, mug shots, personal information of real individuals are scattered on the floor. In analogy, digital biometric data stored in databases can pass into disuse. Central databases are vulnerable to being hacked from outside, but they are also vulnerable from an operator point of view, who has control but is not accountable for misuse (or disuse).

A potential solution is empowering the end-user in selecting, banning and controlling biometrics that would be used in a particular scenario, although at this moment, second generation biometric technologies are not matured enough to offer a great choice, nor an integrated control over the process (see next section) itself.

Fig. 9.2 Pictures of an abandoned police station in the Detroit Area (USA), showing fingerprints and other personal data scattered on the floor

But, difficulties aside, involving the end-user in the process itself has a big advantage over existing scenarios (Schouten and Salah 2008). It allows the rules to be questioned, or even challenged, by removing their a-priori status. The challenge is in preventing the spoon-fed cues of suppression and authority in a world that slowly loses its freedoms and diversity.

9.5 The Use of Biometrics in an Ambient Lifestyle

The ambient lifestyle pertains to a vision of end-user empowerment through technology, in all aspects of everyday existence. To empower the end-user in ID management systems in particular, their meaning and mechanisms must be communicated to the user. One way of leaving control in the hand of individuals is to introduce negotiation into the authentication process. These actualities would make a strong case for decentralized protocol, strong local grounding and an effective use of sensor and actuator technologies, which are emerging in ambient, pervasive, and ubiquitous computing that can be collected under the rubric of the Internet of Things (see www.theinternetofthings.eu). We foresee a future situation where people will carry certain identity tokens (e.g. in a handheld phone, an identity card, or possibly an implanted chip) constituting partial identities by which they would present themselves, enabling them to communicate with their environment through different applications. A very similar representation is valid for digital identities, which are currently in use by millions of people.

The creation of different identities with different levels of security and channels of communication, which do not need to be centralized and organized, is not only a challenging, but also deeply necessary idea. To be accepted, the same technology should come available for local initiatives or certified organizations that can handle different identities, as well as for (local) initiatives where user and end-user are

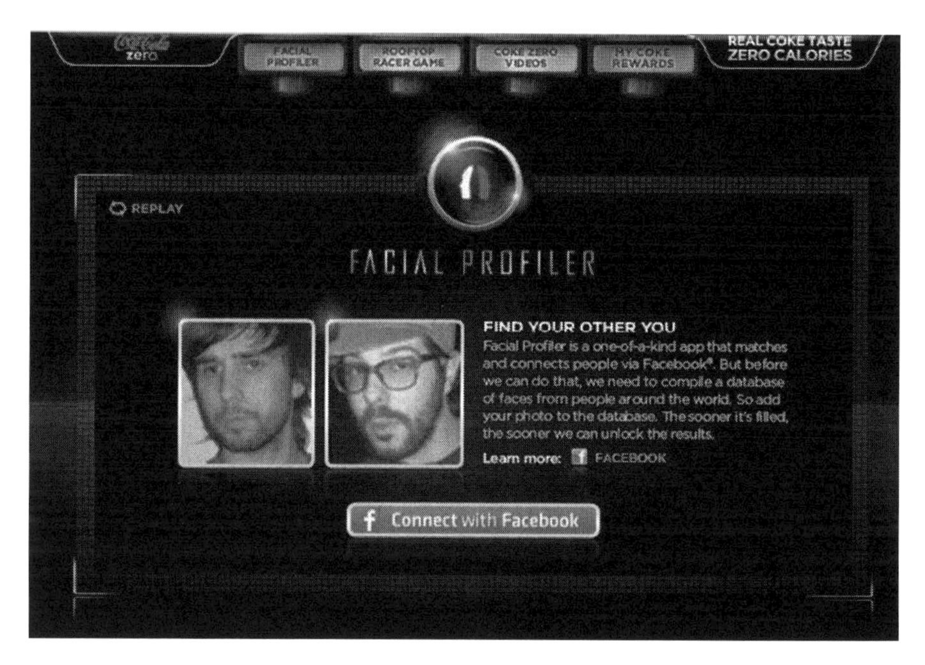

Fig. 9.3 Coke Zero Facial Profiler "…will let you use the same facial recognition software that governments and international security agencies use. But instead of finding criminals or identity thieves, you'll be able to find a person that looks just like you"

the same entity as in the case of Facebook. Biometrics will only be accepted by the public if more commercial and user centred applications will become available (see Fig. 9.3).

There is some interest in the arts community to realize biometric applications. An example is the Bio-Mapping project (see http://www.biomapping.net), which had more than 1,500 participants in 4 years. In this project, participants are equipped with a device which records the participant's Galvanic Skin Response (GSR), which can be used as an indicator of the emotional arousal. This value is linked to the geographical location of the acquired sensor readings. Subsequently, a map is created to visualize emotional arousal levels topographically. Through interpretation and annotation, the community can derive a communal emotion map from its collected biometrics, and visualize one aspect of the social space of the community.

Such examples of constructive use of biometrics are rare, but illustrative. Mordini and Massari (2008) claim in *Body, Biometrics, and Identity*, that "biometric identification technologies may offer a way for individuals to cooperate in the construction of their public identities in a more democratic and polytechnic fashion, and may perhaps eventually replace the current centralized and bureaucratic forms of identification (birth certificates, passports, drivers' licenses, and the like)." In a bold and important step, they offer a turnaround of today's view of biometrics as a tool of control and a system of binaries by stating: "biometric technologies also promise to liberate citizens from the 'tyranny' of nation states and create a new global decentralized, rhyzomatic scheme for personal recognition."

The first step of the authors towards this possible view of a more decentralized yet global system is to ascertain the primary issue of accountability, as there would be "no right, no liberty, without certified personal identities. One can claim her rights, included the right to be left alone, and the right to refuse to be identified, only if she is an identifiable subject, if she has a public identity. …there would be no liberty and no private sphere if there were no public identity" (Mordini and Massari 2008). The crucial issue is however, whether these identities need to be public, or can be private and as they put it, "*the way in which we ascertain public identities.*"[1]

In the current situation, "states hold the power to establish national identities, to fix genders, names, surnames and parental relationships, and to assign rights and obligations to individual subjects according to the names written on their identity documents." All aspects of this identity are contested by the state, regardless of their conformity to the core principles upon which the state is established. To give an example, the Turkish state is constitutionally secular, yet the national identity card includes a field declaring the religion of the citizen, (paradoxically) to ascertain that its religious minorities are treated within the secular framework. In the case of a newborn child born to parents of different religions, father's religion is given to the child by default. To leave this field empty, both parents need to sign a written petition, and personally deliver it to the authorities. The state can thus complicate the procedures that challenge its influence over personal information arbitrarily.

9.6 Conclusions

The future of biometrics involves adapting biometrics to ever more challenging situations. The two main streams in identity management we have previously mentioned (i.e. the technologically driven approach and the sociologically driven approach, in which individuals build and maintain social networks through which they "negotiate" their identities) are in need of a new iteration that brings them together in a different way that bridges the current gap between authorities and end-users. This should naturally favour a trend towards a more liberated and diverse identification bazaar, with munificent commercial implications. Its focus should be long term: educating citizens into socially innovative and inclusive uses of the new technologies that are ever more rapidly being offered to them. Every technology faces the problem of creating its able user groups; biometrics is not an exception in this respect. However, the presence of state edicts necessitates a very wide education in their usage.

[1] "Of course one could argue that this would be a tragedy, and that an ID management solution controlled and operated by governments is absolutely essential in order for government agencies to provide the services citizens expect to receive and to guarantee the survival of the same notion of state. Discussing this question is well beyond the scope of this paper, but there is no doubt that this is one of the main ethical and political challenges raised by biometric technologies." quoted from Mordini and Massari (2008), p. 497.

Some recommendations for further development of ambient biometrics would be, to make biometric generally accepted and part of our culture, to have more user applications and localized (commercial) initiatives, including open-source software in biometrics available for larger user groups (see Fig. 9.4 for example). Artistic projects that are focusing on identity management and democratizing data visualization such as Christian Nold's Bio-Mapping – can open up a public debate.

Fig. 9.4 'Polar Rose', a publicly available identity management system capable of recognising faces in a photobook (Copyright Univ. Lund & Univ. Malmö)

A spin-off company from the Universities of Lund and Malmö proposes simple and straight-forward identity management. Polar Rose is a publicly available application where you can find the identity of person in any photo on any site. The Polar Rose plug-in helps build a local database of identified people by aggregating user input. There is an optional feature to receive a message when someone names the user in a photograph, see http://www.polarrose.com.

With respect to the identity data, it may be argued that data aggregation should be made public; the alternative we would propose is to allow the public to make data. Last but not least, we should have a more fluid notion of identity for which the end-user is empowered to select and control the biometrics to be used in a certain scenario.

We like to end with a quote from Jan Yoors (2004). In his autobiographical story of his life with gypsies, Yoors writes how hard it was for him to be outside in the open for weeks on end. At times he longs for a door and to be able to lock it. The gypsies understand him, but for them privacy is a state of mind: "…privacy was first of all a courtesy extended and a restraint from the desire to pry or interfere in other people's lives. However, privacy must not be the result of indifference to others, but rather a mark of respect for them and of real compassion…."

References

ADABTS, Automatic detection of abnormal behaviour and threats in crowded spaces. http://cordis.europa.eu/fetch?CALLER=FP7_PROJ_EN&ACTION=D&RCN=91158.

Ahmed, A.E., and I. Traore. 2007. A new biometric technology based on mouse dynamics. *IEEE Transactions on Dependable Secure Computing* 4(3): 165–179.

Ananthakrishnan, G., H. Dibeklioğlu, M. Lojka, A. Lopez, S. Perdikis, U. Saeed, A.A. Salah, D. Tzovaras, and A. Vogiannou. 2008. Activity-related biometric authentication. In *Proceedings of the 4th eNTERFACE Workshop*, 56–72. Orsay.

Averill, J.R. 1973. Personal control over aversive stimuli and its relationship to stress. *Psychological Bulletin* 80: 286–303.

Black, J., T. Ellis, and P. Rosin. 2002. Multi-view image surveillance and tracking. In *Proceeding of the IEEE Workshop on Motion and Video Computing*. Orlando.

Boyd, J.E., and J.J. Little. 2005. Biometric gait recognition. In *Lecture notes in computer science*, vol. 3161, 19–42. Berlin: Springer.

Buddharaju, P., I.T. Pavlidis, P. Tsiamyrtzis, and M. Bazakos. 2007. Physiology-based face recognition in the thermal infrared spectrum. *IEEE Transactions on Pattern Analysis and Machine Intelligence* 29(4): 613–626.

Chang, K.I., K.W. Bowyer, S. Sarkar, and B. Victor. 2003. Comparison and combination of ear and face images in appearance-based biometrics. *IEEE Transactions on Pattern Analysis and Machine Intelligence* 25: 1160–1165.

Chen, H., and A.K. Jain. 2005. Dental biometrics: Alignment and matching of dental radiographs. *IEEE Transactions on Pattern Analysis and Machine Intelligence* 27: 1319–1326.

COGNIRON: The Cognitive Robot Companion. [Online] Available: http://www.cogniron.org/Home.php.

Cook, D.J. 2005. Prediction algorithms for smart environments. In *Smart environments: Technologies, protocols, and applications*, Wiley series on parallel and distributed computing, ed. D.J. Cook and S.K. Das, 175–192. Hoboken: Wiley.

Cook, D.J., and S.K. Das. 2005. *Smart environments: Technology, protocols and applications*. Hoboke: Wiley-Interscience.

Damousis, I.G., D. Tzovaras, and E. Bekiaris. 2008. Unobtrusive multimodal biometric authentication – The HUMABIO project concept. *EURASIP Journal on Advances in Signal Processing*. doi:10.1155/2008/265767.

Eagle, N., and A. Pentland. 2006. Eigenbehaviors: Identifying structure in routine. In *Proceedings of Ubicomp'06*. Orange county.

Erickson, T., and W. Kellogg. 2000. Social translucence: An approach to designing systems that Mesh with social processes. In *Transactions on computer-human interaction* (New York: ACM Press) 7(1): 59–83.

Fleuret, F., J. Berclaz, R. Lengagne, and P. Fua. 2008. Multi-camera people tracking with a probabilistic occupancy map. *IEEE Transactions on Pattern Analysis and Machine Intelligence* 30(2): 267–282.

Goffman, E. 1959. *The presentation of self in everyday life*. New York: Anchor Books.

Gökberk, B., A.A. Salah, and L. Akarun. 2005. Rank-based decision fusion for 3D shape-based face recognition. In *Proceedings of the International Conference on Audio- and Video-based Biometric Person Authentication, LNCS 3546*, 1019–1028. Hong-Kong.

Goudelis, G. 2008. Emerging biometric modalities: A survey. *Journal on Multimodal User Interfaces* 2: 217–235. doi:10.1007/s12193-009-0020-x.

Haritaoğlu, S., D. Harwood, and L. Davis. 2000. W4: Real-time surveillance of people and their activities. *IEEE Transactions on Pattern Analysis and Machine Intelligence* 22(8): 809–830.

Hayles, N.K. 1999. *How we became posthuman: Virtual bodies in cybernetics, literature, and informatics*. Chicago: University of Chicago Press.

ISTAG, European Challenges and Flagships 2020 and beyond. 2009. Report of the ICT Advisory Group (ISTAG). ftp://ftp.cordis.europa.eu/pub/fp7/ict/docs/fet-proactive/press-17_en.pdf

Jain, A.K., S.C. Dass, and K. Nandakumar. 2004. Soft biometric traits for personal recognition systems. In *Proceedings of the International Conference on Biometric Authentication*. Hong-Kong.

Kang, J., I. Cohen, and G. Medioni. 2004. Tracking people in crowded scenes across multiple cameras. In *Proceedings of the Asian Conference on Computer Vision*. Korea.

Lamb, R., and E. Davidson. 2002. Social scientists: Managing identity in socio-technical networks. In *Proceedings of the Annual Hawaii International Conference on System Sciences*, 99–99. IEEE Computer Soceity.

Li, S.Z., B. Schouten, and M. Tistarelli. 2009. Biometrics at a distance: Issues, challenges and prospects. In *Handbook of remote biometrics, advances in pattern recognition*, 3–21. London: Springer.

Magnusson, M.S. 2000. Discovering hidden time patterns in behavior: T-patterns and their detection. *Behavior Research Methods, Instruments, & Computers* 32(1): 93–110.

Maltoni, D., D. Maio, A.K. Jain, and S. Prabhakar. 2003. *Handbook of fingerprint recognition*. New York: Springer.

Marcel, S., and J.R. Millan. 2007. Person authentication using brainwaves (EEG) and maximum a posteriori model adaptation. *IEEE Transactions on Pattern Analysis and Machine Intelligence* 29(4): 743–752.

Maurer, D.E., and J.P. Baker. 2007. Fusing multimodal biometrics with quality estimates via a Bayesian belief network. *Pattern Recognition* 41(3): 821–832.

MIKR, The Ministry of the Interior and Kingdom Relations, the Netherlands. 2005. 2b or not to 2b. http://www.minbzk.nl/contents/pages/43760/evaluatierapport1.pdf.

Mittal, A., and L. Davis. 2003. M2tracker: A multi-view approach to segmenting and tracking people in a cluttered scene. *International Journal of Computer Vision* 51(3): 189–203.

Monrose, F., and A.D. Rubin. 2000. Keystroke dynamics as a biometric for authentication. *Future Generation Computer Systems* 16(4): 351–359.

Mordini, E. 2007. Technology and fear: Is wonder the key? *Trends in Biotechnology* 25(12).

Mordini, E., and S. Massari. 2008. Biometrics, body and identity. *Bioethics* 22(9): 488–498. ISSN 0269-9702 (print); 1467-8519 (online) doi:10.1111/j.1467-8519.2008.00700.x.

Newham, E. 1995. *The biometrics report*. SJB Services.

OECD, The Eight Privacy Principles of the Organization for Economic Co-operation and Development 1980. Available online: www.oecd.org.

Pentland, A. 2008. *Honest signals: How they shape our world*. Cambridge: MIT Press.

Poh, N., T. Bourlai, J. Kittler, L. Allano, F. Alonso, O. Ambekar, J. Baker, B. Dorizzi, O. Fatukasi, J. Fierrez, H. Ganster, J.-O. Garcia, D. Maurer, A.A. Salah, T. Scheidat, and C. Vielhauer. 2009.

Benchmarking quality-dependent and cost-sensitive multimodal biometric fusion algorithms. *IEEE Transactions on Information Forensics and Security* 4(4): 849–866.

Rejman-Greene, M. 2005. Privacy issues in the application of biometrics: A European perspective in biometric systems. In *Biometric systems*, ed. J. Wayman, A. Jain, D. Maltoni, and D. Maio. London: Springer.

Ross, A.A., K. Nandakumar, and A.K. Jain. 2006. *Handbook of multibiometrics*. New York: Springer Verlag.

Salah, A.A. 2009. Machine learning for biometrics. In *Handbook of research on machine learning applications*, ed. E. Soria, J.D. Martín, R. Magdalena, M. Martínez, and A.J. Serrano. New York: IGI Global Pub.

Salah, A.A., R. Morros, J. Luque, C. Segura, J. Hernando, O. Ambekar, B. Schouten, and E.J. Pauwels. 2008. Multimodal identification and localization of users in a smart environment. *Journal on Multimodal User Interfaces*. doi:10.1007/s12193-008-0008-y.

Samurai project, www.samurai-eu.org.

Sarkar, S., P.J. Phillips, Z. Liu, I.R. Vega, P. Grother, and K.W. Bowyer. 2005. The humanID gait challenge problem: Data sets, performance, and analysis. *IEEE Transactions on Pattern Analysis and Machine Intelligence* 27(2): 162–177.

Schedler, A. 1999. Conceptualizing accountability. In *The self-restraining state: Power and accountability in new democracies*, ed. A. Schedler, L. Diamond, and M.F. Plattner, 13–28. London: Lynne Rienner Publishers. doi:13. ISBN 1-55587-773-7.

Schouten, B., and A.A. Salah. 2008. Empowering the end-user in biometrics. In *Proceedings of the 10th IEEE International Conference on Control, Automation, Robotics and Vision* (ICARCV). Hanoi.

Socolinsky, D.A., A. Selinger, and J.D. Neuheisel. 2003. Face recognition with visible and thermal infrared imagery. *Computer Vision and Image Understanding* 91: 72–114.

Tabassi, E., C. Wilson, and C. Watson. 2004. NIST fingerprint image quality. NIST Res. Rep. NISTIR7151.

Tangelder, J., and B. Schouten. 2006. Transparent face recognition in an unconstrained environment using a sparse representation from multiple still images. In *Proceedings of the ASCI Conference,* Lommel.

Tavenard, R., A.A. Salah, and E.J. Pauwels. 2007. Searching for temporal patterns in AmI sensor data. In *Constructing ambient intelligence: AmI-07 workshops proceedings*, ed. M. Mühlhauser, A. Ferscha, and E. Aitenbichler. Darmstadt: LNCS.

Tistarelli, M.S., Z. Li, and R. Chellappa (eds.). 2009. *Biometrics for surveillance and security*. London: Springer Verlag.

Torpey, J., C. Arup, and M. Chanock. 1999. *The invention of the passport, surveillance, citizenship and the state*. Cambridge: Cambridge University Press.

van der Ploeg, I. 2005. *The machine-readable body. Essays on biometrics and the informatization of the body*. Maastricht: Shaker.

van Oortmerssen, G. 2009. Evolutie van het Internet (Evolution of the Internet). Inaugural Speech, University of Tilburg, September 9, 2009.

Varela, F.J., E. Thompson, and E. Rosch. 1992. *The embodied mind: Cognitive science and human experience*. Cambridge: MIT Press.

Wren, C., D. Minnen, and S. Rao. 2006. Similarity-based analysis for large networks of ultra-low resolution sensors. *Pattern Recognition* 39: 1918–1931.

Yoors, J. 2004. *The heroic present: Life among the gypsies*. New York: Monacelli Publishers.

Zuo, W., K. Wang, D. Zhang, and H. Zhang. 2004. Combination of polar edge detection and active contour model for automated tongue segmentation. In *Proceedings of the Third International Conference on Image and Graphics*, 270–273. IEEE Computer Society.

Chapter 10
Behavioural Biometrics: Emerging Trends and Ethical Risks

Günter Schumacher

10.1 Introduction

For quite some time, biometrics has been considered as a technology for purely identifying people. We are witnessing a significant change to this perception. *Firstly*, in the context of security, the focus has moved from the question "who you are" towards "how you are". Using an identification technology to prevent threats clearly makes only sense if a certain identity is linked to a certain risk. However, most of the terroristic acts in the past have been committed by before unknown people.[1] Knowledge about their true identity would have hardly prevented their activities. Therefore, the distinction of potential attackers from ordinary people has put emphasise on properties other than identity. It's "suspicious behaviour" which matters.

Secondly, the vision of "Ambient Intelligence (AmI)", introduced in the late 1990s as "next generation IT" in general (Cf. e.g. Aarts and Marzano 2003), implies a technical element able to automatically recognise specific needs of people through observation of behaviour. Four of the five main characteristics of AmI (ibid., p. 14) (namely context awareness, personalisation, adaptiveness, and anticipation) call for technological features able to recognize contemporary states or behaviour of users.

These two trends have created an ever increasing interest in what is now call *behavioural biometrics*. This chapter deals with the emerging trends in this direction and, in particular, with evenly emerging ethical issues.

[1] Apart from the 9/11 event, also the failed attack on Northwest Flight 253 from Amsterdam to Detroit on 25 December 2009 has been committed by a person with very limited suspicion potential. In February 2010, CIA Director Leon Panetta stated that *Al Qaeda is using "clean" recruits with a negligible trail of terrorist contacts* (source: www.nypost.com).

G. Schumacher (✉)
European Commission Joint Research Centre, Institute for the Protection and Security of the Citizen, 1049 Brussels, Belgium
e-mail: Günter-Egon.Schumacher@jrc.ec.europa.eu

E. Mordini and D. Tzovaras (eds.), *Second Generation Biometrics: The Ethical, Legal and Social Context*, The International Library of Ethics, Law and Technology 11,
DOI 10.1007/978-94-007-3892-8_10,

10.2 Behavioural Versus Traditional Biometrics

Behavioural biometric modalities are usually described as "a measurable behaviour trait that is acquired over time versus a physiological characteristics or physical trait".[2] Although categorised as biometrics of "second generation", some types of behavioural biometrics are much older than those referenced as of "first generation". Well-known examples include:

- *Hand signature*: It has been used at times when there was hardly any understanding what biometrics could ever mean. The concept is based on the individual habit how to hold a pencil and write by hand.
- *Voice recognition*: Who ever responded at the telephone by "It's me!" fully relied on the ability to be recognised by a characteristic sound pattern.
- *Lie detector*: Also known as "polygraph", this machine was designed to decide whether someone tells the truth based on physiological changes caused by the *sympathetic nervous system.*

All these examples have in common that through the observation of one or more physical parameters (over time), conclusions about the characteristic behaviour of a person is drawn. Not all these techniques aim to further reveal the identity of that person. Some of them might not even have the discriminatory power to do so. On the other hand, these characteristics may allow conclusions about a particular person far beyond pure identity as the example of the lie detector shows.

As we will see in the following, behavioural biometrics can be roughly distinguished from traditional types of biometrics like fingerprint, face or iris recognition along the dimensions summarised in Table 10.1. We will also see that another way of distinction between the two types is the consideration of 'discrimination power' against 'privacy risks' as depicted in Fig. 10.1. The figure also illustrates the diverging tendencies between traditional and behavioural biometrics. Traditional biometrics aim for the highest possible uniqueness of the identifier at the lowest possible privacy impact (right lower corner). Behavioural biometrics, in contrary, is by nature less discriminatory (because more group specific) but tend to reveal sensitive "hidden" information about a person (left upper corner). It is interesting to note that the combination of these different types of biometrics (in the sense of multimodal biometrics) create identifiers with high discriminatory power at high level of privacy sensitiveness: knowing quite precisely who *and* how a person is.

Finally, the difference to traditional biometrics is reflected also in the level of understanding of related privacy issues. For example, template protection has become a compulsory aspect of consideration for any biometric modality of first generation. For behavioural biometrics, the nature of "templates" (i.e. any kind of abstract representation of the acquired raw data) is different and its proper protection less understood.

---

[2] Cf. Glossary of Terms for Voice Biometrics, Biometric Security Ltd. (2009).

Table 10.1 Traditional biometrics versus behavioural biometrics

	Traditional biometrics	Behavioural biometrics
Measurement of modality	*Direct*: The biometric identifier is directly recorded by the sensor system	*Indirect*: The behavioural modality is indirectly recorded through measurement of one or more physical parameters
Discriminatory power	*High*: The theoretical recognition rate tends to approach 100%	*Low to medium*: A certain profile is more linked to a group of people than to an individual person[a] The modality acts as an indicator rather than an identifier
Scientific basis	*Statistics based on genetics and environmental causes*: The underlying identifier is supposed to be constant through lifetime, except of the problem of disturbances through ageing effects	*Statistics based on unknown causes*: The behavioural parameter may be long lasting or a temporal effect of psychological stress but the rationale how to use it is based on assumptions or statistical observations
Profiling	*Secondary*: The combination with other information can be established but is not necessary to perform identification	*Primary*: The modality (as an indicator only) adds to a profile which is the actual aim for its collection
Risk	*Loss of physical assets*: As used almost exclusively for authentication purposes, the actual risk lies in loosing the asset it protects, i.e. ultimately material values	*Loss of "ethical" assets*: The profile (as a collection of behavioural and other indicators) could reveal very personal details which could result in a loss of respect, reputation, social relations, etc.

[a]However, the combination of a number of "weak" identifiers increase statistically the probability to match with only one person

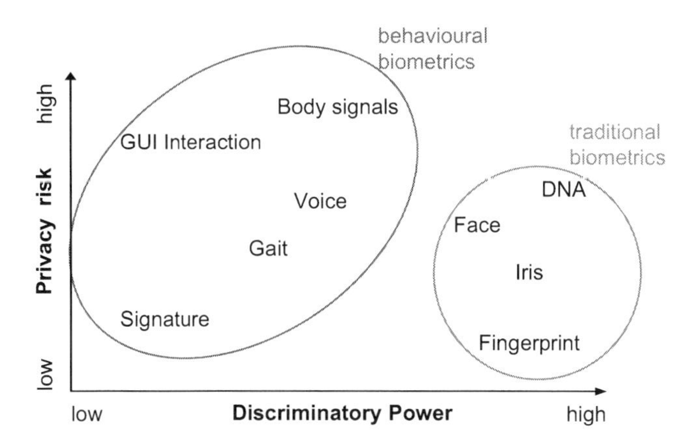

Fig. 10.1 Discrimination power versus privacy protection in the case of first and second generation biometrics modalities

10.3 Categorising Behavioural Biometrics

Before entering in the discussion of risks, we first need to distinguish between certain categories of behavioural biometrics according to the information processed.

In their highly recommendable textbook about behavioural biometrics,[3] Wang and Geng (2009) developed already a comprehensive taxonomy for this type of biometrics but also stated that a "commonly agreed categorisation is yet to be established". Therefore, in the absence of such commonly agreed definition and for the purpose of the consideration about ethical issues, we define and consider only three major categories of behavioural biometrics ENISA (2010); Yampolskiy and Govindaraju (2008):

- *Motor skills*, based on specific muscle control and movements
- *Body signals*, based on the measurement of certain body parameters
- *Machine usage*, based on the complex interaction of various personal elements

All categories have in common that they assume a repeatable "typical" behaviour of the person in question. Whilst the first two categories observe the direct execution of a body function, the third category observes only the final result of an action, usually when performing mentally demanding tasks.

10.3.1 Motor Skills

This category deals with human muscle control and movements such as:

- Signature/Handwriting (the way in which a person writes text by hand, in particular his/her own name)
- Voice (individual sound patterns)
- Speech (individual way of phrasing words and sentences)
- Gait (the way a person walks)
- Dynamic face features (individual way in controlling certain face muscles)
- Key stroke dynamics (individual usage of computer key boards)
- Mouse movements (individual usage of a computer mouse)

The commonly known recognition capabilities for modalities like handwriting and voice demonstrate that humans tend to have typical motor skills. This fact becomes operational when the movements can be recorded and the records be compared. Except for the case of (non-dynamic) consideration of signatures, all profiles generated by these modalities consist of the measurement of one or more physical parameters over time.

---

[3] Chapter 1 of Wang and Geng (2009) deals in particular with "Taxonomy of Behavioural Biometrics" in which also the historical roots and the related science areas are discussed. The authors introduce the five categories "authorship", "direct human computer interaction", "indirect human computer interaction", "motor skills", and "purely behavioural". However, most of the modalities can be mapped to more than one of these categories.

An additional dimension of these modalities comes with the praxis of interpreting certain movements (or results of these movements) as belonging to a certain population group regardless whether or not this can be scientifically substantiated. For example, some people claim to distinguish a female handwriting from a male or to recognise the gait of a homosexual. Other conclusions could be with respect to certain assumed inabilities like spelling problems (handwriting, voice), stuttering (voice), walking impairment (gait), computer handling (key stroke dynamics) or with respect to a certain ethnicity (speech, face features). Regardless of the actual "recognition rate" when comparing such profiles, the possibility of generating this type of indicators for certain groups and their usage cannot be neglected. We will discuss this further in Sect. 10.4.

10.3.2 Body Signals

The modalities considered under this category have been traditionally exclusively used by medicine:

- Heart beat and electrocardiogram (ECG)
- Body temperature profile (in particular of the face)
- Brain activity and electroencephalogram (EEG)
- Breathing frequency
- Eye blinking
- Transpiration
- Trepidation (level of trepidation of the body as a changing parameter under psychological stress)

The idea behind the measurement of these parameters is the indirect conclusion on psychological states through the interaction with the sympathetic nervous system. This was exactly the rational of the polygraph ("lie detector"), although the possibility of defining the profile of a "liar" is still not sufficiently proven (see National Research Council (2003)). Nonetheless, that basic idea has now again been adopted by a project funded by the US Department of Homeland Security under the acronym FAST (Future Attribute Screening Technology).[4] The project aims to detect "malintent" prior to its execution through the combination of body signal and motor skills biometrics. A group of US research and development organisations investigates the feasibility of developing risk profiles of persons about "the intent or desire to cause harm" (see Fig. 10.2).

[4] http://www.dhs.gov/files/programs/gc_1218480185439.shtm

Fig. 10.2 FAST project (illustration)

According to information made available to the public,[5] the modalities used for the entire system include:

- the temperature of a face, measured by a thermal imaging device, in order to observe temperature changes as a person is asked questions;
- an eye tracker which follows a person's gaze, checks the amount of blinking and measures pupil dilation;
- heartbeat and respiratory rate and volume;
- fidgeting, monitored by a Wii™ balance board, a device on which people normally stand on to interact with certain Nintendo™ Wii™ video games.[6]

The combination of various modalities should on one hand increase the reliability of the analysis conclusions, and on the other hand prevent attackers to get used to the system.

Less ambitious (though not less potentially useful) applications of biosensing modalities range from alerting people about decreasing attention (e.g. truck drivers) up to real identification devices based on heart beat sound (Phua et al. 2006).

Due to its very nature as medical data, also body signals allow for further conclusion beyond the indented original purpose. However, unlike motor skills biometrics where some of the diverted profiling is based on subjective reasoning, there exists an enormous reservoir of scientific literature interpreting certain profiles involving the mentioned modalities. More precisely, such profiles can be linked to certain diseases, both of physical and mental nature. Again, this can be the source of potential discrimination as we will see in Sect. 10.4.

---

[5]See e.g. CNN.com/Technology: http://edition.cnn.com/2009/TECH/10/06/security.screening/index.html

[6] See http://www.nintendo.com/wii/what/accessories/balanceboard

10.3.3 Machine Usage

This ever growing group of modalities consists of interactions with a machine, mostly computers. Here, it is not directly motor skill which counts but e.g. the time in which decisions are taken or strategic aspects. Typical examples with computers are:

- GUI interaction (individual way of interacting with the graphical user interface of a program)
- Game strategy (individual way of interacting with a certain computer game)
- Audit logs (tracking of the activities performed by a computer user)
- Email behaviour (individual way of phrasing and dealing with emails)
- Credit card usage (tracking of purchases)

Although not so much expected as for the other two categories, even machine usage modalities can achieve high recognition rates (Wang and Geng 2009), i.e. imply even the possibility to track specific indicators down to a single person.

A characteristic feature of this category of modalities is the ability to acquire the necessary information completely from remote. This makes it interesting for businesses as well as for criminals. It is important to note that collecting this type of behavioural information (including computer related motor skills) is not covered by relevant legislation such as the European Directive 94/46 on data protection. Thus a company might well and completely legally collect information how a user is interacting with the web interface provided by that company. On the other hand, the same type of observations can help to identify system intrusion and other illegal computer activities.

10.4 Ethical Risks in the Usage of Behavioural Biometrics

We have seen that an important aspect of behavioural biometric modalities is the recording of data over time, resulting in a characteristic profile of dynamic behaviour. Literally, the profile describes how a person has "behaved" in a certain context, either consciously or unconsciously. Therefore, "profiling" as defined in the context of data protection[7] is unavoidably inherent to the application of behavioural biometrics. Behavioural modalities create as "raw data" profiles because their main purpose is usually to develop a set of "classifiers" rather than establishing a real identifier.

At first glance, behavioural biometrics seems to lack many of the risks of traditional biometrics (like fingerprint or face recognition) due to less stringent links to a particular

[7] Hildebrandt and Gutwirth (2008) and Rannenberg et al. (2009) defines profiling as "*the process of 'discovering' patterns in data in databases that can be used to identify or represent a human or nonhuman subject (individual or group) and/or the application of profiles (sets of correlated data) to individuate and represent an individual subject or to identify a subject as a member of a group (which can be an existing community or a 'discovered' category).*"

person. Identity theft and personal tracking seems to be less likely. But at second glance, behavioural biometrics goes obviously deeper "under the skin" than traditional modalities do. Not physical properties are of interest but psychological states or habits which are by nature more privacy sensitive than ridge lines of a fingerprint.

This sensitiveness becomes more critical when modalities are combined to deliver more complex profiles. The predictability of risks becomes almost impossible if other types of information are mixed with those delivered by behavioural modalities. The combination with data like age, sex, size, weight, ethnicity, etc. (sometimes also referred to a soft biometrics) could allow for a whole range of indicators about a person in question.

Probably the most comprehensive analysis related to the profiling aspect of behavioural biometrics has been performed by the FIDIS project. In FIDIS Project (2009), the authors identified a couple of major risks:

- *Discrimination*: As the examples given for motor skills modalities indicate, there is a huge risk that such information can be used to exclude persons from certain areas. For example, employers or online stores could keep away interested people with "risky profile" by letting them run through an online questionnaire while their computer interaction is recorded and analysed, together with other information provided. This is also a consequence of a particular *de-individualisation.* It does not matter any more who you are as a person. You are just one of "those". A further consequence is a changed attitude towards the data subject, including a *reduced or limited information supply.* Once a service provider has classified a person according to the recorded profile, it could limit any further information flow to that person to the extent necessary for just doing business.
- *Stigmatisation*: This is the risk of longer term "sticky" profiles with negative interpretation. For instance, privacy advocates speculated during the first presentations of the FAST project (see above) that the technology would misidentify innocent people who might then be branded as potential terrorists – as long as the corresponding risk profile is kept.[8] Stigmatisation is clearly dependent on the way the sensitive profiling data is treated. Any security breach or other kind of data loss can create dramatic damages to the reputation of a person. But even under strict data protection rules, the negative consequences for the data subject can be severe and long lasting.
- *Unwanted confrontation*: Right to privacy implies the right of not being informed. In particular, body signals can indicate certain diseases for which medical treatment is unlikely or even impossible. Such information might completely upset the life of this person and of others. On the other hand, it is hard to predict which information could be harmful or unwanted (Cf. e.g. Sherlock and Morrey 2002, p. 444).

[8]Source: The Boston Globe: http://www.boston.com/news/local/massachusetts/articles/2009/09/18/spotting_a_terrorist/

Though the described risks apply to any profiling activity, they become an issue *for all* applications related to behavioural biometrics due to the inherent need for profiling. In other words, the described risks are unavoidable side effects when using behavioural biometric modalities.

10.5 Risk Mediators

As almost any new technological development, behavioural biometrics has opportunities and risks. This implies that risk mediation should not result in a complete show-stopping in order to preserve the opportunities for further exploration. In the context discussed here, the mentioned risks are completely data related and thus a matter of suitable data protection. Without entering into a general discussion about modalities, benefits and problems of data protection, we can summarize the situation as follows:

- Existing data protection legislation and guidelines provide a comprehensive framework but do not guarantee the respect of these rules. On the other hand, emerging new technologies like biometrics call continuously for a particular interpretation and application of these rules.
- Efficient data protection is usually build on three pillars:
  - Technological and organisational measures
  - Awareness about issues
  - Control and audit

We concentrate here on two particular aspects: technological measures and awareness. The other measures are complementary and can be organised as soon as the potential of the highlighted two measures has been fully implemented.

10.5.1 Technological Measures

Technical solutions to tackle privacy issues (apart from pure IT security) are relatively new to biometrics in general and are even more immature for second generation biometrics. The focus is mainly to avoid the abuse of the acquired data by certain encryption techniques (template protection) but some research is also devoted to address unavoidable communication of biometric data. For the latter, there a currently two major concepts discussed for securing the data exchange between the data subject (the one who gives biometric data) and the data controller (the one who collects and processes the biometric data), namely Privacy Enhancing Technologies (PET) and Transparency Enhancing Technologies (TET). The following paragraphs shall provide a brief summary of these concepts.

10.5.1.1 Template Protection

Biometric templates can be considered as an abstraction of the originally measured raw data which also reduces the amount of data to the actual distinctive points. For fingerprints, the template consists mainly of the found so-called minutiae points which are ridge endings or bifurcation of ridge lines; for face recognition, the focus is on landmarks and other geometric information. Templates for behavioural biometric modalities can be even more complex, depending on the number of aspects to relevant. For example, in gait recognition, quantitative information about the following aspects are recorded (Cf. Wang and Geng 2009): amount of arm swing, the rhythm of the walker, bounce, length of steps, vertical distance between head and foot, distance between head and pelvis, maximum distance between the left and right foot.

Apart from data reduction, the mapping to a template has also the advantage that the "richer" raw data can be eliminated. "Richer" means that the original data (e.g. a video stream) may also contain information which would allow further conclusions about the person in question beyond the scope of the considered biometric modalities. This could also help to prevent replay attacks (Ratha et al. 2001) to spoof the involved sensor system since it is very difficult (in some cases almost impossible) to generate from the template the original raw data.

Nevertheless, ongoing research is concentrating to further protect the templates as well, both from any attempt to reconstruct the raw data and from any kind of abuse. The already existing techniques for traditional biometrics exploit for this purpose the relatively static nature of the raw data in combination with cryptographic techniques (Jain et al. 2008). This does not equally apply to behavioural biometrics. First attempts to overcome this problem have been published only recently (Argyropoulos et al. 2010).

10.5.1.2 PET (Privacy Enhancing Technologies)

The basic idea of PETs is to provide "*a system of ICT measures protecting informational privacy by eliminating or minimising personal data thereby preventing unnecessary or unwanted processing of personal data, without the loss of the functionality of the information system*" (Van Blarkom et al. 2003). Therefore, PET aims to reduce the personal data as much as possible and does not (and cannot) control its later usage (or abuse). A major application scenario consists of an online transaction with a service provider. PETs make sure that the user discloses only information necessary to conclude the transaction. However, the focus is on static data like name, address, payment credential etc. and does not include any information on dynamic behaviour as described in the previous sections. In any case, PETs only work if introduced in the technological elements of both sides, i.e. service provider and user, established through a particular communication protocol ("privacy by design").

10.5.1.3 TET (Transparency Enhancing Technologies)

Conceptually introduced by the FIDIS project,[9] TETs aim to overcome the limits of PETs as described above. Such tools should somehow anticipate the creation of certain profiles and the resulting consequences for the data subject. Even though a precise definition of TETs is still to be established, FIDIS defined and distinguished between two types of TETs (FIDIS Project 2009):

- *Type A: legal and technological instruments that provide (a right of) access to (or information about) data processing, implying a transfer of knowledge from data controller to data subjects.*
- *Type B: legal and technological instruments that (provide a right to) counter profile the smart environment in order to 'guess' how one's data match relevant group profiles that may affect one's risks and opportunities, implying that the observable and machine readable behaviour of one's environment provides enough information to anticipate the implications of one's behaviour.*

While PETs build on data minimisation, TETs build on minimisation of information asymmetry. "Type B" TETs would come close to the actually desired "golden bullet" but exist so far only as a concept. "Type A" TETs have already been further analysed and are available for specific type of web service applications. Nevertheless, both types have in common that increased transparency makes only sense if the additional information provided to the user are understandable. Otherwise, the gain in information symmetry would be of very limited use.

Again, similar to PETs, the implementation of any TET requires cooperation with the data controller in order to understand the nature of data which is acquired in a certain context.

Apart from these early considerations about TET, further research will be necessary to introduce this concept into system developments involving behavioural biometrics.

10.5.2 Awareness and Training

If technology cannot yet solve the obvious problems with behavioural biometrics, users need to be informed and even trained how to best use this technology. This clearly requires a certain level of knowledge about biometrics and profiling techniques in general. It also requires the openness of technology providers both for security and non-security application. Whilst for the latter this could be normally assumed (for reasons of trustworthiness), the situation for security related applications is – for obvious reasons – different. However, once a technology has been developed, it can (and will be) exploited for all applicable purposes – including those unwanted.

---

[9] http://www.fidis.net

Training is also needed to grasp any additional information provided by TETs described in the previous paragraph. Otherwise, such techniques would serve only experts who are probably the least group to need protection. A minimum of knowledge is required for everybody deliberately or undeliberately exposed to this kind of technology. Awareness raising and training should therefore draw as much as possible on latest research results.

10.6 Conclusions

The realisation of an "Orwellian future" can still not be excluded, in particular if security remains the major driver for this technological area. It is unlikely that new methods to look "underneath the skin" can be restricted to protective or purely supportive applications. Once the knowledge is created, it will be used for all those purposes we might be afraid of – except technical or other conceptual barriers would prevent this. Yet, the current technical measures to avoid an almost arbitrary invasion of privacy are still insufficient and require further intensive research. The concept of TET is promising – though in its infancy – but require a minimum of knowledge on the user side in order to be really beneficial. Hence, increased awareness as well as specific training about the relevant aspects are indispensible to avoid any dark scenario.

References

Aarts, E., and S. Marzano. 2003. *The new everyday – Visions of ambient intelligence*. Rotterdam: 010 Publishers.

Argyropoulos, S., et al. 2010. Biometric template protection in multimodal authentication systems based on error correcting codes. *Journal of Computer Security* 18(1): 161–185.

Biometric Security Ltd. 2009. http://www.voicebiogroup.com/glossary.html.

ENISA. 2010. ENISA briefing: Behavioural biometrics. European Network and Information Security Agency, document URL: http://www.enisa.europa.eu/act/rm/files/deliverables/behavioural-biometrics/at_download/fullReport.

FIDIS Project. 2009. Behavioural biometric profiling and transparency enhancing tools. Deliverable D7.12, FIDIS Project (Future of Identity in the Information Society), document URL: http://www.fidis.net/fileadmin/fidis/deliverables/fidis-wp7-del7.12_behavioural-biometric_profiling_and_transparency_enhancing_tools.pdf.

Hildebrandt, M., and S. Gutwirth (eds.). 2008. *Profiling the European citizen*. Berlin/Heidelberg/New York/Tokio: Springer.

Jain, A.K., et al. 2008. Biometric template security. *EURASIP Journal on Advances in Signal Processing* 2008(12): 1–18.

National Research Council. 2003. *The polygraph and lie detection, committee to review the scientific evidence on the polygraph*. Washington, DC: National Academy of Science.

Phua, K., T.H. Dat, J. Chen, and L. Shue. 2006. Human identification using heart sound. Paper presented at the Second International Workshop on Multimodal User Authentication, Toulouse, France.

Rannenberg, K., et al. (eds.). 2009. *The future of identity in the information society*. Berlin/Heidelberg/New York/Tokio: Springer.
Ratha, N.K., et al. 2001. Enhancing security and privacy in biometrics-based authentication systems. *IBM Systems Journal* 40(3): 614–634.
Sherlock, R., and J.D. Morrey. 2002. *Ethical issues in biotechnology*. Lanham: Rowman & Littlefield.
Van Blarkom, G.W., et al. 2003. *Handbook of privacy and privacy-enhancing technologies*. The Hague: College bescherming persoonsgegevens, TNO-FEL.
Wang, L., and X. Geng. 2009. *Behavioral biometrics for human identification: Intelligent applications*. Hershey: Medical Information Science Reference.
Yampolskiy, R., and V. Govindaraju. 2008. Behavioural biometrics: A survey and classification. *International Journal of Biometrics* 1(1): 81–113.

Chapter 11
Facial Recognition, Facial Expression and Intention Detection

Massimo Tistarelli, Susan E. Barrett, and Alice J. O'Toole

11.1 Introduction

Visual perception is probably the most important sensing ability for humans to enable social interactions and general communication. As a consequence, face recognition is a fundamental skill that humans acquire early in life and which remains an integral part of our perceptual and social abilities throughout our life span (Allison et al. 2000; Bruce and Young 1986). Not only faces provide information about the identity of people, but also about their membership in broad demographic categories of humans (including sex, race, and age), and about their current emotional state (Hornak et al. 1996; Tranel et al. 1988; Calder et al. 1996; Humphreys et al. 1993; Calder and Young 2005). Humans "sense" this information effortlessly and apply it to the ever-changing demands of cognitive and social interactions.

Face recognition has now also a well established place within the current biometric identification technologies. It is clear that machines are becoming increasingly accurate at the task of face recognition, but not only at this. While only limited information can be inferred from a fingerprint or the ear shape, the face itself conveys far more information than the identity of the bearer. In fact, most of the times humans look at faces not just to determine the identity, but to infer other issues which may or may not be directly related to the other person's identity. As an example, when speaking to someone we generally observe his/her face to either better understand the meaning of words and sentences, but mostly to understand the emotional

M. Tistarelli (✉)
Computer Vision Laboratory, University of Sassari, Sassari, Italy
e-mail: tista@uniss.it

S.E. Barrett
Lehigh University, Bethlehem, PA, USA

A.J. O'Toole
University of Texas at Dallas, Richardson, TX, USA

E. Mordini and D. Tzovaras (eds.), *Second Generation Biometrics: The Ethical, Legal and Social Context*, The International Library of Ethics, Law and Technology 11,
DOI 10.1007/978-94-007-3892-8_11,

state of the speaker. In this case, visual perception acts as a powerful aid to the hearing and language understanding capabilities (Hornak et al. 1996).

Even though most of past and current research has been devoted to ameliorate the performances of automatic face identification in terms of accuracy, robustness and speed in the assignment of identities to faces (a typical classification or data labeling problem) (Mika et al. 1999; Lucas 1998; Roli and Kittler 2002), the most recent research is now trying to address more abstract issues related to face perception (Humphreys et al. 1993; Haxby et al. 2000). As such, age, gender and emotion categorization have been addressed with some level of success (again, as a broader labeling or classification problem), but with limited application in real world environments. This paper analyzes the potential of *current* and *future* face analysis technologies as compared to the human perception ability. The aim is to provide a broad view of this technology and better understand how it can be beneficially deployed and applied to tackle practical problems in a constructive manner. Airports constitute a typical and interesting example where identification and personal categorization is often required. The adoption of the new electronic passport, encompassing biometric data, such as the face and fingerprint, constitutes a challenge for the use of biometric data without compromising the traveller's privacy. A simple solution is to use the data to reveal ancillary information about the traveller without even attempting to identify him/her.

Emotions are an integral and unavoidable part of our life. As such, should we be scared if machines acquire the ability to discern human emotions? Would it be an advantage or a limitation for the user if a Personal Computer or a Television is capable of sensing our mood and provide appropriate services accordingly? Should we perceive a "trespassing" of our privacy if a hidden camera in an airport lounge recognizes our tiredness and activates a signal to direct us to the rest facilities?

Nowadays, there is a considerable concern for biometric technologies breaking the citizen's privacy. These concerns have yet hindered the adoption of these technologies in many environments, thus depriving many of potential service improvements and benefits. Unfortunately, as in the past, also today misinformation and politically driven views impair technological advances which can be adopted for the wellness of the citizens. This chapter analyzes the impact of *current* and *future potential* face analysis technologies as compared to the human perception ability. The aim is to provide a broad view of this technology and better understand how it can be beneficially deployed and applied to tackle practical problems in a constructive manner. Yet, as any other technology, the way it is used, either for the good or bad, entirely depends on the intentions of the men behind its application.

11.2 Human Face Perception and Recognition

Human face perception seems effortless. We recognize people by their faces and make instantaneous judgments about their emotional state and intent scores of times per day. The human ability to extract this information quickly and efficiently from a face is an important pre-requisite for our survival as a species. In the next few sections, we discuss human face recognition and facial expression perception abilities.

We do this in a way that keeps in mind the question of whether or not machines are "up to the task" of stepping in for humans in automated surveillance applications. Most of us are entirely comfortable presenting an identification card to a border guard or customs official for verifying the identity of our face against the picture on the card. We are sometimes even comforted by the sight of a police officer or security guard in a deserted subway station or parking lot late at night. And yet, we are concerned at the possibility of handing our identification card over to a machine and standing in front of a camera, before being allowed to proceed on our way.

Indeed, some ethical concerns about allowing machines to aid or replace humans in security and surveillance tasks come from not knowing answers to the following questions. "How well do machines perform security and surveillance tasks?" "Are they going to make mistakes that humans would not make?" "If they do make a mistake, are we going to be able to talk to a person about it, to challenge the decision?" For the question of accurate face recognition, the first of these questions has been addressed continuously over the past decade with large scale evaluations of state-the-art face recognition algorithms (e.g. Phillips et al. 2005). We will discuss the results of some of these studies shortly. To the best of our knowledge, there have been no formal objective tests for the evaluation of machine-based expression recognition systems.

Implicitly, the questions we ask about trusting machines to do these tasks, depends on understanding how well humans recognize faces. Psychologists have long studied human strengths and weaknesses of human face recognition and expression perception. On this question, we will argue that computer scientists, engineers, and the policy makers, who make decisions about the application of automated surveillance systems, need to understand the competition for their machines–human beings.

First, we begin with a discussion of the basic issues involved in relating human and machine abilities to face recognition and expression perception. Second, we discuss human abilities to perceive facial expressions. We start with the influential categorical models of expression and emotion, discussing the universal human facial expressions defined by Ekman and Friesen in 1976 and how they are produced using specific sets of coordinated facial muscle movements. Third, because the ability to produce and perceive facial expression develops early in life, we briefly sketch out the course of facial expression development in infants and young children. This developmental course comes surprisingly close to offering a "to-do list" for machine recognition of expression. What do automatic expression recognition systems need to master before they are to be trusted in important applications? Fourth, expression and identity information coexist on the face and must be accessed simultaneously, with minimal cross-interference. Thus, we discuss the support for long-standing theories in the psychological and neuroscience literatures that posit a strong separation of the processes that underlie facial expression and face identity perception (Bruce and Young 1986; Calder and Young 2005). Although the idea of *strictly* separated systems has been challenged in recent years, the basic tenant of perceptual and neural separability in facial expression and identity processing remains a fundamental part of our understanding of human facial analysis. Finally, we will sum up by linking our understanding of human abilities to the factors that must be considered when we apply machines to the problems.

11.2.1 Face Recognition: Issues Relating Humans and Machines

Human performance on the task of face recognition has generally been considered the "gold standard" against which the performance of algorithms is judged. In fact, the available psychological and computational data suggest human face recognition performance is superior to even the best state-of-the-art algorithms in some, but not all (O'Toole et al. 2007b), cases. What is certain is that our ability to recognize the faces of people we know well (i.e., friends and family), displays a remarkable robustness to photometric variables such as changes in illumination, viewpoint, and resolution (cf., Burton et al. 1999). Face recognition algorithms have not yet achieved the level of robustness typical of human face recognition for highly familiar faces. Photometric variables, especially changes in viewpoint, remain highly challenging for face recognition algorithms and have a substantial negative impact on the performance of the algorithms (e.g., Phillips et al. 2005).

Less clear is the comparison between face recognition algorithms and humans on the task of recognizing relatively unfamiliar faces (Hancock et al. 1991). By unfamiliar faces, we mean those that we have limited experience with (e.g., have seen only once or twice from a photograph or live encounter). In this case, humans, like algorithms, perform less accurately when there are photometric changes between learning and test images (e.g., illumination Braje et al. (1999); Hill and Bruce (1991), viewpoint Moses et al. (1996); Troje and Bülthoff (1996)). In general, the kinds of face recognition and verification tasks done by security personnel are carried out almost exclusively with unfamiliar faces. Because human face perception skills are not at their best for unfamiliar faces, the difference between human and machine accuracy in real world applications might be less pronounced than one would fear.

In fact, under reasonably controlled viewing conditions, when there is a good match between image characteristics on viewpoint, the best face recognition algorithms now compete favorably with humans. In the Face Recognition Grand Challenge (FRGC), a recent U.S. Government sponsored test of face recognition algorithms, three out of the seven algorithms under evaluation surpassed humans matching identity in pairs of images that differed in illumination (O'Toole et al. 2007b). A follow-up experiment showed that there were qualitative differences between human and machine recognition strategies. This was demonstrated by fusing human and machine identification judgments, which can improve decisions only when the pattern of errors differs enough to combine the independent strengths of both strategies. In fact, the fusion of human judgments with the judgments of face recognition algorithms from the FRGC produce better performance than any of the systems operating alone (O'Toole et al. 2007a).

Combined, one important implication of this work is that we may over-estimate human performance on the task of face recognition. This suggests that the confidence we have in our own ability to recognize faces accurately may be misplaced. Ample evidence exists for the human ability to mistakenly identify someone with high confidence, a finding that is now well-established in the context of DNA-based exonerations of unjustly convicted individuals in the United States over the past few decades.

Mistaken identifications by witnesses figure prominently in many cases (Saks and Koehler 2005). When face recognition technologies are applied in surveillance situations, policy makers must make an effort to understand how these technologies can *compensate* for human weaknesses, while also understanding the limits of the technologies.

On this topic, there are analogous issues about the fairness of both human and machine recognition systems. For humans, it is well known that face recognition is not equally accurate for all faces. For example, *typical* faces are recognized less accurately than unusual faces (Light et al. 1979; Valentine 1991). The practical consequence of this finding is that people with typical faces, (i.e., few distinctive features) are more often mistakenly recognized than are people with more distinctive faces. It is unclear, to date, if machines show a similar bias, though it is highly likely. Certainly, the performance of machines on biometric recognition tasks has been characterized in terms of the recognizability of individuals (Doddington et al. 1998). One framework for understanding this issue in machine performance is referred to as the *Doddington Zoo*, which posits four categories of individuals based on the statistical pattern of recognition success. *Sheep* are individuals who are easy to recognize. *Goats* are difficult to recognize. *Lambs* are easy to imitate and *wolves* are good at imitating others. Recognition schemes can be adjusted to take into account aspects of a face that might make it a wolf or a lamb. This Doddington zoo characterization suggests potential inequities in the application of face recognition systems across individuals. The important point, however, is that these inequities may not differ substantially between humans versus machines.

More systematic inequities exist for humans recognizing faces of different races. The *other-race effect* has been established and replicated in many human face recognition studies. This effect refers to the finding that humans recognize faces of their own race more accurately than faces of other races (Malpass and Kravitz 1969; Meissner and Brigham 2001). The critical false identification problems that arise from this phenomenon have been well-documented in the eyewitness literature. Indeed, the recognition of a suspect by a witness of another race is less likely to be accurate than the recognition of the suspect by a person of the same race, though prosecutors, juries, and judges are often unwilling or unable to factor this into their confidence about an identification. Although there is still some controversy regarding the cause of the other-race in humans, most evidence relates this effect with the amount of experience we have with faces of our own race versus faces of other races (Bryatt and Rhodes 1998). Moreover, there is recent work suggesting the particular importance of learning faces early in life (Kelly et al. 2007). The idea in this case is that learning in infancy contributes to an optimization of the feature sets used to encode faces. Generally, this optimization is based on the statistical variations in the set of faces we learn first, usually faces of our own race.

Do automated face recognition systems show an other-race effect? In other words, does the geographic origin of an algorithm (i.e., where it was developed) affect how well it performs on faces of different races. Presumably, if the cause of the other-race effect is a bias in the selection of the features used to encode faces, there is reason to be concerned that algorithms will fare best with faces of the race(s) used to train them.

A recent study tested whether state-of-the-art face recognition systems show an other-race effect (O'Toole et al. 2008). The U.S. Government's Face Recognition Vendor Test 2006 (FRVT 2006) provided the face recognition algorithms used for the test. In the FRVT 2006, five algorithms from East Asian countries (China, Japan, and Korea) were compared with eight algorithms from Western Countries (France, Germany, and the United States) on the task of recognizing East Asian and Caucasian faces. This test revealed a clear other-race effect that was especially pronounced in the range of low false alarm rates required for most security applications. In other words, under the operating conditions that would be used for most applications, the algorithms developed in East Asia recognized East Asian faces more accurately than Caucasian faces and the Western algorithms recognized Caucasian faces more accurately than East Asian faces. In a direct comparison with humans of East Asian and Caucasian ethnicities (on a subset of the faces used in the first comparison), an other-race effect was seen for both the humans and algorithms. Humans performed somewhat more accurately than machines overall. Of note, however, the major difference between machine and human accuracy was that human performance was more stable across changes in the race of a face than was machine performance. Indeed, the face recognition algorithms showed much larger variations in performance across changes in the face race than did humans.

The cautionary note in the airport is that Big Brother is already there in human form, with his/her strengths and weaknesses. Regardless of whether a machine or human stands guard at the border, it is important to understand, anticipate, and mitigate the kinds of recognition errors that are likely to occur.

11.2.2 Facial Expression Perception: Human Perspectives That May Inform Machine Vision

An important difference between face identity and facial expression perception, is that identity has ground truth. Facial expressions are produced in response to felt emotions, or are in some cases, pretended emotions. Remarkably little is known about how accurately humans perceive the felt emotions of others from their facial expressions. Although it seems as if expressions provide us with direct access to the a person's emotional state, classic examples of the mismatch between felt emotions and perceived expressions abound (e.g., tears during the World Cup Finals and at weddings can be more than a bit difficult to interpret out of context). Thus, it's worth bearing in mind that facial expressions, which consist of non-rigid deformations of facial structure, provide only partial information about felt emotions. The rest of the information comes from the cognitive and social context of the episodes or events that elicit the expressions and emotions. More formally, psychologists have tried to put the definition of facial expressions onto solid ground by quantifying the patterns of facial muscle movements used to create (or portray) expressions that are universally recognizable across cultures. In the next section, we discuss ideas about the structure of facial expressions.

11.2.2.1 Universal Facial Expressions

Over the past several decades, the most influential theories have advocated a categorical view of expressions (Ekman and Friesen 1976). According to this view, each instance of an emotional expression is treated as a variation within a set of basic facial configurations. Facial expressions are part of a biologically based communication system for signaling emotions. Cross-cultural studies have looked at how people in different societies and with different cultural backgrounds perceive expressions (Ekman et al. 1969, 1972; Izard 1971). These studies have been used to argue that basic emotions are universally expressed and recognized. There is a striking degree of similarity in what elicits certain emotions and in how these emotions are expressed on the face. Although agreement about the perception of particular facial expressions is well above chance, it is not perfect. Notwithstanding, most psychologists agree that there are *universals* in emotional expression. Anger, disgust, fear, happiness, sadness and surprise have been offered as emotions that are associated with specific facial expressions and are basic to all human emotional life.

It is hard to overstate the importance of cross-cultural work in laying down the foundations of psychological and anthropological understandings of emotion. Darwin's insights gave rise to systematic studies of how emotions are expressed in people from widely different cultural backgrounds (Darwin 1898). Both Ekman and Izard and their colleagues drew on their own cross-cultural observations as they developed theories and coding systems for human facial expression. For example, Ekman and Friesen's Facial Action Coding System (FACS) defines a *prototypical* version of each of the six basic emotions (anger, disgust, fear, happiness, sadness and surprise). To do this, the system uses 40 separate and functionally independent muscle action units to classify a face as exemplifying a particular emotional expression. Timing, intensity, and laterality are also considered important variables for distinguishing among emotional expressions.

The status of these *prototypical expressions* as "ground truth" for the associated emotion, however, is unclear. Although Izard (1997) laments the "false notion that there is one full face prototypical expression for each of several basic emotions" (Izard 1997, p. 60), emotion prototypes are a driving force in both Ekman and Matsumoto's view: "When we perceive emotions in others, the process is analogous to template matching with the universal facial prototypes of emotion. Before a judgment is rendered, however, that stimulus is also joined by learned rules [that] may differ according to stable sociocultural dimensions such as IC [individualism vs. collectivism] and SD [behavioral differentiation based on status differences]" (Matsumoto 2001, p. 185).

One critical aspect of these rules is whether it is appropriate to show specific emotions in different contexts. Ekman and Friesen provided an early and powerful demonstration of how culture shapes emotional expression (Ekman et al. 1972). They found that both Japanese and American observers readily displayed negative emotions when viewing a stressful film in isolation but that Japanese observers attempted to conceal these negative emotions, often with a smile, when others were present. Smiles are popular masks and even children use social smiles to convey an emotional message they may not feel but are required to show. Feigning a negative emotion is more difficult as most individual are less experienced at falsifying emotions such as fear (Ekman 2009).

Although social conventions may lead individuals to try to cover up their feelings, Ekman believes that in most cases the message leaks through, especially if the feeling state is strong. Hence, with practice, someone can become skilled at detecting true expressions, which may be masked because of social conventions or because the expresser is trying to mislead an audience."The stronger the emotion, the more likely it is that some sign of it will leak despite the liar's best attempt to conceal it. Putting on another emotion, one that is not felt, can help disguise the felt emotion being concealed. Falsifying an emotion can cover the leakage of a concealed emotion" (Ekman 2009, p. 31–32).

When an emotion is masked, the original expression may be covered up after it is appears. Full facial expressions that last for only a fraction of their usual time are referred to as "micro expressions." Micro expressions, however, do not occur very often. Emotions are often squelched, that is, the facial expression may be covered up before it is fully revealed in the face. Detecting deception may hinge on detecting full or partial expressions before they are masked by another emotion. Training programs have been developed to heighten sensitivity to "squelched emotions" and "micro expressions." Differences in timing can also be used to discriminate feigned and felt emotions.

Individuals may be especially adept at reading emotions, or interpreting the "emotional dialect" of their culture. Matsumoto attributes these differences to learned cultural display rules, but perceptual and attentional factors may also play a role. Cultural differences in display rules affect which regions of the face are most informative (Matsumoto 2001). The mouth is likely to be a particularly informative region when individuals are emotionally demonstrative, whereas felt emotions are likely to be evident in the eyes when the emotion is masked. Whereas Western Caucasian observers distribute their gaze evenly across a face; East Asian observers focus on the eye region, at the expense of mouth region, and have difficulty when the eyes do not disambiguate the expression (Jack et al. 2009). Consequently, East Asian observers tend to confuse fear with surprise and disgust with anger. When asked to choose a label for less clearly expressed emotions, East Asian observers also have a tendency to choose the more socially desirable label. Hence, social desirability, cultural concepts, and gaze patterns (or sampling biases) all play an important role in how facial expressions are perceived (Jack et al. 2009; Russell 1991; Schyns 1998).

11.2.2.2 Development of Facial Expression Perception

In the past few decades, a revolution of sorts has taken place in the developmental literature as infancy researchers have uncovered hitherto unimagined competencies in young infants. Psychologists studying infants are faced with challenges similar to those one might encounter when testing expression recognition systems. Infants, like machines, can differentially respond to stimuli. But, we cannot ask them to explain how they reach their decisions and it is not obvious which aspects of the stimulus are driving their performance. A careful analysis of what infants can and cannot do shows that emotion perception, broadly defined, occurs on many levels. Infants may be able to discriminate among facial expressions, but this does not mean that they understand the "meaning" of the message. Infants are surprisingly competent social partners, even though they are not born with a system for recognizing basic universal emotions. Children are sensitive to emotional valence (positive or negative) and intensity, long before they are able to classify "prototypical examples" of the facial expressions that characterize the basic emotions). This suggests a potentially different path for developing emotion recognition systems and facial expression recognition systems.

The steps that infants go through in becoming competent perceivers of the emotions of others provide a kind of blue print for machine recognition systems. First, there is a need to discriminate the patterns of muscle movements that convey different expressions and to generalize these patterns across all faces. Next, natural displays of emotion include changes to facial expression, voice, and body posture. Being able to discriminate these displays reliably is a first step in understanding what they mean. Finally, emotional displays make us react with empathy or with an otherwise adaptive response. These adaptive responses are the ultimate measure for our understanding of the emotional displays of other people. Through these responses, emotional expression serves its function as a system of communication.

A prerequisite for identifying and labeling facial expressions is an ability to reliably discriminate different expressions (e.g., fear from anger, happy from surprise). Infancy researchers use simple behaviors to test how infants perceive the world. Changes in responses, such as increases in looking time, are taken as evidence that infants notice a stimulus change. In the habituation-dishabituation paradigm, infants are given the opportunity to inspect a stimulus (e.g., a happy face) until there is a decrease in looking time. If a new stimulus (e.g., a sad face) is presented, and infants detect the change, looking time increases. But does this increase in looking time really mean that infants discriminate facial expressions? Perhaps the infant only notices that the mouth has changed: tightly compressed lips have replaced bright teeth. Discriminating facial expressions and discriminating mouths are two different things. One way around this problem is to show the infant the same emotional expression by different people. Individual features (e.g., the "toothiness" of the smile) vary across individuals and infants do generalize smiles across individuals (Walker-Andrews and Bahrick 2001). If infants perceive facial expressions as wholes or "gestalts," the orientation of the face should matter, and it does. Infants smile at dynamic upright faces, but fail to engage emotionally with inverted faces (Muir and Nadel 1998). Thus, there seems good evidence from visual presentations of facial expressions that infants can reliably distinguish most facial expressions, and to generalize these expressions over the individuals who display them.

To ask the more difficult question about whether the ability to discriminate expressions is accompanied by an understanding of their meanings (i.e., the emotional states that underlie the expressions), researchers have made use of cross modal matching tests as a first step. These cross-modal tests are used to assess the extent to which infants can integrate emotionally similar information across different sensory modes of presentation. In studies of this kind, the modes are defined usually as visual presentations of faces and auditory presentations of voices. Indeed, caregivers communicate emotions with their voices and faces, as well as their bodies. Data from cross modal matching studies suggest that infants are sensitive to connections between facial and vocal expressions of emotion. In a cross-modal matching study, the infant might be presented with a happy voice while two silent films are presented on adjacent television screens. If infants are sensitive to the affective information conveyed in each modality, we would expect that they would look at the display that matches the sound track. Five to seven months old look at the matching display, and they make this connection even when the voice and video are not synchronized (see Walker-Andrews (1997) for a review). These findings suggest that infants are sensitive to the concordance of affective messages across sensory modalities.

Although cross-modal matching studies suggest that infants are sensitive to information that conveys the emotional meaning of the messages, these studies were not designed to test whether infants *react* in meaningful ways to specific facial expressions. Social-interactional theorists argue that if we are interested in whether infants recognize the meaning of facial expressions, we should focus on how infants respond to these messages (Lewis and Goldberg 1969; Stern 1985). Detailed observations of face-to-face interactions provide compelling evidence that infants and their caregivers are engaged in emotional dialogues (Sroufe 1995). At first, it is the parents who shoulder the responsibility for "meaning-making" as they interpret the infants actions as intentional and respond in meaningful ways to changes in the infants facial expressions. Infants engage and reengage with their partner's face. Surprisingly, which expression is posed does not seem to matter. A happy expression is nearly as disruptive as a sad or neutral one (D'Entremont and Muir 1997), although happy faces do elicit slightly more smiling, perhaps suggesting that infants expect their interactive bids will be more successful if they are directed at a smiling partner.

Stronger evidence that infants understand the emotional meaning of facial expressions comes from studies of social referencing. In social referencing studies, the infant is presented with an ambiguous situation, for example a "scary" visual cliff with a moderately steep drop-off. A visual cliff is made by juxtaposing two steps or ledges with a "drop-off" between the two. A clear plastic sheet of flooring is placed overtop to connect the ledges and to provide a safe bridge-like connection over the "drop-off". Babies can see the drop-off and are usually too frightened to cross over. An adult, often the mother, stands at the far side of the drop-off and poses a specific facial expression. If the mother poses a happy face, the infant will venture across the cliff; the infant is not likely to do so if the mother poses an angry, fearful, or sad face (Sorce et al. 1985).

One reason developmental psychologists are careful not to over interpret the infant data is that they are well aware of the difficulties older children have recognizing emotions. Toddlers can label emotions at above chance levels, but their performance is far from error-free. Although 2 year olds rarely place a happy face in the "angry" box, they place similar numbers of sad, fearful, angry, disgusted faces in this box (Widen and Russell 2008a). Preschoolers continue to have difficulty with sorting tasks even when the verbal requirements are minimized and the boxes are labeled with pictures. Children's emotion categories and their ability to recognize facial expressions have a rather protracted developmental time course (Markham and Adams 1992).

For both humans and machines, ultimately, there is a need to label emotional expressions. Two-year-olds use a variety of emotion words both to describe their own feelings and to talk about other's feelings (Dunn et al. 1987; Wellman et al. 1995). Children clearly differentiate positive feelings (happy, better, okay) from negative feelings (afraid, sad, angry) but they use these terms in a much less specific way than adults (Harris 1989). Children acquire emotion labels in a predictable order, although mastery of these labels is a gradual process. The first three emotion words children use are "happy", "sad", and "angry." "Scared" and "surprised" are added next with "disgust" being added a bit later (Widen and Russell 2003). The frequency with which these labels are used also differs, the earliest acquired labels are also the most frequently used. This is a reflection of both the fuzzy nature of emotion categories and a pull toward more accessible labels (Widen and Russell 2008a).

One assumption that has been pervasive across the psychological literature on facial expression perception is that human expressions are categorically generated and perceived. In more recent theorizing, this assumption has been re-evaluated in a way that counters Ekman's influential view of basic emotions that can be represented by discrete labels and conveyed with prototypical facial expressions. Specifically, dimensional models, such as the circumplex model, view affective experiences along a continuum. These experiences themselves are often ambiguous, or at least, best described by multiple labels that convey the interrelations among affective states. Variations in affective states are best captured by two dimensions, the first corresponds to something akin to valence and the second to arousal. Widen and Russell (2008b) argue that these two dimensions serve as a starting point for building emotional knowledge. Data from both free labeling and classification tasks support their view. Children use emotion words, such as angry, but these words map onto a broad fuzzy categories that only partially overlaps with what adults means when they use the label.

Widen and Russell (2008b) argue that the infant data do not make a compelling argument for the discrete-category account of emotional development. More specifically, they argue that while it is clear that infants distinguish between happy and negative expressions, there is little evidence that they differentiate among specific negative expressions, such as fear and anger. The problem, in their view, is that most studies focus on emotion pairs that differ in valence. Evidence for differential responding to specific negative emotion is scant, at best. For example, infants tend to avoid the apparent drop-off of the visual cliff when the mother poses both a fearful and angry face. In some studies, infants do respond differently to negative expressions, for example, sad expressions generate less interest than angry expressions (Haviland and Lelwica 1987) and infants will look in the appropriate direction when angry and sad faces are presented side by side in a cross-modal matching task (Soken and Pick 1999). But differences in infants' responses to sad and angry faces are taken as evidence that they are sensitive to second dimension, intensity, as well as valence.

Emotional labels together with the child's attempts to actively make sense of emotionally charged events, including antecedents and behavior consequences, drive emotional understanding. As children's perspective taking skills improve, they become increasingly tuned to the central role that appraisals play in emotion (Harris 1989). And as parents of preschoolers can attest, children ask many questions as they try to figure out the "why's" of emotion. Children face numerous challenges as they try to make connections between these explanations and experiences. Words, such as "angry" and "scared," map onto broad fuzzy categories and negative facial expressions are often confused, which can lead to incorrect inferences about causes and consequences. And as adults, we know that even if we could perfectly recognize and classify emotions, explaining the *why's* is never simple.

11.2.3 Separable Neural Processing Systems for Perceiving Facial Identity and Expression

Information about facial identity and facial expression coexist on a face. The simultaneous access of identity and expression from a face is computationally, but not perceptually, challenging. The computational difficulties occur because facial expressions consist of a set of complex non-rigid deformations of the features and shape of the face. Thus, the information that specifies identity is distorted to produce expressions. Concomitantly, facial expression movements are made with many different facial identities. In understanding how expression and identity analyses are coordinated neurally and psychologically, a look at the entire face processing system is helpful.

One of the first comprehensive models of human face processing, proposed by Bruce and Young in 1986, posited a separation of expression and identity processing. This was supported by evidence from neuropsychological case studies that indicated a *double dissociation* between expression and identity processing in brain-injured individuals. Indeed, evidence for the importance of particular local brain areas for processing face identity has been available for over a half of a century (Bodamer 1947). This evidence comes in the form of case studies of selective impairment of face recognition following brain damage to the inferior temporal area of the brain. Damage to this region of the brain can produce *prosopagnosia*, a neuropsychological impairment in which patients lose the ability to identify people by their faces, while retaining general visual object recognition skills.

Evidence for the first part of the double dissociation of face identity and expression processing comes from case studies of prosopagnosics who have impaired face recognition abilities, but spared expression perception (e.g., for a review Young et al. (1993)). The second part of the double dissociation comes from documented cases of brain damaged individuals who fail to perceive facial expressions, but have no difficulty recognizing faces (e.g., Adolphs et al. 1994). These individuals generally have damage to the amygdala, a sub-cortical brain area with importance for processing emotional stimuli. The amygdala connects to other subcortical structures that generate emotional responses, including increases in heart rate and blood pressure.

With the advent of more sophisticated functional brain imaging technologies, it has been possible to view the neural activity in intact brains in a way that highlights the importance of brain regions for specific perceptual and cognitive tasks. Functional magnetic resonance imaging (fMRI), for example, yields a high resolution three-dimensional map of brain activity. The application of this technology to human subjects actively engaging in face processing tasks has refined our knowledge of the face recognition and expression perception systems beyond what was known from neuropsychological case studies.

Based on functional neuroimaging data, neuropsychological case studies, and electrophysiological recordings from neurons in non-human primates, Haxby et al. (2000) proposed an updated functional model of the human face processing system. They proposed three brain regions as the core of a distributed neural network for processing faces: (a) the lateral fusiform gyrus (in the inferior temporal lobe), (b) the superior temporal sulcus (pSTS); and (c) the inferior occipital gyrus. According to this model, the lateral fusiform gyrus, sometimes called the fusiform face area (FFA) (Kanwisher et al. 1997), processes the invariant information from faces useful for identification. The pSTS processes the changeable information in faces that specifies facial expression, facial speech, and eye gaze. This separation between FFA and pSTS functions is linked to low level visual processing differences for static and dynamic (moving) stimuli. The FFA receives input from visual processing streams associated with high resolution form information and is thought to analyze the features and configuration of a face. The pSTS, on the other hand, receives input from motion sensitive neurons in the visual cortex and is thought to be involved in the perception of dynamic facial movements (i.e., for expression, gaze, speech). Haxby et al. note that a secondary set of brain structures continues processing the information from the core network. Among these structures is the amygdala, which is involved in the activation of emotional responses, including the increase in blood pressure.

One additional insight of the Haxby et al. model is that the pSTS as a processor of facial expression, facial speech, and gaze/head movement, by its nature, processes the *motions* of the face and body. A facial expression is created through movements of the face. Eye gaze shifts and head movement indicate changes in our focus of attention. Moreover, the areas of the pSTS that process facial motion are contiguous with areas that process gait (i.e., the way people walk) and a host of other body part motions including motions of the hands. It is through these movements that people express their intent to act.

In summary, a look at the neural processing of moving and static faces and bodies, is consistent with both a structural and functional separation of the information. The function of a moving face and body is often to communicate social or affective intent. The brain areas that process these messages are closely connected to the parts of the brain that can act and react appropriately. The function of the static information in faces and bodies may largely be more useful to cognitive processes that recognize and categorize people. Thus, the neural systems may suggest a logic for dividing the processing of people and their expressions/intents into components that focus on the invariant and changeable parts of the signal.

11.2.4 Lessons Learned: Human Facial Expression and Identity Processing

The primary lesson we learn from considering human face recognition and expression perception is that, like machines, we are not perfect. Indeed, there seems to be a bias in the machine recognition literature to believe that "If only, my machine could perform as well as humans, all would be well". A look at human performances suggests that we should beware of such simplistic goals. The fact that there is, as yet, no perfect system for these tasks should motivate us to set more realistic goals for developing machines that are aimed at identifying people and their intentions. The machine and human literatures seem to be moving toward a broader inclusion of multiple biometrics in establishing identity. The psychology literature on expression likewise points to the need to understand the full socio-cognitive and affective context to determine the intent of an individual. More basic research is needed to elucidate the many factors that may play a role in establishing this context. It is likely however that the interest of machine vision researchers in the problem of understanding human intent, will open the door to an interesting and productive dialog on these issues.

11.3 Face Recognition Technologies

First generation biometrics has been puzzled in devising a 100% error-free face recognition system. The large variations implied in face appearance make it rather difficult to be able to capture all relevant features in faces regardless of the environmental conditions. Therefore, the most successful face recognition systems are based on imposing some constraints on the data acquisition conditions. This is the case for the CASIA F1 system (Li et al. 2007), which is based on the active illumination by infrared illuminators and a near-infrared camera sensor. This is also the case for commercial products like the automated portal for automated border check developed by L1 Identity Solutions. This is not the way face recognition is going to revolutionarize today's use of identification technologies. We expect that face recognition systems will be more and more able to work in "open air", detecting and identifying people remotely (from a distance), in motion and with any illumination. The main breakthrough is expected to come from the use of high resolution images and exploiting the time-evolution of data rather than single snapshots.

In general, face recognition technologies are based on a two step approach:

- An off-line enrollment procedure is established to build a unique template for each registered user. The procedure is based on the acquisition of a pre-defined set of face images, selected from the input image stream, or a complete video, and the template is build upon a set of features extracted from the image ensemble;
- An on-line identification or verification procedure where a set of images are acquired and processed to extract a given set of features. From these features, a face description is built to be matched against the user's template.

Regardless of the acquisition devices exploited to grab the image streams, a simple taxonomy can be based on the computational architecture applied to: extract distinctive and possibly unique features for identification and to derive a template description for subsequent matching.

The two main algorithmic categories can be defined on the basis of the relation between the subject and the face model, i.e. whether the algorithm is based on a subject-centered (eco-centric) representation or on a camera-centered (ego-centric) representation. The former class of algorithms relies on a more complex model of the face, which is generally 3D or 2.5D, and it is strongly linked with the 3D structure of the face. These methods rely on a more complex procedure to extract the features and build the face model, but they have the advantage of being intrinsically pose-invariant. The most popular face-centered algorithms are those based on 3D face data acquisition and on face depth maps.

The ego-centric class of algorithms strongly relies on the information content of the gray level structures of the images. Therefore, the face representation is strongly pose-variant and the model is rigidly linked to the face appearance, rather than to the 3D face structure. The most popular image-centered algorithms are the holistic or subspace-based methods, the feature-based methods and the hybrid methods. Over these elementary classes of algorithms several elaborations have been proposed. Among them, the kernel methods greatly enhanced the discrimination power of several ego-centric algorithms, while new feature analysis techniques, such as the local binary pattern (LBP) representation, greatly improved the speed and robustness of Gabor-filtering based methods. The same considerations are valid for eco-centric algorithms, where new shape descriptors and 3D parametric models, including the fusion of shape information with the 2D face texture, considerably enhanced the accuracy of existing methods.

11.3.1 Face Analysis from Video Streams

When monitoring people with surveillance cameras at a distance it is possible to collect information over time. This process allows to build a rich representation than using a single snapshot. It is therefore possible to define a "dynamic template". This representation can encompass both physical and behavioral traits, thus enhancing the discrimination power of the classifier applied for identification or verification.

The representation of the subject's identity can be arbitrarily rich at the cost of a large template size. Several approaches have been proposed to generalize classical face representations based on a single-to-multiple view representations. Examples of this kind can be found in Lucas (1998), Lucas and Huang (2004) and Raytchev and Murase (2003, 2002, 2001), where face sequences are clustered using vector quantization into different views and subsequently fed to a statistical classifier. Recently, Zhou et al. (2003, 2004) proposed the "video-to-video" paradigm, where the whole sequence of faces, acquired during a given time interval, is associated to a class (identity). This concept implies the temporal analysis of the video sequence with dynamical models (e.g., Bayesian models), and the "condensation" of the tracking and recognition problems. Other face recognition systems, based on the still-to-still and multiple stills-to-still paradigms, have been proposed (Li et al. 2001, 2000; Howell and Buxton 1996). However, none of them is able to effectively handle the large variability of critical parameters, like pose, lighting, scale, face expression, some kind of forgery in the subject appearance (e.g., the beard). Other interesting approaches are based on the extension of conventional, parametric classifiers to improve the "face space" representation. Among them are the extended HMMs (Liu and Chen 2003), the Pseudo-Hierarchical HMMs (Bicego et al. 2006; Tistarelli et al. 2008) and parametric eigenspaces (Arandjelovic and Cipolla 2004), where the dynamic information in the video sequence is explicitely used to improve the face representation and, consequently, the discrimination power of the classifier.

11.4 Facial Expression and Emotion Recognition

Rosalind Picard in (2000) defined the notion of *affective computing* as the computational process that relates to, arises from, or deliberately influences emotions. This concept may be simply formulated as giving a computer the ability to recognize and express emotions (which is rather different from the question of whether computers can have 'feelings') and, consequently, develop the ability to respond intelligently to human emotion. Most human interactions are conveyed through emotional expression. Therefore it should not be surprising if computers are enabled to discern and interpret emotions. This should be regarded as an added feature which improves man machine interaction, and may facilitate the fruition of goods and services.

The most expressive (and the easiest to capture) way humans display emotion is through facial expressions (Sebe et al. 2005). Humans detect and interpret faces and facial expressions in a scene with little or no effort. Still, developing an automated system that accomplishes this task is rather difficult. There are several related problems: detection of an image segment as a face, extraction of the facial expression information, and accurate classification of the facial expression within a set of pre-assessed emotion categories. The achievement of this goal has been studied for a relatively long time with the aim of achieving human-like interaction between human and machine. Since the early 1970s Paul Ekman and his colleagues have performed extensive studies of human facial expressions (Ekman 1994). They found evidence for "universal facial expressions" representing happiness, sadness, anger, fear, surprise and disgust. They studied facial expressions in different cultures, including preliterate cultures, and found much commonality in the expression and recognition of emotions on the face. However, they observed differences in expressions as well and proposed that facial expressions are governed by "display rules" within different social contexts. In order to facilitate the analytical coding of expressions, Ekman and Friesen (1978) developed the Facial Action Coding System (FACS) code. In this system movements on the face, leading to facial expressions, are described by a set of action units (AUs). Each AU has some related muscular basis. This system of coding facial expressions is done manually by following a set of prescribed rules. The inputs are still images of facial expressions, often at the peak of the expression. Ekman's work inspired many researchers to analyze facial expressions by means of image and video processing. By tracking facial features and measuring the amount of facial movement, they attempt to categorize different facial expressions. Past work on facial expression analysis and recognition has used these "basic expressions" or a subset of them. Fasel and Luettin (2003), Pantic and Rothkrantz (2000) and more recently Gunes et al. (2008) and Zeng et al. (2009) provide an in-depth review of much of the research done in automatic facial expression recognition over the years since the early 1990s.

Among the different approaches proposed, Chen (2000) and Chen and Huang (2000) used a suite of static classifiers to recognize facial expressions, reporting on both person-dependent and person-independent results. Cohen et al. (2003) describe classification schemes for facial expression recognition in two types of settings: dynamic and static classification. In the static setting, the authors learned the structure of Bayesian network classifiers using as input 12 motion units given by a face tracking system for each frame in a video. For the dynamic setting, they used a multilevel HMM classifier that combines the temporal information and allows one not only to classify video segments with the corresponding facial expressions, as in the previous works on HMM-based classifiers, but also to automatically segment an arbitrary long sequence to the different expression segments without resorting to heuristic methods of segmentation (Sebe et al. 2005). These methods are similar in that they first extract some features from the images, then use these features as inputs into a classification system, and the outcome is one of the preselected emotion categories. They differ mainly in the features extracted from the video images and in the classifiers used to distinguish between the different emotions. An interesting feature of this approach is the commonality with some methods for identity verification from video streams (Shan and Braspenning 2009). On the other hand, a recurrent issue is the need to either implicitly or explicitly determine the changes in the face appearance. The facial motion is thus a very powerful feature to discriminate facial expressions and the correlated emotional states.

Recently Gunes et al. (Gunes and Piccardi 2006, 2009; Gunes et al. 2008) extended the concept of emotion recognition from face movements to a more holistic approach, including also the movements of the body. In this work also the range of detected emotions is increased, adding most common emotions such as Boredom and Anxiety. The integration of face and body motion allowed to improve the recognition performances as compared to the analysis of face alone.

11.4.1 Computation of Perceived Motion

The motion of 3D objects in space induces an apparent motion on the image plane. The apparent motion can either be computed on the basis of the displacement of few image features points or as a dense field of displacement vectors over the entire image plane. This vector field is often called optical flow. In general, there is a difference between the displacement field of the image points and the actual projection of the 3D motion of the objects on the image plane (the velocity field). This difference is due to the ambiguity induced by the loss in dimensionality when projecting 3D points on the image plane. A simple example is the motion of a uniformly colored disc, parallel to the image plane, rotating around its center. As there is no motion on the image plane the optical flow is zero, but the 3D motion of the object, and consequently its projection, the velocity field, is non-zero. Nonetheless, apart from some degenerate conditions, the optical flow is generally a good approximation of the velocity field. The optical flow can be computed by processing a set of image frames from a video. The process is based on the estimation of the displacement of the image points over time. In general terms, the displacements can be computed either explicitly or implicitly. In the first case, the motion of image patterns is computed by matching corresponding patches on successive image frames. In the latter case, the instantaneous velocity is determined by computing the differential properties of the time-varying intensity patterns. This generally implies the computation of partial derivatives of the image brightness, or filtering the image sequence with a bank of filters properly tuned in the spatial and temporal frequencies, and the fulfillment of some constraints on the apparent motion (Tretiak and Pastor 1984; Tistarelli 1996; Horn and Schunck 1981). The combination of multiple gradient-based constraints allows to compute the value of the flow vector, best approximating the true velocity field, at each image point. Some examples of flow fields computed from different image streams and different actions are presented in Figs. 11.1 and 11.2

11.4.2 Performances of Expression and Emotion Recognition Algorithms

Several algorithms to detect facial expressions and infer emotions have been developed and tested. The datasets used and the reported performances are quite varying, but all share some common denominators.

Firstly, no single system is capable of achieving 100% correct categorization, even with a very limited number of subjects (Sebe et al. 2006; Gunes and

Fig. 11.1 (*Left*) Optical flow from an image sequence of a person jumping. (*Right*) Optical flow from an image sequence of two persons walking

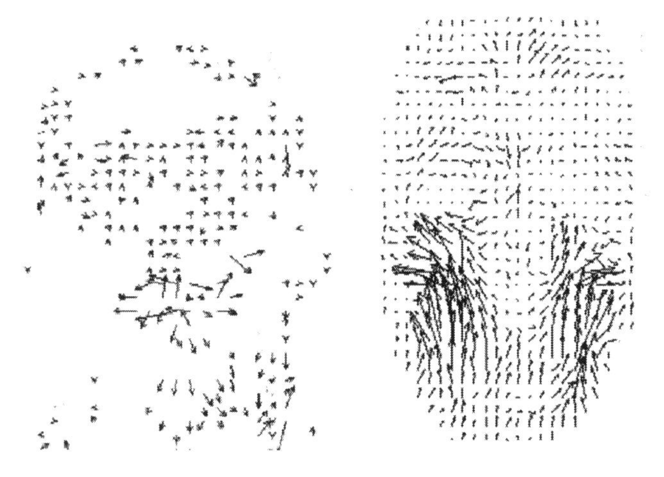

Fig. 11.2 (*Left*) Optical flow from an image sequence of a talking face. (*Right*) Optical flow from an image sequence of a face changing expression from neutral to hanger

Piccardi 2006). The datasets employed for the experiments are quite varied and include from 10 up to a maximum of 210 subjects. The Cohn Kanade AUCoded facial expression database (Kanade et al. 2000), being the largest data collection with 2105 videos from 100 subjects. Yet, the number of captured expression is not totally uniform and, more importantly, it is not fully representative of the variety of emotions which can be experienced and displayed by humans in different environments. The datasets and relative experimental results are limited to six basic expressions

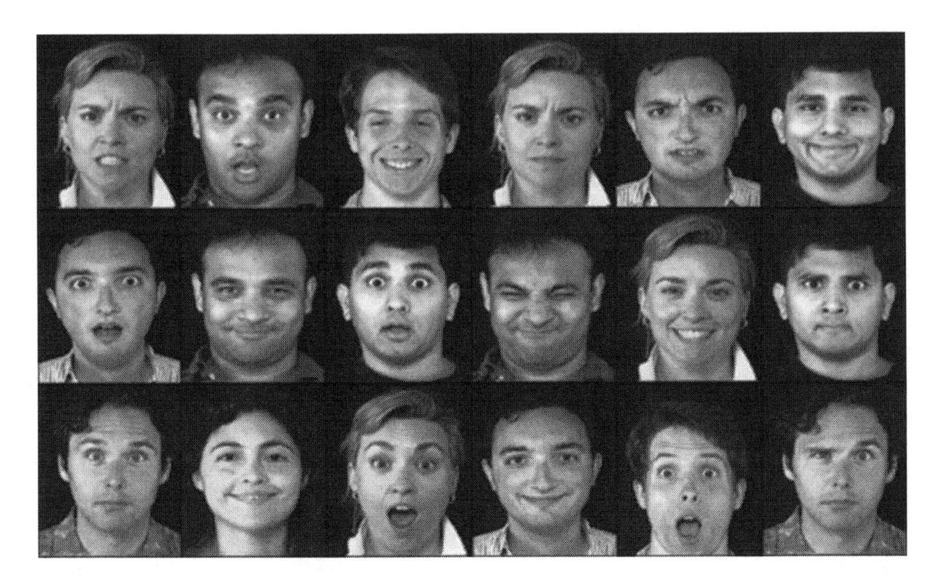

Fig. 11.3 Sample snapshots from a standard video database for facial expression analysis (Essa and Pentland 1995)

(Fear, Anger, Disgust, Happiness, Sadness, Surprise). More recently, these expression have been augmented to a more complete set including also Uncertainty, Anxiety and Boredom. These last three being most important to detect uncomfortable states of the subject, for example, while waiting in a long queue at the post office or at a ticket counter.

The second common denominator is the acquisition condition. All the datasets have been collected in indoor, controlled environments (generally research labs). As a consequence, most of expressed emotions are purposively constructed, rather than grabbed from a real scenery (see Fig. 11.3). Therefore, the datasets display a number of "expected" expressions, rather than genuine expressions as caused by either external (physical) or internal (emotional) factors. In a few cases the data was acquired from spontaneous emotions derived from external stimuli (Zeng et al. 2009). Even though, in most cases, humans can perceive a difference between the real and posed (or exaggerated) facial expression, devising an algorithm to automatically determine this difference requires an extensive test set.

Another similarity is the variation in the classification performances across different subjects. Not everybody displays the emotions exactly in the same way. Depending on the cultural background, there are several differences in the visible effects of emotions. Sometimes differences among emotions are just subtile movements of the eyes or the mouth. In other cases emotions are expressed by largely evident motions of the face parts. As a consequence, it is rather difficult to conceive a single system which is able of performing equally well over a large variety of subjects, especially in real world conditions.

Remarkably several systems have been recently proposed to improve the classification performances by using multiple visual and non visual cues (Zeng et al. 2009; Busso et al. 2004; Castellano et al. 2008; Datcu and Rothkrantz 2008; Ratliff and Patterson 2008; Paleari et al. 2009). The work by Sebe et al. (2006) and by Gunes et al. (Gunes and Piccardi 2006, 2009) are examples of how speech and the body motion can be exploited to improve the inference of displayed emotions. If emotional states are detected to facilitate the human computer interactions, as it is in many research projects, the available datasets and emotion categorizations are generally sufficient to describe the range of possible emotional states to be discovered by a computer to start the necessary actions. In the case of non-cooperative subjects, where a surveillance camera grabs images from freely moving people in the environment, the range of possible emotions and consequent motions and expressions is quite larger. Therefore, it is necessary to better understand the target environmental conditions first, and then understand the range of emotions which may be observed.

11.5 Emotion and Intention Recognition

There is strong relation between emotional states and intentions. Anxiety status can easily lead to an unpredictable action or a violent reaction. As discussed in the preceding sections, current technology can only discern among basic emotions through the categorization of facial motions. Still, tomorrow's technology may allow to discern more subtile expression changes to capture a wider spectrum of emotions, which may lead to detect dangerous intentions or possible future actions.

Nowadays, there is a considerable concern for biometric technologies breaking the citizen's privacy. The public view of the widespread adoption of surveillance camera is that the use of face identification systems may lead to a "state of control" scenario where people are continuously monitored against their supposed intentions. Science fiction and action movies greatly contributed to shape the public view of intelligent surveillance. Many people may see a "preemptive police corp" into action within the next few years. Even though we can't reasonably predict what the advance of technology will be in the next few years, these concerns have yet hindered the adoption of these technologies in many environments, thus depriving of many potential service improvements and benefits. Unfortunately, as in the past, also today misinformation and politically driven views impair technological advances which can be adopted for the wellness of the citizens. Yet, as any other technology, the way it is used, either for the good or bad, entirely depends on the intentions of the men behind its application. As stated above, the research on facial expression recognition has been largely lead by the aim to develop more intuitive and easy to use human computer interfaces (Picard 2000; Fasel and Luettin 2003). The overall idea being to "instruct" computers to adapt the delivered services or informations according to the "mood" and appreciation level of the users (Gunes et al. 2008). On the other hand, if security is the target application for an "emotion detection" surveillance system, at the basis there is a tradeoff between security and freedom. We already traded much of our personal freedom for air travel security.

Nobody can access any airport's gate area without being checked. We can't even carry a number of items onboard. Exceptions are quite limited and very hard to achieve. Still, millions of people travel by air every day (apparently) without any complaint. A more realistic scenario, where intelligent camera networks continuously check for *abnormal* and *potentially dangerous* behaviors in high risk areas or crowded places, may on the contrary increase the "sense of protection" and security of citizens. The same cameras may then be used in shops and lounges to determine the choice and satisfaction level of users to provide better customized services. In reality, there is no need to really impair personal privacy by coupling personal data with the identity of the bearer. This is far from being useful in the real practice.

We don't need to wait for the massive introduction of intelligent "emotionally driven" surveillance systems. Real privacy threats are already a reality. The widespread use of social networks and e-services, coupled with the misuse of personal data cross-check and fusion, is already a potential danger which may impair our privacy. This, unless the proper ruling actions are taken to hinder the mis-use (rather than the use) of personal data. Better spread of knowledge, at all levels, is the first step, then followed by an appropriate development of pan-national rules supported by technology standards is the only solution to defeat the privacy threats.

11.6 Conclusion

There is always a deep concern about the introduction of new technologies which may impact on the personal privacy. For this reason, all technologies which are devoted to either identify individuals or even for understanding some personal features, are looked suspiciously by the large public. Nonetheless, there are research projects which are now seriously looking not only into the identification of individuals, but indeed at the understanding of potentially hostile intentions. An example is the FAST (Future Attribute Screening Technologies) project of the US Department of Homeland Security.[1] A popular web magazine addresses this project in this way: *The Department of Homeland Security has been researching a sensor system that tries to predict "hostile thoughts" in people remotely for a while, but it's just spoken up about developments and renamed the system "Future Attribute Screening Technologies" FAST, which sounds really non-intimidating. It was called "Project Hostile Intent." But check out the technology's supposed powers for a rethink on how intimidating it sounds: it remotely checks people's pulse rate, breathing patterns, skin temperature and momentary facial expressions to see if they're up to no good.*

As any other technology, biometrics can be used for the good or bad of the citizens, but this is not a motivation to ban the technology per se. Rather this is a

[1] A description of the project can be found under the "Screening Technologies to detect intentions of humans" on the web page http://www.dhs.gov/xabout/structure/gc_1224537081868.shtm

strong incentive to consider better the potential of these technologies at the *service* of the citizens and to ameliorate the quality of life. Of course, the prevention of terroristic attacks is the first reference which comes to one's mind, and probably the most heavily advertised, but this is not the only nor the most important one. Many potential applications exist, related both to secure or not secure areas, which are worth deploying. Probably everybody is familiar with surveillance cameras placed in public areas, such as parks, shopping centers, squares, metro and train stations and others, devoted to detect potentially dangerous situations after the sunset. There are millions of such cameras located in every city throughout the world. Unfortunately it's simply impossible to make a good use of the tremendous amount of data constantly delivered by these cameras. No human operator could constantly look at these videos and reliably identify dangerous situations or suspects. Still, several criminals have been convicted over time, thanks to the videos recorded by these cameras. But how many lives could have been saved if the same cameras would have been able to automatically detect a malevolent behavior in the same environment, and just send an alarm to a human operator? The answer is obvious, and we believe the actions to be taken, at several levels, to promote the proper introduction of these technologies in everyday life are obvious as well.

Acknowledgements Funding from the European Union COST action 2101 "Biometrics for Identity Documents and Smart Cards" and from the Regional Research Authority are acknowledged.

References

Adolphs, R., D. Tranel, H. Damsio, and A.R. Damasio. 1994. Impaired recognition of emotion in facial expressions following bilateral damage to the human amygdala. *Nature* 372: 669–672.

Allison, T., A. Puce, and G. McCarthy. 2000. Social perception from visual cues: Role of the STS region. *Trends in Cognitive Sciences* 4: 267–278.

Arandjelovic, O., and R. Cipolla. 2004. Face recognition from face motion manifolds using Robust Kernel resistor-average distance. In *Proceedings of FaceVideo04*, 88.

Bicego, M., E. Grosso, and M. Tistarelli. 2006. Person authentication from video of faces: A behavioral and physiological approach using Pseudo Hierarchical Hidden Markov Models. In *Proceedings first International Conference on Biometrics 2006*, Lectures Notes in Computer Science, vol. 3832, 113–120. Hong Kong, January 2006.

Bodamer, J. 1947. Die prosopagnosie. *Archiv fur Psychiatrie und Nervenkrankheiten* 179: 6–54.

Braje, W.L., D.J. Kersten, M.J. Tarr, and N.F. Troje. 1999. Illumination effects in face recognition. *Psychobiology* 26: 371–380.

Bruce, V., and A.W. Young. 1986. Understanding face recognition. *British Journal of Psychology* 77(3): 305–327.

Bryatt, G., and G. Rhodes. 1998. Recognition of own-race and other-race caricatures: Implications for models of face recognition. *Vision Research* 38: 2455–2468.

Burton, A.M., V. Bruce, and P.J.B. Hancock. 1999. From pixels to people: A model of familiar face recognition. *Cognitive Science* 23: 1–31.

Busso, C., Z. Deng, S. Yildirim, M. Bulut, C.M. Lee, A. Kazemzadeh, S. Lee, U. Neumann, and S. Narayanan. 2004. Analysis of emotion recognition using facial expressions, speech and multimodal information. In *Proceedings of the 6th international conference on multimodal interfaces, (State College, PA, USA, October 13–15, 2004). ICMI '04*, 205–211, ACM, New York. doi:http://doi.acm.org/10.1145/1027933.1027968

Calder, A., and A. Young. 2005. Understanding the recognition of facial identity and facial expression. *Nature Reviews Neuroscience* 6: 641–651.

Calder, A., A. Young, D. Rowland, D. Perrett, J. Hodges, and H. Etcoff. 1996. Facial emotion recognition after bilateral amygdala damage: Differentially severe impairment of fear. *Cognitive Neuropsychology* 13: 699–745.

Castellano, G., L. Kessous, and G. Caridakis. 2008. Emotion recognition through multiple modalities: Face, body gesture, speech. In *Affect and emotion in human-computer interaction*, Lectures Notes in Computer Science, vol. 4868, ed. C. Peter and R. Beale. Heidelberg: Springer.

Chen, L.S., and T.S. Huang. 2000. Emotional expressions in audiovisual human computer interaction. In *Proceedings of ICME 2000*, 423–426. New York, July–August 2000.

Chen, L.S. 2000. Joint processing of audio-visual information for the recognition of emotional expressions in human-computer interaction. PhD Thesis, University of Illinois at Urbana-Champaign.

Cohen, I., N. Sebe, A. Garg, L.S. Chen, and T.S. Huang. 2003. Facial expression recognition from video sequences: Temporal and static modeling. *Computer Vision and Image Understanding* 91(1–2): 160–187.

Darwin, C. 1898. *The expression of emotions in man and animals.* New York: D. Appleton and Company.

Datcu, D., and L. Rothkrantz. 2008. Automatic bi-modal emotion recognition system based on fusion of facial expressions and emotion extraction from speech. In *IEEE face and gesture conference FG2008*. September 2008 ISBN 978-1-4244-2154-1.

D'Entremont, B., and D.W. Muir. 1997. Five-month-olds' attention and affective responses to still-faced emotional expressions. *Infant Behavior and Development* 20(4): 563–568.

Doddington, G., W. Liggett, A. Maartin, M. Pryzbocki, and D. Reynolds. 1998. Sheep, goats, lambs, and wolves: A statistical analysis of speaker performance in the NIST 1998 Speaker Recognition Evaluation. In *International conference on spoken language processing*. Sydney: Sydney Convention Centre.

Dunn, J., I. Bretherton, and P. Munn. 1987. Conversations about feeling states between mothers and their young children. *Developmental Psychology* 23(1): 132–139.

Ekman, P. 1994. Strong evidence for universals in facial expressions: A reply to Russell's mistaken critique. *Psychological Bulletin* 115(2): 268–287.

Ekman, P. 2009. *Telling lies: Clues to deceit in the marketplace, politics, and marriage*. New York: Norton.

Ekman, P., and W.V. Friesen. 1976. *Pictures of facial affect.* Palo Alto: Consulting Psychologists Press.

Ekman, P., and W.V. Friesen. 1978. *Facial action coding system: Investigator's guide*. Palo Alto: Consulting Psychologists Press.

Ekman, P., R.E. Sorenson, and W.V. Friesen. 1969. Pan-cultural elements in facial displays of emotion. *Science* 164(3875): 86–88.

Ekman, P., W.V. Friesen, V. Wallace, and P. Ellsworth. 1972. Emotion in the human face: Guidelines for research and an integration of findings. Oxford: Pergamon.

Essa, I., and A. Pentland. 1995. Facial expression recognition using visually extracted facial action parameters. In *Proceedings of international workshop on automatic face and gesture recognition*. Zurich.

Fasel, B., and J. Luettin. 2003. Automatic facial expression analysis: A survey. *Pattern Recognition* 36: 259–275.

Gunes, H., and M. Piccardi. 2006. A bimodal face and body gesture database for automatic analysis of human nonverbal affective behavior. In *Proceedings of the 18th international conference on pattern recognition (ICPR'06)*, vol. 1, 1148–1153. Hong Kong.

Gunes, H., and M. Piccardi. 2009. Automatic temporal segment detection and affect recognition from face and body display. *IEEE Transactions on Systems, Man, and Cybernetics – Part B, Special Issue on Human Computing* 39(1): 64–84.

Gunes, H., M. Piccardi, and M. Pantic. 2008. From the lab to the real world: Affect recognition using multiple cues and modalities. In *Affective computing, focus on emotion expression, synthesis and recognition*, ed. Jimmy Or, 185–218. Vienna: I-Tech Education and Publishing. ISBN 978-3-902613-23-3.
Hancock, P.J.B., V. Bruce, and A.M. Burton. 1991. Recognition of unfamiliar faces. *Trends in Cognitive Sciences* 4(9): 263–266.
Harris, P.L. 1989. *Children and emotion: The development of psychological understanding*. Oxford: Blackwell.
Haviland, J.M., and M. Lelwica. 1987. The induced affect response: 10-week-old infants' responses to three emotion expressions. *Developmental Psychology* 23(1): 97–104.
Haxby, J.V., E.A. Hoffman, and M.I. Gobbini. 2000. The distributed human neural system for face perception. *Trends in Cognitive Sciences* 20(6): 223–233.
Hill, H., and V. Bruce. 1991. Effects of lighting on the perception of facial surface. *Journal of Experimental Psychology: Human Perception and Performance* 4(9): 263–266.
Horn, B.K.P., and B.G. Schunck. 1981. Determining optical flow. *Artificial Intelligence* 17(1–3): 185–204.
Hornak, J., E. Rolls, and D. Wade. 1996. Face and voice expression identification in patients with emotional and behavioral changes following ventral frontal lobe damage. *Neuropsychologia* 34: 173–181.
Howell, A.J., and H. Buxton. 1996. Towards unconstrained face recognition from image sequences. In *Proceeding of the IEEE international conference on automatic face and gesture recognition (FGR'96)*, 224–229. Killington.
Humphreys, G., N. Donnelly, and M. Riddoch. 1993. Expression is computed separately from facial identity, and it is computed separately for moving and static faces: Neuropsychological evidence. *Neuropsychologia* 31: 173–181.
Izard, C.E. 1971. *The face of emotion*. East Norwalk: Appleton-Century-Crofts.
Izard, C.E. 1997. Emotions and facial expressions: A perspective from Differential Emotions Theory. In *The psychology of facial expression: Studies in emotion and social interaction*, ed. J.A. Russell and J.M. Fernndez-Dols, 57–77. New York: Cambridge University Press.
Jack, R.E., C. Blais, C. Scheepers, P.G. Schyns, and R. Cadara. 2009. Cultural confusions show that facial expressions are not universal. *Current Biology* 19: 1543–1548.
Kanade, T., J. Cohn, and Y. Tian. 2000. Comprehensive database for facial expression analysis. In *Proceedings of the 4th IEEE international conference on automatic face and gesture recognition (FG'00)*, 46–53. Grenoble, March 2000.
Kanwisher, N., J. McDermott, and M. Chun. 1997. The fusiform face area: A module in human extrastriate cortex specialized for face perception. *The Journal of Neuroscience* 17: 4302–4311.
Kelly, D.J., P.C. Quinn, A.M. Slater, K. Lee, L. Ge, and O. Pascalis. 2007. The other-race effect develops during infancy: Evidence of perceptual narrowing. *Psychological Science* 18: 1084–1089.
Lewis, M., and S. Goldberg. 1969. Perceptual-cognitive development in infancy: A generalized expectancy model as a function of the mother-infant interaction. *Merrill-Palmer Quarterly* 15(1): 81–100.
Li, Y., S. Gong, and H. Liddell. 2000. Support vector regression and classification based multiview face detection and recognition. In *Proceedings of the IEEE international conference on automatic face and gesture recognition (FGR'00)*, 300–305. Grenoble.
Li, Y., S. Gong, and H. Liddell. 2001. Modelling faces dynamically across views and over time. In *Proceedings IEEE international conference on computer vision*, 554–559. Vancouver, July 2001.
Li, S.Z., R.F. Chu, S.C. Liao, and L. Zhang. 2007. Illumination invariant face recognition using near-infrared images. *IEEE Transactions on PAMI* 29(4): 627–639.
Light, L., F. Kayra-Stuart, and S. Hollander. 1979. Recognition memory for typical and unusual faces. *Journal of Experimental Psychology: Human Learning and Memory* 5: 212–228.
Liu, X., and T. Chen. 2003. Video-based face recognition using adaptive hidden Markov models. In *Proceedings CVPR03 (I)*, 340–345. Madison

Lucas, S.M. 1998. Continuous n-tuple classifier and its application to real-time face recognition. *In IEE Proceedings-Vision Image and Signal Processing* 145(5): 343.
Lucas, S.M., and T.K. Huang. 2004. Sequence recognition with scanning N-tuple ensembles. In *Proceedings ICPR04 (III)*, 410–413. Cambridge.
Malpass, R.S., and J. Kravitz. 1969. Recognition for faces of own and other race faces. *Journal of Personality and Social Psychology* 13: 330–334.
Markham, R., and K. Adams. 1992. The effect of type of task on children's identification of facial expressions. *Journal of Nonverbal Behavior* 16(1): 21–39.
Matsumoto, D. 2001. Culture and emotion. In *The handbook of culture and psychology*, ed. D. Matsumoto. New York: Oxford University Press.
Meissner, C.A., and J.C. Brigham. 2001. Thirty years of investigating the own-race bias in memory for faces: A meta-analytic review. *Psychology, Public Policy, and Law* 7: 3–35.
Mika, S., G. Ratsch, B. Scholkopf, A. Smola, J. Weston, and K.-R. Muller. 1999. Invariant feature extraction and classification in kernel spaces. In *Advances in neural information processing systems*, vol. 12, ed. M.I. Jordan, Y. LeCun, and Sara, A. Solla. Cambridge, MA: MIT.
Moses, Y., S. Edelman, and S. Ullman. 1996. Generalization to novel images in upright and inverted faces. *Perception* 25(4): 443–461.
Muir, D.W., and J. Nadel. 1998. Infant social perception. In *Perceptual development: Visual, auditory, and speech perception in infancy*, ed. A. Slater, 247–285. Hove: Psychology Press.
O'Toole, A.J., H. Abdi, F. Jiang, and P.J. Phillips. 2007a. Fusing face recognition algorithms and humans. *IEEE Transactions on Systems, Man and Cybernetics* 37: 1149–1155.
O'Toole, A.J., P.J. Phillips, F. Jiang, J.J. Ayadd, N. Penard, and H. Abdi. 2007b. Face recognition algorithms surpass humans matching faces across changed in illumination. *IEEE Transactions on Pattern Analysis and Machine Intelligence* 29: 1642–1646.
O'Toole, A.J., P.J. Phillips, A. Narvekar, F. Jiang, and J. Ayyad. 2008. Face recognition algorithms and the 'other-race' effect [Abstract]. *Journal of Vision* 8(6): 256–256.
Paleari, M., R. Benmokhtar, and B. Huet. 2009. *Evidence theory based multimodal emotion recognition MMM 2009, 15th international multiMedia modeling conference*, 435–446. Sophia Antipolis, January 7–9, 2009.
Pantic, M., and L.J.M. Rothkrantz. 2000. Automatic analysis of facial expressions: The state of the art. *IEEE Transactions on PAMI* 22(12): 1424–1445.
Phillips, P.J., P.J. Flynn, W.T. Scruggs, K.W. Bowyer, J. Chang, K. Hoffman, J. Marques, J. Min, and W.J. Worek. 2005. Overview of the face recognition grand challenge. In *Proceedings of the IEEE conference on computer vision and pattern recognition*, vol. 1, 947–954. San Diego: IEEE Computer Society.
Picard, R.W. 2000. *Affective computing*. Cambridge: MIT.
Ratliff, M.S., and E. Patterson. 2008. Emotion recognition using facial expressions with active appearance models. In *Proceedings of IASTED human computer interaction 2008*, 92–138. Anaheim.
Raytchev, B., and H. Murase. 2001. Unsupervised face recognition from image sequences. In *Proceedings ICIP01(I)*, 1042–1045. Thessaloniki.
Raytchev, B., and H. Murase. 2002. VQ-Faces: Unsupervised face recognition from image sequences. In *Proceedings ICIP02 (II)*, 809–812. Rochester, New York.
Raytchev, B., and H. Murase. 2003. Unsupervised recognition of multi-viewface sequences based on pairwise clustering with attraction and repulsion. *Computer Vision and Image Understanding* 91(1–2): 22–52.
Roli, F., and J. Kittler (eds.). 2002. *Multiple Classifier Systems*, Lectures Notes in Computer Science, vol. 2364. Berlin: Springer.
Russell, J.A. 1991. Culture and the categorization of emotions. *Psychological Bulletin* 110(3): 426–450.
Saks, M.J., and J.J. Koehler. 2005. The coming paradgm shift in forensic identification science. *Science* 309: 892–895.
Schyns, P.G. 1998. Diagnostic recognition: Task constraints, object information, and their interactions. *Cognition* 67(1–2): 147–179.

Sebe, N., I. Cohen, F.G. Cozman, T. Gevers, and T.S. Huang. 2005. Learning probabilistic classifiers for human-computer interaction applications. *Multimedia Systems* 10(6): 484–498.
Sebe, N., I. Cohen, T. Gevers, and T.S. Huang. 2006. Emotion recognition based on joint visual and audio cues. In *Proceedings of the 18th international conference on pattern recognition (ICPR'06)*, vol. 1, 1136–1139. Hong Kong.
Shan, C., and R. Braspenning. 2009. Recognizing facial expressions automatically from video. In *Handbook of ambient intelligence and smart environments*, ed. H. Nakashima, J. Augusto, and H. Aghajan. New York: Springer.
Soken, N.H., and A.D. Pick. 1999. Infants' perception of dynamic affective expressions: Do infants distinguish specific expressions? *Child Development* 70(6): 1275–1282.
Sorce, J.F., R.N. Emde, J.J. Campos, and M.D. Klinnert. 1985. Maternal emotional signaling: Its effect on the visual cliff behavior of 1-year-olds. *Developmental Psychology* 21(1): 195–200.
Sroufe, L.A. 1995. *Emotional development: The organization of emotional life in the early years*. New York: Cambridge University Press.
Stern, D. 1985. *The interpersonal world of the infant: A view from psychoanalysis and developmental psychology*. New York: Basic Books.
Tistarelli, M. 1996. Multiple constraints to compute optical flow. *IEEE Transactions on Pattern Analysis and Machine Intelligence* 18(12): 1243–1250.
Tistarelli, M., M. Bicego, and E. Grosso. 2008. Dynamic face recognition: From human to machine vision. In *Image and Vision Computing: Special issue on Multimodal Biometrics*, M. Tistarelli and J. Bigun eds. doi:10.1016/j.imavis.2007.05.006.
Tranel, D., A. Damasio, and H. Damasio. 1988. Intact recognition of facial expression, gender, and age in patients with impaired recognition of face identity. *Neurology* 38: 690–696.
Tretiak, O., and L. Pastor. 1984. Velocity estimation from image sequences with second order differential operators. In *Proceeding of 7th IEEE international conference on pattern recognition*, 16–19. IEEE.
Troje, N.F., and H.H. Bülthoff. 1996. Face recognition under varying pose: The role of texture and shape. *Vision Research* 36: 1761–1771.
Valentine, T. 1991. A unified account of the effects of distinctiveness, inversion, and race in face recognition. *Quarterly Journal of Experimental Psychology* 43A: 161–204.
Walker-Andrews, A.S. 1997. Infants' perception of expressive behaviors: Differentiation of multimodal information. *Psychological Bulletin* 121(3): 437–456.
Walker-Andrews, A.S., and L.E. Bahrick. 2001. Perceiving the real world: Infants' detection of and memory for social information. *Infancy* 2(4): 469–481.
Wellman, H.M., P.L. Harris, M. Banerjee, and A. Sinclair. 1995. Early understanding of emotion: Evidence from natural language. *Cognition and Emotion* 9(2–3): 117–149.
Widen, S.C., and J.A. Russell. 2003. A closer look at preschoolers' freely produced labels for facial expressions. *Developmental Psychology* 39(1): 114–128.
Widen, S.C., and J.A. Russell. 2008a. Children acquire emotion categories gradually. *Cognitive Development* 23(2): 291–312.
Widen, S.C., and J.A. Russell. 2008b. Young children's understanding of others' emotions. In *Handbook of emotions*, 3rd ed, ed. M. Lewis, J.M. Haviland-Jones, and L.F. Barrett, 348–363. New York: Guilford Press.
Young, A.W., F. Newcombe, de Haan, M. Small, and D.C. Hay. 1993. Face perception after brain injury: Selective impairments affecting identity and expression. *Brain* 116(4): 941–959.
Zeng, Z., M. Pantic, G.I. Roisman, and T.S. Huang. 2009. A survey of affect recognition methods: Audio, visual, and spontaneous expressions. *IEEE Transactions on PAMI* 31(1): 39–58.
Zhou, S.K., V. Krueger, and R. Chellappa. 2003. Probabilistic recognition of human faces from video. *Computer Vision and Image Understanding* 91(1–2): 214–245.
Zhou, S.K., R. Chellappa, and B. Moghaddam. 2004. Visual tracking and recognition using appearance-adaptive models in particle filters. *Image Processing* 13(11): 1491–1506.1 A description of the project can be found under the "Screening Technologies to detect intentions of humans" on the web page http://www.dhs.gov/xabout/structure/gc_1224537081868.shtm

Chapter 12
The Transparent Body: Medical Information, Physical Privacy and Respect for Body Integrity

Emilio Mordini and Holly Ashton

12.1 Introduction

Scientific literature on the ethical and privacy implications of biometrics is becoming increasingly important. A sharp debate is emerging about whether biometric technology offers society any significant advantages over conventional forms of identification, and whether it constitutes a threat to privacy and a potential weapon in the hands of authoritarian governments. The main issues at stake concern large scale applications, biometric databases, remote and covert biometrics, respect for fair information principles, enrolment of vulnerable and disabled groups, information sharing and system interoperability, technology convergence, and behavioural biometrics, surveillance. These have been covered elsewhere in this book. One area which has not been greatly considered however, relates to the potential of biometrics to detect medical information about the persons enrolled in a system and the implications of this on future policy concerning these technologies.

Biometrics are constantly 'sold' to the general public in terms of their potential for enhancing personal security. However, the data used to authenticate an individual – measurable physical and behavioural attributes – could also potentially be used for studying patterns in individuals and populations, for looking for change or consistency over time, and in different contexts. Decisions about attributes to be measured and the algorithm to be adopted are never neutral, in the sense that they are crucial decisions which are influencing the way in which the whole biometric system works and its degree of "informational intrusiveness". Not all biometrics are good for all purposes, yet no single biometric is suited for one purpose alone. In other words,

E. Mordini (✉) • H. Ashton
Centre for Science, Society and Citizenship, Piazza Capo di Ferro 23,
Rome, 00186, Italy
e-mail: emilio.mordini@cssc.eu; holly.ashton@cssc.eu

E. Mordini and D. Tzovaras (eds.), *Second Generation Biometrics: The Ethical, Legal and Social Context*, The International Library of Ethics, Law and Technology 11,
DOI 10.1007/978-94-007-3892-8_12,

there are no biometrics that are completely "clean", as is exemplified by the fact that the same biometrics can be used in a variety of ways, say, for population statistics and genetics research, for medical monitoring, for clustering human communities and ethnic groups, and for individual identification. Some attributes are more appropriate for individual recognition (e.g. fingerprint, iris), others are more suited for group categorization (e.g. face geometry, body shape), and still others are more apt to medical monitoring (e.g. thermal images, voice), yet all biometrics may generate data that are not strictly relevant only to personal identification. This data surplus runs the risk of becoming available for "unintended or unauthorized purposes". If a biometric system generated only those details which are indispensable for making the system work, function creep – at least in the sense of a malicious use of biometric applications – would be hardly feasible.

Sensors – which are the most important part for capturing data within biometric systems – target physical properties of body parts, or physiological and behavioural processes, which are called 'biometric characteristics'. The output of the sensor(s) is an analogical, or digital, representation of the biometric characteristic, this representation is called a 'biometric sample'. Sensors unavoidably generate data about time and location, say, when and where the sample was captured. They may also collect shadow information, for instance any system for facial recognition inescapably ends up collecting extra information on people's age, gender, skin colour and eye colour, and – given that facial expressions are topological configurations that can be measured – they have also the potential to detect people's emotional states, as reflected in their expressions.

Sensors can also elicit details on the medical history of the identifying person. Medical details can be elicited in various ways (Zhang 2008). First, injuries or changes in health can prevent someone from being enrolled by the system and might then be recorded.[1] Although most current technologies have no capability for determining the causes of recognition failure, no one can exclude the possibility that future applications may be designed to identify these causes. Second, medical information can be deduced by comparing selected biometric characteristics captured during initial enrolment and subsequent entries.[2] Indeed, using biometrics to search for consistency over time – as for recognition/authentication purposes – is not so different from using biometrics to look for patterns of change as medical doctors do. Third, biometric characteristics could directly disclose health information[3]

[1]E.g. some eye diseases could prevent iris scanning, arthritis could prevent hand geometry, finger burns can prevent fingerprinting, etc.

[2]E.g. facial geometry taken in different periods of time can reveal some endocrinopathies. Also, a bitonal voice is a result of a paralysed recurrent laryngeal nerve – this can be a side effect of lung cancer.

[3]E.g. certain chromosomal disorders – such as Down's syndrome, Turner's syndrome, and Klinefelter's syndrome – are known to be associated with characteristic fingerprint patterns. In another example, Wong, M., Choo, S.P. & Tan, E.H. (2008) reported that several cancer patients on the drug Capecitabine have reported loss of fingerprint as a side-effect of taking the medication (Wong et al. 2008).

(Wong et al. 2008). Additionally, some sensors may detect surgical modifications to the body.[4] Finally, by knowing that certain medical disorders are associated with specific biometric patterns, researchers might actively investigate such questions as whether biometric patterns can be linked to behavioural characteristics or predispositions to medical conditions. As a consequence, biometric characteristics could become a covert source for prospective medical information, allowing people to be profiled according to their current and potential health status. Mitigating this state of affairs is not easy, because – as we shall discuss below – it is partly due to the very nature of the human body. However there are two critical measures that should be adopted. First, biometric characteristics should always be chosen with their potential for disclosing health details taken into consideration. Medical doctors should be involved in biometric capture device design, and sensors with the effective capability to detect medical conditions should always be used proportionately to risk (i.e. such sensors would be more acceptable used within high security infrastructures than say, a workplace canteen). Second, people should always be aware of the presence of biometric sensors. This is a very difficult goal, not only because some applications are by definition covert (e.g. screening and surveillance), but also because biometric sensors are more and more embedded in ambient intelligence environments. According to the EU Data Protection Directive (art.7 par.1) (Directive 95/46/EC of the European Parliament and of the Council of 24 October 1995) no data collection can go unnoticed by the subject that is being monitored. The goal is that the individual is aware of all *types* of data about him/her that are collected. Yet this is exactly what embedded biometrics would prevent. The loophole to escape from this legal dilemma comes from art.7 par.2, which states that par.1 is not applied in the case of '*processing of data relating to offences, criminal convictions or security measures*'. Yet is it legitimate to extend the concept of 'security measures' to any technology used in any context? The growing proliferation of embedded biometrics cannot be justified by a never-ending growth of an indistinct 'security area'. Nor does warning people that they are entering into a biometric controlled area seem to be sufficient to ensure respect for privacy rights. Research from as far back as the early 1990s showed that warning labels tend not to be perceived any longer as people become familiar with them (Stewart and Martin 1994).

Thus we have established that almost any working phase biometric system might generate extra information, which has the potential to be further used for unintended, unauthorized, purposes. This state of affairs can certainly be mitigated but it is highly improbable that it can be totally prevented. Generating extra information is not due to imperfect technologies, or to procedures still to be refined, but depends on the very nature of the human body. In real life, communication and recognition are but two sides of the same coin. When sender and receiver exchange messages

---

[4]Infrared cameras can easily, and covertly, detect dental reconstruction and plastic surgery (e.g. added or subtracted skin tissue, added internal materials, body implants, scar removal, skin resurfaced by laser, removed tattoos, etc.) because the temperature distribution across reconstructed and artificial tissues is different from normal.

they unavoidably produce details that can be used to identify both. A totally anonymous sender or receiver does not exist either in the real world or in cyberspace. Distinguishing and individualizing are two different things, although they are both forms of recognition. Communication always introduces a distinction, but does not necessarily individualize. Similarly, processes of recognition channel messages that go well beyond mere personal identification, think for instance how, even moments after birth, the newborn seeks out the mother's eyes and face in an intricate and insoluble mix between recognition, auto-identification, and non-verbal messages. Thus it cannot be realistic to assume that systems designed with the intention of recognition (and thus, authentication) will not also entertain the possibility for the flip-side, communication. The collection of biometric information may consequently raise privacy concerns among the target audience. The concern is, what information could be *communicated* about them if the data finds itself in the wrong hands? User's perception of the degree of a particular device's or system's intrusiveness, is a vital element to be considered and must not be ignored by those developing and advocating the use of these systems.

12.2 The Importance of Medical Privacy

As far back as Ancient Greece, the importance of respecting medical privacy was acknowledged, as can be seen from its inclusion in the Hippocratic Oath, an oath of good practice to be taken by members of the medical profession. The Oath has lasted as an important element of medical practice for over 2,000 years and though it has been through modernising processes, the need to maintain a section on respecting the privacy of patients has never been overlooked. Translation of the original text includes the line;

> What I may see or hear in the course of the treatment or even outside of the treatment in regard to the life of men, which on no account one must spread abroad, I will keep to myself, holding such things shameful to be spoken about. (Edelstein 1943)

The origins of medical privacy go back to the origin of the feeling of "sacredness" and are interlaced with the charm related to the human body. The notion of "body" was hardly present in the Homeric language, where the two polar terms *psyche* and *soma* were to connote the corpse (*soma*) and the vital breath (*psyche*) (Wright and Potter 2000). When Homer speaks of the living human body, he does not distinguish between physical and mental processes. On the contrary he projects mental contents on to certain organs and physiological functions, as though mental and bodily operations were the same. For him, mental activity is "distributed" in the body and manifested in the activities of the organs. The lack of distinction between body and mind is also present in the Bible, where for example, the term *nefes* (literally "throat") alludes, through reference to nutrition, to the man as being indigent, and a wishful being. The word *ruach* (breath) indicates the state of emotional agitation, and *basar* is both the flesh and human impotence and futility. In these ancient cultures the body is "holy" because it is inherently a "living body", say, because it is animated by the god. As a consequence the body belongs to the god(s) and trespassing body boundaries is violating a "holy area". This is well illustrated by the *Lapis Niger*, which is a large piece of black marble which covers, according to legend, the tomb of Romulus, or the place of Romulus' assumption to heaven (the place however is strictly linked with Romulus' body). Under the Lapis Niger, there is the oldest known Latin inscription, dating from the sixth century BC. The writing is extremely damaged, so only one third to one half of each line survive, but the meaning is fairly clear. The first line declares the sanctity of the site (*sakros esed* – it is holy). The second line mentions the king (*recei* — to the king). The third line mentions the king's herald (*kalatorem*) and beasts of burden (*iouxmenta*). The general meaning is that the place is declared sacred, and all those who enter are threatened with terrible punishments (Coulston and Dodge 2000). Body integrity (Romulus' body) was thus assimilated to a sacred area.

The secularization of the ancient Greek culture in the democratic Athens eventually led to a conception of body integrity closer to the modern notion of physical privacy. Pericles (1916) said that one of the reasons behind the greatness of Athens was the respect and distinction made between public and private spheres, and such a cultural atmosphere also generated the Hippocratic Oath (which still conserves, however, a religious aura). The medic, is a stranger to the most private and sacred sphere of an individual, their own body, and is invited only through the necessity of the need for a cure. The relationship between patient and doctor is based around the patient disclosing and entrusting personal knowledge of themselves to their doctor which creates an uneven and one-sided relationship,[5] thus it is the doctors' duty to honour the private sphere of the individual they are assisting by keeping the knowledge secret and using it only for the benefit of their patient. The importance of respecting this privacy is protected by law and in rare instances where doctors are found to have failed in this responsibility, they face medical councils and tribunals and lose their right to practice.

Within this context, it seems strange to consider that the world is entering a phase of technology progression involving technology which could be capable of disclosing the very same private data that doctors are sworn to such secrecy over. These biometric devices, and the companies developing them, are not sworn to any kind of Hippocratic oath, and the privacy of the individuals enrolled in these systems lies in the hands of data protection laws which may not even be developed enough to cover the influx of new technologies. It is vital therefore that serious consideration is given to the potential of these technologies to detect medical findings and to contemplating whether policy changes are required to further safeguard the personal sphere which was recognized as so important all those years ago in Ancient Greece.

There is no point entering this debate without first considering what medical information could feasibly be gleaned from biometric devices. In the next section,

[5]http://www.enotes.com/everyday-law-encyclopedia/doctor-patient-confidentiality

we present various second generation biometrics and consider what medical information they could potentially disclose about an individual enrolled into a system that used them.

12.3 The Potential for Disclosure of Medical Information

Second Generation Biometrics refers to the new wave of biometric technologies which aim to inch closer to being able to mimic the way in which humans identify and authenticate one another. That is, via recognition of individual features and continuous analysis of unique body dynamics which can be captured in 'real time' and unobtrusively in the sense of not necessarily requiring an individual to stop at a machine to allow a scanning process to occur. These biometrics tend to focus on less persistent (weaker) traits than standard biometrics (such as fingerprinting or iris scanning) and they are often prone to changing over time. Although second generation biometrics address traits that are unstable and with low discriminating content, they can be fruitfully used for any kind of recognition, from authentication, identification, to screening – particularly when several biometrics are combined into a multimodal system which takes into account several different biometrics at a time. They can also be effectively combined with semantic knowledge (such as a pin code) in order to add another level of security to procedures. In this way, an individual must not only perform an action 'correctly' (in their unique way suitable for authentication), but they must also know the required knowledge. This makes opportunities for faking or spoofing a system even harder.

Although it isn't possible to provide an exhaustive list of second generation biometrics within this chapter, current ones which are most often referred to include; behavioural biometrics such as gait, face dynamics, signature dynamics, human computer interfacing, voice and even odour; electrophysiological biometrics and soft biometrics. These broad categories will be explored in more depth below.

The possibility for second generation biometrics to disclose incidental medical findings may vary depending on which technologies are used and of course on what types of data protection measures are in place. Certainly strong arguments could be made that with the correct measures and technology in place, no biometric system would be able to detect medical information – the fact remains however that usually data cannot be filtered to support only one application, which might raise further ethical and privacy concerns. We shall refer to this again later in the chapter. For now however, it seems pertinent to consider exactly what medical information various types of biometric *could* disclose (in a worst-case scenario) in terms of the information and data they detect and use.

12.3.1 Behavioural Biometrics

As most sensors for second generation biometric systems are designed to be non-intrusive, the subject may not even be aware of them, meaning that behaviour is

Table 12.1 Classification of actions

1. Voluntary
Intentional (planned, self-initiated)
Responsive (induced by external stimulus)
2. Semivoluntary (unvoluntary)
Induced by an inner sensory stimulus
Induced by an unwanted feeling/compulsion
3. Involuntary
Non suppressible (reflex, seizures, myoclonus)
Suppressible (tics, dystonia, stereotype...)
4. Automatic
Learned, without conscious effort (gait, speech, facial expressions, voice prosody, etc.)
Learned, with conscious effort (cycling, playing a musical instrument, typing, etc.)

likely to be natural and authentic. As such, the amount of personal information that can be elicited by behavioural biometrics is huge. The standard neurophysiological classification (Jankovic and Tolosa 2002) of behaviour separates it into four categories: voluntary, semi-voluntary, involuntary, and automatic (Table 12.1).

Voluntary actions are actions carried out by a conscious agent, according to some intentions or as a response to external events. Semi-voluntary actions are actions that the subject can stop but that are usually performed under the normal level of awareness. They are originated by inner sensory stimuli (e.g. most unaware body postures and gestures like crossing legs, stretching arms, etc.) or unwanted feelings and compulsions (e.g. nail biting, scratching the head, creasing the eyes, etc.). Involuntary actions concern both non suppressible (e.g. seizures, tremors, etc.) and suppressible (e.g. tics) actions provoked by nervous reflexes. Automatic actions are actions performed by the subject following an action sequence that has previously been learned but has since become totally unconscious. They include essential human actions such as walking or speaking that are learned in childhood and become completely automatic with age, and other action sequences learned in any other moment of life (e.g. cycling, driving, playing a musical instrument, etc.). These categories of actions have different biometric profiles, which correspond to their physiology.

Voluntary actions are a very refined and complex set of movements that imply the functioning of the higher mental functions. They imply the existence of belief, desires and intentions. They are possible thanks to a special kind of memory that is called "short-term" or "working" memory. Short term memory allows us to plan and perform voluntary actions. Everyone occasionally experiences a failure in the normal integration of memories, perceptions, and consciousness – typically, such a failure does not disrupt everyday activities. However, sometimes individuals become partly incapable of performing voluntary actions. This can be due to psychiatric conditions (e.g. dissociative disorders) or to progressive neurological disorders (e.g. dementia).

Voluntary actions may present a high degree of uniqueness, be easily collectable and be universal, but they rarely remain stable over time and they require a collaborative subject. They can easily be used for intention detection, for early

Table 12.2 Medical conditions potentially detected by behavioural biometrics

Condition	Biometric feature	Sensors
Psychiatric conditions		
Dissociative disorders	Gait, body dynamic, HCI biometrics	Cameras, sound recorders, HCI sensors
Acute anxiety, panic	Gait, body dynamic, HCI biometrics, signature dynamic	Cameras, sound recorders, HCI sensors
Major depression	Gait, body dynamic, HCI biometrics	Cameras, sound recorders, HCI sensors
Neurological conditions		
Movement disorders	All	All
Sleep and wakefulness disorders	HCI biometrics	HCI sensors (keyboard, mouse, haptic, etc.)
Muscolo-skeletal and articular disorders		
Hand disorders	HCI biometrics, signature dynamics	HCI sensors
Foot and ankle disorders	Gait, body dynamic	Cameras, sensing floors, sensing seats
Joint disorders	Gait, body dynamic	Cameras, sensing floors, sensing seats
Other medical conditions		
All those conditions that can generate symptoms similar to those of previous disorders	All	All

detection of certain medical conditions (notably psychiatric and neurological conditions), and for ethnic and cultural categorization (the way in which a subject performs a voluntary action is highly influenced by his cultural background), and also for detecting minor particularities, which could present the risk of stigmatisation (e.g. left-handed people). As a biometric signature, voluntary actions can present a high degree of specificity and uniqueness, yet they tend to be very unstable and influenced by several confounding factors.

Semi-voluntary Actions are expressions of emotions, feelings, compulsions and responses to inner sensory stimuli such as proprioceptive stimuli, posture reflexes, and balance movements. They include a series of subconscious movement patterns, partly governed by the brainstem. Their presence implies the functioning of the whole mental and neuromotor apparatus, although not necessarily of the higher mental functions (for instance, some semi-voluntary movements are kept in vegetative states). They have never been studied as biometric signatures, and there is no data about their capacity for recognizing individuals. They are however important confounding factors in behavioural biometrics because they can alter patterns of all voluntary and automatic movements, which are used as biometric signatures (Table 12.2).

Involuntary Actions are movements that are governed by subcortical and peripherical nervous centres. They don't require the presence of higher, cortical, activities. Although different from semi-voluntary actions, they can represent confounding factors in behavioural biometrics (they have not been studied as biometric signatures of themselves). Involuntary actions are almost always a sign of a medical condition. They include the main movement disorders, those with decreased or slow purposeful movements (hypokinesia) and those with excessive voluntary or abnormal involuntary movements (hyperkinesia).

12.3.2 *Human Computer Interfacing*

Technically classed as behavioural biometrics, we mention these separately as they are in a slight class of their own. Human Computer Interfacing biometrics use the interactions between humans and computers, and the individual ways that each person has for doing this as its basis. HCI could be used to infer medical information about an individual. In particular, movement abnormalities resulting from such illnesses as arthritis, Parkinson's Disease and osteoporosis could be detectable.

12.3.3 *Voice Biometrics*

Although voice biometrics are technically classed as behavioural biometrics, as with HCI, we deal with them separately here as they do not seem to fit with conventional behavioural biometrics. In this instance, 'voice' does not constitute a specific behaviour (when considered as 'speech' in general and combined with face dynamics it fits the definition more closely). Here we consider 'voice' in terms of the numerical model of the sound, rhythm and pattern of an individual's voice. Such sounds and patterns are unique as the physical components and vibrations of vocal chords involved in producing them are different in each person.

Voice is perhaps one of the easiest biometrics to work with and one of the easiest to incorporate in existing security systems as the technology for using them (e.g. microphones and speakers) is already widely in use.

Voices are capable of giving away some medical information about the speaker. 'Gravelly' speech is a common side-effect of colds or laryngitis and a person may lose their voice altogether for a period of time. Vocal cord nodules and polyps may cause a hoarse and breathy voice over a longer period of time. Voice abnormalities may also be a result of neurological disease or a problem ensuing from, for example, a stroke, throat cancer, or even developmental abnormalities. One study found that patients with short stature have a high incidence of voice and laryngeal abnormalities (Heuer et al. 1995).

12.3.4 *Electrophysiological Biometrics*

Different systems of the human body (such as the heart and the nervous system) naturally emit electric signals that can be recorded (Revett 2008) in various ways.

Genetic and environmental influences in early life interact to shape fine details of the bio-electric signals, which are unique and consistent to every individual (Prance et al. 2008). The medical literature has hardly considered the possibility of using electro physiological signals as a method for personal identification, partly because until a few years ago collecting these signals was a complex operation which demanded wired sensors and the absence of ambient noise (e.g. power line artifacts, muscular artifacts, badly positioned electrodes, etc.). Today we are able to use both wireless sensors and effective filters which substantially reduce noise. Electrophysiological biometric signatures, which amplify, measure and record changes of electrical potentials on skin are increasingly studied (Harland et al. 2002). They include:

Electrocardiogram (ECG) records electrical activity of the heart as measured on skin. Medical doctors use 12 standard electrode positions and 6 standard electrical connections. For identification purposes the number of electrodes and electrical connections can be substantially reduced. ECG signals consist of periodic waves which follow the periodic heart beat. There are six kind of different waves, P, Q, R, S, T and U waves, each one with its own characteristics and patho-physiological meaning.

Electroencephalogram (EEG) records electrical activity of the brain as measured on the scalp. The standard is made up by 21 electrode positions (but for biometric purposes far fewer are used). EEG recorded in the absence of an external stimulus is called spontaneous EEG; EEG generated as a response to external or internal stimulus is called an event-related potential (ERP). Signals include typical spontaneous oscillation in 4–20 Hz frequency range. Types of signals which can be observed in EEG include different frequency bands, Delta, Theta, Alpha, and Beta.

Electrooculogram (EOG) records eye movements. Signals record low changes of potential related to these movements.

Electromyogram (EMG) records muscle activity as measured by using electrodes. Signals are typically high frequency electrical spikes closely related to muscle activity.

Sensory Evoked Potentials (SEP)[6] are electrical potentials recorded following presentation of a stimulus, and are distinct from spontaneous potentials as detected by EEG, ECG, EMG. They include Visual and Auditory SEPs. Ordinarily, these small potentials are lost in EEG background noise, but computer processing cancels out the noise to reveal a waveform.

Event-Related Potentials (ERPs) are a non-invasive method of measuring brain activity during cognitive processing. ERPs occur as small fluctuations in Electroencephalogram (EEG) in response to the presentation of a single stimulus,

[6]Other kind of evoked potential are Somatosensory Evoked Potentials (SSEPs) and Motor Evoked Potentials (MEP), which are not relevant to biometrics.

either sensory, motor or a mental event. By measuring the time it takes for an ERP to occur after presentation of the stimulus, and by taking recordings from several areas of the brain, it is possible to determine the sequence and timing of the specific areas activated within the brain. While evoked potentials reflect the processing of the physical stimulus, event-related potentials involve memory, expectation, attention, or changes in the mental state.

Originally all electrophysiological signals were biometrics used in medicine to detect anomalies in normal organ physiology (e.g. brain, nerves, heart, muscles). They have also been extensively studied as a tool for investigating perceptual, motor, emotional and cognitive processes across the lifespan. Electrophysiological biometric signatures are universal (they are present in all living human beings), highly accurate, very difficult to fake, cost effective to implement, easy to integrate in electronic devices (e.g. into a mobile phone, into a lap top, etc.), small in size (and technology promises to make them smaller and smaller), and have low power consumption. As Revett (2008) points out;

> The deployment of biological signals – either directly or through the process of biofeedback – provides an innovative and practical way of user authentication/identification. Users will no longer have to sign in, enter a password or select images from a gallery. They may just be able to think of their authentication details and these will be transmitted faithfully to the authentication device [...] This technology, coupled with a VRE, may provide an unique opportunity to implement multimodal biometrics within a single framework, allowing fusion to occur at all levels. These ideas are already in place – and the components are being developed in laboratories around the globe. It is really up to government agencies and related legislative bodies to decide how this technology will be utilized.

What he fails to appreciate is that there is no proportion at all between the goal of this technology and its degree of intrusiveness. To be sure, electrophysiological biometrics can become the building block of several Brain Computer Interfaces (BCIs) and many of these applications are largely justifiable. For instance people with very severe neurological conditions – like the so called "locked-in" syndrome, where a person is completely paralyzed and unable to speak, which affects half a million people worldwide – can use altered biometric devices as a non-muscular channel of control for computers and other devices. Patients learn to master their mental activity in order to produce minute electrophysiological changes in brain signals, which have been used to restore communication. What is instead untenable is that a technology, which is in principle able to detect mental contents, is used to access control or people authentication in the working place. However, Revett correctly goes on to point out that the main barriers to the development of electrophysiological biometrics for automated human recognition are ethical, legal, and political ones.

The main reasons for ethical and privacy concerns with electrophysiological biometrics are twofold. First of all, they may prove to be psychologically intrusive. Electrophysiological signals can also be used for emotion and lie detection. Lie detectors record not lies but physiological changes. When using a lie detector, the idea is to elicit emotions by questioning. In today's lie detectors, physiological parameters monitored are (1) blood pressure; (2) pulse changes; (3) respiratory changes; (4) skin conductivity. This information is recorded, classified, and compared in order to detect incongruous emotional reactions, which can reveal a deception. Physiological changes, which are investigated by standard polygraph testing, could be better investigated by using electromagnetic signals like those captured by EEG, ECG and ERPs. It is likely then that next generation lie detectors will largely overlap with biometric devises. In high security environments it would make sense to use the same technology both to authenticate persons and to detect imposters. What makes these considerations ethically and politically worrisome is that emotion and lie detection can be carried out covertly, completely unnoticed by users. Of course this would be against all good privacy practices and it would not be respectful of personal dignity and psychological integrity. Yet no one could exclude the possibility that it will happen soon or later, or that it has already happened, given the current security oriented political climate.

The second key area of concern, which is more aligned to the thread of this chapter, relates to additional medical, information collected by the sensors. Electrophysiological signals are employable for authentication purposes as long as they are normal. Abnormal signals, like an ECG showing cardiac irregularities, prevent individual's pattern recognition (Agrafioti and Hatzinakos 2008). This also holds true for other electrophysiological signals. Whilst it is true that the number of electrodes and type of analysis used within biometric utilisations of electrophysiology vastly differ from the level required for medical analysis and diagnosis, it remains the case that it would still be possible to infer some kind of abnormality of function from the biometric data. With this in mind, it is irrelevant that the level of information would be insufficient for medical diagnosis, as the merest hint of unusual function could be enough to invade someone's privacy.

The risk of function creep for this technology is partly prevented, because recognition is in opposition with detection of medical conditions. In other words, particularly combined with the difference in the standard of the technology used for medical diagnosis, it is quite improbable that Electrophysiological Biometrics is used for covert medical diagnosis or medical screening. Yet it is unarguable, as previously mentioned, that the *potential* of this technology for disclosing at least some medical information is extremely high, also because many conditions addressed by electrophysiological studies may go unnoticed (for instance many cardiac disturbances such as arrhythmias and ischemic disorders) till the moment in which an ECG is recorded. The same holds true with EEG, which can detect unknown neurological conditions by detecting electrical charges associated with seizure disorders, sleep disorders, and metabolic or structural encephalopathies. Also, Evoked Potentials may detect unknown medical conditions, such as identifying pre-clinical deficits in multiple sclerosis, discovering unsuspected optic nerve damage, demonstrating the presence of deficits in patients suspected to suffer from hysteria or to be imposters. In a working environment, employers could be tempted to use this technology to covertly verify employees' medical conditions and to screen and sort employees according to their medical conditions and risks to develop specific diseases. Another risk is that people may unwillingly receive a medical diagnosis or get referred to their GPs. On the other hand it is also evident that a good deal of legal controversies could be generated by missed or wrong medical referrals, or by a lack of communication about incidental medical findings. Moreover it seems unreasonable to request system controllers to include in their teams cardiologists or/and neurologists as a rule. If electrophysiological biometrics is used, it is then paramount to provide exhaustive and effective information to the data subject and to involve him in the decision about the policy to be followed in case of incidental medical findings. Finally it is crucial to ensure that no discrimination against the subject will be carried out because of the disclosure of previously unknown medical conditions.

12.3.5 Soft Biometrics

All biometric sensors also produce ancillary information, which is rarely stored and used during automatic recognition. These soft biometrics can be collected in order to create a system with thresholds and weighting for different classes of users. This can create a further biometric module, which can be either a pre-identification module, which filters subjects (when soft biometrics is used for filtering a large biometric database) or a secondary biometric system, which is fused with the primary biometric system at scoring or decision levels. Thus, soft biometrics are traits like gender, height, weight, age and ethnicity which are used to complement the identity information provided by the primary biometric identifiers (Ratha et al. 2001). Although soft biometric characteristics lack the distinctiveness and permanence to recognize an individual uniquely and reliably, and can be easily faked, they provide some evidence about the users identity that could be beneficial. In other words, despite the fact they are unable to individualize a subject, they are effective in distinguishing between people. Combinations of personal attributes like gender, race, eye colour, height and other visible identification marks can be used to improve the performance of traditional biometric systems. Most soft biometrics can be easily collected and are actually collected during enrolment. However, this information is not currently utilized during the automatic identification/verification phase. The soft biometric traits can either be continuous (e.g. height and weight) or discrete (e.g. gender, eye colour, ethnicity, etc.).

Some biometric systems may perform better, given a target audience with a majority that possess (or don't possess) a certain feature or characteristic related to, for example, ethnicity, gender, occupation, age, or colour of eyes. It implies that some soft biometrics can be indirectly deduced from difficulties in enrolment or more false rejections than average, as follows:

Age (elderly): more false rejections than average (any biometrics), also they have more difficulties in enrolment for many reasons (e.g., they tend to have very dry skin, which can make adequate contact with certain types of fingerprint capture devices difficult, etc.)

Age (children): more false rejections than average (any biometrics)

Disability (wheelchair): difficulties in enrolment (height of the device)

Disability (nanism, genetic conditions, inherited disorder): difficulties in enrolment (height of the device)

Disability (amputation): difficulties in enrolment (unable to use hand or finger based systems)

Disability (arthritis): difficult to use hand or finger based systems, more false rejections than average in these systems, difficulty in gait recognition

Disability (blindness): difficulties in enrolment (unable to use iris or retina based systems, and also affects user positioning for other systems)

Medical conditions (any): difficulties in enrolment and more false rejections than average (e.g., COLDS, LARYNGITIS: temporary effect on voice; CONTUSIONS: temporary effect on face or hand images; CRUTCHES: may make it difficult to stand steadily; SWELLING: temporary effect on face, body profile or hand images; etc.)

Physiognomy (race & ethnicity, religion, cultural area, political beliefs): BEARDS & MOUSTACHES (facial recognition, more false rejections than average); BALDNESS (facial recognition, more false rejections than average); EYELASHES (iris recognition, difficulties in enrolment); FINGERNAIL GROWTH (fingerprinting, difficulties in enrolment); VERY TALL OR VERY SHORT PEOPLE (any biometric, difficulties in enrolment); IRIS COLOUR INTENSITY (iris recognition, difficulties in enrolment); SKIN TONE AND COLOUR (iris and facial recognition, difficulties in enrolment because of increased difficulties in correctly locating faces or irises);

Look and Appearance (ethnicity, religion, cultural area, political beliefs): BANDAGES/BANDAID/CLOTHING (any biometrics, difficulties in enrolment); HATS, EARRINGS, SCARVES, PIERCINGS/TATTOOS (facial recognition, difficulties in enrolment); SLEEVES/RINGS (hand based systems, difficulties in enrolment); HEEL HEIGHT (body profile and gait recognition, change apparent user height); TROUSERS/SKIRTS/SHOES (gait recognition, difficulties in enrolment and more false rejections than average); CONTACT LENSES (iris recognition, difficulties in enrolment and more false rejections than average); COSMETICS (facial recognition, difficulties in enrolment and more false rejections than average); GLASSES, SUNGLASSES (iris and facial recognition, difficulties in enrolment); FALSE FINGERNAILS (fingerprint, difficulties in enrolment); HAIR STYLE/COLOUR (facial recognition, difficulties in enrolment).

Behaviour (ethnicity, religion, cultural area, political beliefs, medical conditions): DIALECT, ACCENT, and NATIVE LANGUAGE (voice recognition, difficulties in enrolment); EXPRESSION, INTONATION, and VOLUME (voice recognition, difficulties in enrolment); FACIAL EXPRESSIONS (facial recognition, difficulties in enrolment); MISPOKEN OR MISREAD PHRASES (voice recognition, difficulties in enrolment); OUT OF BREATH (voice recognition, difficulties in enrolment);

INVOLUNTARY MOVEMENTS (any biomctrics, difficultics in cnrolment); STRESS, TENSION, MOOD, DISTRACTIONS (any biometrics, difficulties in enrolment and more false rejections than average).

We shall now consider a number of different soft biometrics which may be capable of disclosing medical information about an individual.

12.3.5.1 Age

Age has three different components:

Chronological age: age measured by the time (years and months) that someone has existed

Biological age: measurable changes that take place in us all as we age. They cover a wide variety of measurements. Some are simple, such as the changes in eye-sight, hand grip strength or the thinning of skin. Others are more complex, such as measuring lung capacity and force or blood work to see the changes in various hormone levels. These measurements are called biomarkers.

Social age: a way of grouping people based on particular roles, cultural experience, performances, social networks of belonging, etc. All societies possess some tacit norms for determining social functions. These age rules are often described as "social clocks," the culturally determined moment in time for specific social activities to be carried out. School attendance, weddings, employment, childbirth, grandparenthood, and so on are some of the human activities ruled by social clocks.

Chronological age is usually considered personal data when it is included in health records or in other medical documents, while it is usually considered non sensitive when it is included in identity documents. Biometric sensors are not able to detect chronological age but they can target characteristics that are also used to evaluate biological and/or social age. Some biomarkers are in fact also biometric characteristics (e.g. skin luminescence depends on dermal thickness and subcutaneous layers, which are in their turn age related; gait dynamic can reveal details about age; etc.) and in their turn some biometric characteristics may indirectly provide details on age (e.g. fingerprint quality tends to decrease with ageing and after 65 years is often very poor). Likewise some "social clocks" can be detected by using biometric sensors (e.g. any biometric system that collects recorded images of an individual in a social context is also collecting details on her way of dressing and interacting with other humans). Extra data, which are particularly sensitive, can emerge from comparing these three age components. In particular, by comparing chronological age with biological age, it is possible to infer sensitive medical data without the subject being aware of this. In practice one could evaluate the level of "wear" of one's body and – consequently – one could also infer some details on her lifestyle and major diseases. This could be misused in various ways. Health insurers could be interested in this kind of information and one can also imagine, for instance, an employer who is interested in determining the biological age of his employees in order to exclude those who are older than their chronological age because they are likely to be less fit.

Despite some good research, there is still a lack of detailed understanding about how biometric data and biometric templates behave with ageing, especially specifically within an elderly population. Many human features change with age, and current biometric technologies have not been able to overcome some of these impacts to date. Accuracy of facial recognition technologies, which are the primary biometrics in use for travel/border security applications, does not hold up well over time. Fingerprint biometrics, which are inexpensive and fast to implement, and which are broadly deployed (e.g. payment systems, logical access to computer systems), have very high failure rates for older persons, whose fingerprints have become worn over a lifetime (also an issue for manual labourers, no matter what their age) (Modi and Elliott 2006). In addition, where biometric applications are designed to rely upon a particular physical characteristic for identification, they will result in exclusion of those individuals who may, due to injury or disability, not possess such a characteristic (e.g. injuries to eyes, fingers, limbs, etc.). Usually, systems are designed to offer fallback alternatives, but these are often time-consuming to pursue and can create some perception of stigma at "failing" the system. A 2005 trial for biometric passports held in the UK found that individuals over the age of 60 had more difficulty in enrolling in biometrics than their younger counterparts (UK Passport Service 2005). As previously mentioned, one suggested approach to remedy these issues is multimodality, which uses more than one personal trait or characteristic to identify a person. If one sample does not register, such as a scarred or faded fingerprint, another sample could be used, such as facial or signature recognition, which would lessen false rejection rates (Khan 2006).

12.3.5.2 Gender

Gender includes four different components:

1. *Genetic sex*: the gender as determined by chromosomes. Chromosomes are arranged in pairs and one particular pair govern gender. XX denotes female, and XY denotes male.
2. *Genital gender*: the gender as determined by genitalia, which include external and internal genitalia. Genitalia are mostly directly determined by genetic sex and are usually described as "primary sexual characters". The male external genitalia consists of the penis, scrotum, and pubic hair. The internal genitalia include the epididymis, vas deferens, prostate gland, and testicles. The female external genitalia include vulva, vagina, hymen, clitoris , labia minora and majora, and pubic hair. The internal genitalia include the ovaries, fallopian tubes and uterus.

3. *Biological gender*: biological gender results from the combination of primary and secondary sexual characters. Secondary sexual characters are those anatomical and physiological characteristics (e.g. body size and shape, breast development, pitch of the voice, body hair and musculature, etc.) that are strongly associated with one sex relative to the other and are chiefly due to sexual hormones.
4. *Social gender*: the gender as a distinct social group. "Social Gender" is the term used to refer to those ways in which a culture redefine biological sexes and channel them into different roles in various culturally dependent ways. Males and females in each culture have their different roles and respect different norms in order to adhere to those roles.

Each of these gender components may or may not be consistent with all others, either through free choice of an individual or for several medical reasons. Discrepancies between genetic, genital, biological and social genders determine cases of gender ambiguity, hermaphroditism, transsexualism, transgenderism, etc.

Most biometric characteristics could provide important clues to determine gender components. All biometric sensors which imply any bodily contact collect tiny biological samples which might theoretically allow the determining of chromosomal sex. Genital gender is hardly detected by biometric sensors (some new body scanners – such as Millimeter Wave Holographic Body Scanners – can actually detect genitalia[7] but they are not biometric sensors). On the contrary, secondary sexual characteristics are easily detected by most sensors but those more in use, i.e. fingerprint and iris sensors, are targeting characteristics that are hardly influenced by sexual hormones (yet it must be noted that some indirect details can be deduced from fingerprint and iris, e.g., nails and eyelash length). Face geometry and dynamics have high capacity for biological and social gender recognition as psychological studies have well illustrated (Bruce and Burton 1993; Burton et al. 1993). Artificial intelligence applied to biometric facial recognition has been effectively used for gender classification (Golomb et al. 1991; Brunelli and Poggio 1992; Moghaddam and Yang 2000).

Details about gender are considered sensitive data when they are included in medical records and in any circumstance in which they can be used to discriminate against an individual. Details about gender can also be used for people profiling. Gaps between genetic sex, genital gender, biological gender, and social gender can reveal highly sensitive personal details (e.g. several medical conditions including genetic diseases, genital gender reassignment surgery, cosmetic surgery, cultural and religious mutilations, sexual orientation, transgenderism, etc.) – similar to 'age', it

[7]The scanners bounce harmless "millimeter waves" off passengers who are selected to stand inside a portal with arms raised after clearing the metal detector. A screener in a nearby room views the black-and-white image and looks for objects on a screen that are shaded differently from the body. Finding a suspicious object, a screener radios a colleague at the checkpoint to search the passenger. Privacy is partly protected by blurring passengers' faces and deleting images right after viewing. Yet the images are detailed, clearly showing a person's gender.

is when conflicts appear between the different definitions of gender that medical information could be gleaned. For example, a biometric system that identified an individual as biologically male but socially female could be used to infer private information that a person did not wish to disclose. Another key fact to consider regarding gender as a soft biometric concerns the fact that often, only social gender is really checked by biometrics. As social gender is a construct determined by cultural expectations, it becomes unreliable anyway.

12.3.5.3 Race and Ethnicity

"Race" is probably one of the least measurable and most often used features used for the categorization of human beings. The term "race" is used to denote the differences between groups of people but is now regarded by many social and natural scientists as highly problematic and contentious. The modern concept of race is far removed from the typological concept adopted by nineteenth and partly twentieth Century biologists, which classified the human species in four main races identifiable by certain physical features – in particular skin colour, defined as "white", "black", "yellow" or "red" – and which were considered to have originated in different continents.

Modern genetics recognizes the existence of an almost infinite number of sub-populations within the human species which may be classed as "races", and which are characterized by various different genetic elements. The key finding of modern genetics as far as the concept of race is concerned is summed up by Lewontin et al. (1984) as follows: "*Of all human genetic variation known for enzymes and proteins (…) 85 per cent turn out to be between individuals within the same local population, tribe or nation. A further 8 per cent is between tribes or nations within a major "race", and the remaining 7 per cent is between major "races". That means that the genetic variation between one Spaniard and another, or between one Masai and another, is 85% of all human genetic variations, while only 15 per cent is accounted for by breaking people up into groups*" (1987, p. 206–7). Theoretically any number of races may be classified, be it 4, 16, 64, or 256 etc. depending on the criteria adopted in their definition. The geneticist Theodosius Dobzhansky (1962), having stated beforehand that "*differences in race can be ascertained objectively, however the number of races an individual chooses to acknowledge is merely a question of convenience*", defined a classification of races which is extremely indicative of the approach of modern genetics, and which defines 34 races in all, some of which (for example, neo-Hawaiians, Ladins and coloured North Americans) may seem quite disconcerting to the layman. Dobzhansky's classification implies acknowledgement of the fact that races, far from being strictly defined "types", are in reality Mendelian populations[8] which change through time, and which the scientist must approach as

[8] A Mendelian population is an interbreeding group of organisms that share a common gene pool. As defined by Michael Allaby; "Mendelian population." A Dictionary of Zoology. 1999. Encyclopedia.com. 2 Feb. 2010 http://www.encyclopedia.com

such, independently of their hybrid origin and more or less recent formation. As Dobzhansky himself stressed, “race” is a process rather than a fact (Dobzhansky 1973). Physical differences between groups certainly exist, but these differences are comparatively trivial in genetic terms. The vital point to bear in mind has been that there is much greater variation between individuals within any designated “racial group” than there is between such designated racial groups. In large measure, “race” is a *social construct* that is capable of exercising great force. However, race and ethnicity classification are used in public health surveillance for epidemiological purposes in order to identify differences in health status among racial/ethnic minorities.

Some biometric characteristics include physical features that have been used in the past to classify races and ethnic groups (e.g. skin, eye and hair colour, face and body shape, etc.). Although these features hardly inform about any biological characteristic of the individual, and hence are unlikely to disclose any sort of medical information, they keep on being relevant in social sorting and people profiling. The seemingly innocuous act of assigning people to “races” still sets them sociologically and biologically apart in a way that scientists and anthropologists have long since rejected. More recently a different approach, based on binary face recognition (e.g. ‘Asian’ v. ‘non-Asian’), has shown to be effective in creating a binary classification of face images such as gender and race classification without much modification (Ou et al. 2005). Of course such binary classification is still implicitly based on geography and social constructs, whereas in reality, a person who physically looks Asian may be just as likely in these times to be a European (albeit of Asian descent) and hence the racial classification may be irrelevant.

12.3.5.4 Skin Colour

Skin colour is due primarily to the presence of a pigment called melanin. Both light and dark people have this pigment. Two forms are produced, pheomelanin , which is red to yellow in colour, and eumelanin, which is dark brown to black. People with light skin mostly produce pheomelanin, while those with dark coloured skin mostly produce eumelanin. In addition, individuals differ in the number and size of melanin particles. In lighter skin, colour is also affected by red cells in blood flowing close to the skin. To a lesser extent, the colour is affected by the presence of fat under the skin and carotene, a reddish-orange pigment. Skin colour is rather variable in humans. It ranges from a very dark brown among some Africans, Australians, and Melanesians to a near yellowish pink among some Northern Europeans. There are no people who have true black, white, red, or yellow skin. These are commonly used colour terms that do not reflect biological reality. Skin colour differs from person to person, and it is perceived differently according to illumination conditions.

The main biometrics which detect skin colour are hand and face biometrics. There are various face detection algorithms based on skin colour because colour processing is faster than processing other facial features. Moreover localizing and tracking patches of skin-coloured pixels is a tool used in many face and gesture tracking systems, because, under certain lighting conditions, colour is orientation invariant. Skin colour detection can provide details on race and ethnicity and on some medical conditions. For instance, melanin is produced thanks to a protein called the P protein codified by the OCA2 gene. This gene is associated to three known genetic diseases;

Oculocutaneous albinism, people with this form of albinism often have light yellow, blond, or light brown hair, creamy white skin, light-coloured eyes, and problems with vision.
Angelman syndrome, a disorder characterized by intellectual and developmental delay, seizures, erratic movements especially hand-flapping, frequent laughter or smiling.
Prader-Willi syndrome, a disorder characterized by polyphagia, small stature and learning difficulties.

Skin colour might also suggest other medical issues such as vitiligo, post-inflammatory hyperpigmentation (the darkening of skin) occuring after an injury or burn, or after certain skin disorders such as acne or eczema.

Visible and distinguishable skin marks such as tattoos, scarification, piercing, ritual body painting, branding, scarring (surgical or different origin), can also disclose details about socio-cultural categories. Skin analysis also implies detection of cosmetic surgery, sub dermal implants, silicon injection. Skin luminescence depends on dermal thickness and subcutaneous layers, which are age related, and thus it may also provide details about age.

12.3.5.5 Eye Colour

Eye colour comes from different combinations of melanin in the iris. There are many base eye colours:

- ***Albino****, lack of brown pigment lets the red colour of blood cells show through.*
- ***Amber/yellow****, more commonly found in animals and known commonly as cat's eyes.*
- ***Blue****, contain no melanin in the front part of the iris. Found in Europeans or areas populated by people of European descent, such as Canada, The USA or Australia. 8% of the world's population has blue eyes.*
- ***Brown****, contain a large amount of pigment which can sometimes appear to be black and is the most dominant of eye colours.*
- ***Grey****, a variation of blue but considered more dominant*
- ***Green****, most commonly found in Germanic, Slavic or Celtic descent.*
- ***Hazel****, a medium brown which sometimes appears to be changing, particularly when the light catches it or different coloured clothing is worn.*

Eyes can change colour due to age or disease – most usually fading but sometimes darkening. Many babies are born with blue eyes which then darken to their natural colour within the first few months after birth. This is due to exposure to light which triggers the production of melanin in the iris. Some people have one brown eye and one green eye. This abnormality can be caused by either a trauma in the womb, faulty developmental pigment transport or a benign genetic disorder. Additional information involved in eye colour recognition concerns some medical conditions (e.g. albinism, jaundice, Kayser-Fleischer rings, due to copper deposition as a result of particular liver diseases, etc.) and ethnicity. Eye colour details can be produced as metadata of iris and facial recognition and it is used to improve the performance of these traditional biometrics. Information can be binary (e.g. blue or brown eye) or based on multiple parameters.

12.3.5.6 Beards and Hair

Hair growth in both men and women is regulated by androgens. Testosterone stimulates hair growth in the pubic area and underarms. Dihydrotestosterone stimulates beard hair growth and scalp hair loss. Whether hair growth is considered appropriate may differ depending on ethnic background, cultural interpretation, and individual opinion. A person's age, sex, racial and ethnic origin, as well as hereditary factors, determine the amount of body hair. Beard is a male secondary sex characteristic consisting of terminal facial hair on the cheeks, lips, chin, and neck. Some biometric sensors, in particular those related to facial recognition, may detect the presence of beard and hair. The presence of beard is hardly considered personal data. Yet beard and body hair can reveal religious affiliations (e.g. Hasidic and in general Orthodox Jews have distinctive hair – men don't shave their beards, and let their sideburns grow long and curl them), ethnicity, gender; and some medical conditions, which include alopecia, hypertrichosis and hirsutism.

Beards and Medical Conditions

Unusual (because of age or gender) beard presence can reveal various medical conditions. Likewise total absence of facial hair, in particular when coupled with absence of eyebrows, can reveal that the person is undergoing anti-cancer chemotherapy.

Hirsutism is the excessive growth of thick or dark hair in women in locations that are more typical of male hair patterns like beards. Hirsutism is most often the result of a familial tendency, particularly among people of Mediterranean or Middle Eastern ancestry. It can also depend on Polycystic Ovary Syndrome (PCOS). Sometimes, hirsutism is caused by tumours or other disorders of the pituitary gland, adrenal glands, or ovaries that cause levels of male hormones (androgens) to increase abnormally. Anabolic steroids, which may be abused by female athletes and bodybuilders, are androgens.

Hypertrichosis is an increase in hair growth anywhere on the body in men and women. Hypertrichosis is usually caused by a body-wide (systemic) illness or a drug. Illnesses include the following: Dermatomyositis, General systemic illness (such as advanced HIV infection), Hypothyroidism or other endocrine disorders, Malnutrition, Porphyria cutanea tarda and some central nervous system disorders. Drug causes of hypertrichosis include minoxidil, phenytoin, cyclosporine, and anabolic steroids.

Alopecia is defined as loss of hair. Alopecia can be classified as focal or diffuse and by the presence or absence of scarring. Nonscarring alopecia results from processes that reduce or slow hair growth without irreparably damaging the hair follicle. The alopecias comprise a large group of disorders with multiple and varying etiologies, including, genetic, psychological, medication side-effects and even environmental factors.

12.3.5.7 Hair Colour

Similarly to 'presence of hair', details about hair colour are usually got through facial recognition systems. Hair colour is usually implicated in race classification and details about age can also be elicited from it.

Some medical conditions can also be disclosed by hair colour details, such as some genetic conditions (e.g. albinism); skin diseases (e.g. alopecia, psoriasis, etc.) and drug intake (e.g. anti-cancer chemotherapy).

12.3.5.8 Height

Height in our society is perceived as a plus. People tend to see tall stature as fitness, health, vigour, and ultimately authority and success. Stature is easily visible and measurable, permanent and a factor upon which people discriminate, even if only unconsciously. Height, especially for men, has historically connected with power. In the business world, several studies show that the executive ranks are over-represented by tall men (Herpin 2005; Judge and Cable 2004).

Some biometrics produce metadata about height. Height can be implicated in race classification. Details about age and gender can also occasionally be indirectly elicited from height. Some medical conditions (e.g. thalassemia) are accompanied by short stature. Gigantism and dwarfism have various causes, mainly genetic.

12.3.5.9 Weight

Weight is hardly considered sensitive personal information, unless it is part of a medical record. However biases against people who are overweight are largely diffused. *"We live in a culture where we obviously place a premium on fitness, and fitness has come to Symbolize very important values in our culture, like hard work and discipline and ambition. unfortunately, if a person is not thin, or is overweight or obese, then they must lack self-discipline, have poor willpower, etc., and as a result they get blamed and stigmatized"* (Puhll and Heuerl 2009).

In no European state are there legislations which explicitly prohibit discrimination on the basis of weight. As a result, weight discrimination is increasing and ends up by impairing also the obesity epidemic. Weight discrimination occurs in employment settings and daily interpersonal relationships virtually as often as race discrimination, and in some cases even more frequently than age or gender discrimination. It has been found that even medical professionals demonstrate implicit negative biases against overweight people (Brownell and Puhl 2003). Overweight women are twice as vulnerable as men, and discrimination strikes much earlier in their lives. Compared to other forms of discrimination, weightism is the third most prevalent cause of perceived discrimination among women (after gender and age) and the fourth most prevalent form of discrimination among all adults (after gender, age and race) (Puhll and Heuerl 2009). ibid.

Biometric metadata about weight can hardly be produced. More frequently weight is directly measured during enrolment. Weight details can include information on socio-cultural status, ethnicity, race and of course, medical conditions. It can also be used as a possible indicator of future ill health.

12.3.5.10 Glasses

Details on wearing glasses are hardly considered sensitive information, unless this piece of information is part of a medical record or is related to any risk of working discrimination. Although there is no regular discrimination against people with glasses, glasses are often a sign of a correctable disability, which may disqualify the wearer from performing a variety of jobs.

Wearing glasses can provide indirect information on some medical conditions (eye diseases in general, eye disorders, and drug use and in specific cases of mental illness) or disabilities (e.g. albinism, blindness). Facial and iris recognition can work less well with people wearing glasses and this can cause some discomfort in people who want (or need) to be enrolled in the system. In some cultural areas the use of sunglasses is more frequent in certain socio-cultural groups and can be an indirect clue of socio-cultural categories.

12.4 A Real Risk?

In the previous sections we outlined why the potential for biometric systems to detect incidental medical findings could be problematic from an ethical and privacy perspective, and we went on to show exactly what findings specific types of second generation biometrics could disclose. The questions to consider now though are, what is the actual likelihood that;

(a) biometric systems would be capable of detecting this kind of information in reality and
(b) that this information could be used against people?

As we have seen, there are a number of medical problems that could potentially be detected or inferred from the data captured by second generation biometrics. This is because the nature of the biometrics – using characteristic signature 'traits' of an individual as identifiers/authenticators, relies on information disclosed by the human body in a similar way to medical diagnosis. In such systems, 'non-average' people (the 'outliers') might be exposed to discrimination. Even if such discrimination was not intended malevolently, it could occur in the form of 'failure to enrol' thus immediately setting aside a class of people made conspicuous by their inability to be authenticated by biometric technologies.

If such systems were genuinely capable of detecting so much medical information then the privacy invasion would be huge. However, in reality, are these second generation biometrics really more invasive than current systems for ensuring security? In the case of soft biometrics for example, many of these are already captured on passport photographs, and similarly, behaviour can already be observed on CCTV cameras. Perhaps the difference comes in the amount of control that the individual has about disclosing their data and how much they feel their data is subsequently at risk of being abused in some way. Showing a passport at a border control post (or indeed any form of identity card used in enhancing security in a certain area) involves a fleeting presentation of your document which is observed by another person. As soon as this procedure has occurred, the data itself remains with the individual (on their passport or ID document) rather than the authority performing the check, and thus is perhaps less 'intrusive' for the general public. Similarly, although behaviour can be observed via CCTV cameras, it does not automatically involve the recognition or authentication of the individuals caught on camera. Thus, there remains some level of anonymity. Although the current systems certainly allow for individuals to be traceable and would perhaps allow for some level of medical scrutiny to someone determined enough, the information is not all presented on a plate to those who might be interested in performing such invasions of privacy – the fear is that the nature of second generation biometrics systems could make such data much easier to obtain. If we consider what damage could have been done had a totalitarian regime had access to such amounts and types of data readily at hand, the case for safeguarding privacy at all costs becomes even clearer.

There are two great risks relating to potential medical disclosures for the future users of biometric systems. They relate to the individual's place in society in terms of their public and private spheres. At a private level, there is the simple issue of the Right to Privacy. Any individual should have the right to keep private any information on themselves that they do not wish to share. Thus there should be no concerns for people enrolling in biometric systems, that the technology will be able to disclose private, personal information about them to anyone else. It could be interpreted almost that the individual cooperating with a security system involving biometrics is providing a gift to the authority who requires them to comply, and as pointed out by Murray in 1987 (Murray 1987), *"It would be wrong to treat something dear to the donor in an undignified manner, as merely a commodity; likewise it would be wrong to use it in a way the donor would disapprove of"* (p. 32). The second risk concerns the public sector and relates to practical issues such as work and obtaining health insurance. Policies need to be in place to ensure that these technologies are not used as a tool for stigmatising individuals or for covertly obtaining information which an individual does not wish to disclose – though obviously within the relevant legal framework – for instance certain health insurance policies require disclosure of certain medical conditions and may be invalid if not correctly stated. This disclosure is still however at the discretion of the individual – the fear is that biometric technologies could take such disclosures out of the individuals hands. Careful thought needs to be applied to this as at first it may not always be apparent that infringements may occur. There could be temptation, particularly amongst scientific communities, to petition to use such vast amounts of information in statistical analyses and the performing of correlations to keep track of medical outbreaks. In situations approaching mass hysteria, such as were witnessed to some extent during the recent Avian and Swine Flu outbreaks, members of the public could be coerced into agreeing to privacy intrusions that they would normally instinctively resist – there needs to be clarity concerning the protection rules to ensure that relevant decisions are in place before hysteria and fear sets in. The extent to which we could reasonably expect rational decisions to come from a market largely driven by scare-mongering and fear about security issues needs to be addressed. Such decisions need to be considered from a rational and responsible viewpoint with the genuine interests of the general public in mind.

Overall, the greater fear concerning systems using second generation biometrics, probably relates more to the nature of data storage than the potential of the systems to detect medical findings. A human controller at a check point would be just as capable of detecting certain medical irregularities as a machine processing biometric details, however, the concern relating to this technology relates to the possibility for systems to collect and store data. This would mean that data could be re-visited and analysed, making the possibility of inferring and using medical information much higher. Thus surely one of the safeguards to be built into these systems should concern the way data is captured and disposed of. For security purposes, the need for the data is instantaneous and thus technically, the second after data has been captured and approved by the system, it can be disposed of (aside from the stored biometric template against which it is being measured of course). Policy safeguards should be in place to ensure not only that all biometric security systems of the future are used in such a way as to make data retaining and scrutiny difficult, but also that it is technically *impossible* for them to be used in such a way. Thus at design and implementation level, these issues must be taken into account. In this way, the need for protecting people from incidental medical findings becomes another issue under the general data protection and privacy umbrella.

Bearing all this in mind however, it seems pertinent to also consider the deeper issues raised by these technologies. Merely battling to protect privacy might not be an adequate response, and indeed, to enter such a battle we also need to clarify exactly who we are battling against and how strongly we believe in what we are fighting for. The fact that technologies are being created which can potentially have such an intrusive impact on our lives must lead to some soul-searching questions. Perhaps in an era of z-list celebrities and television shows aimed at giving members of the public their 15 min of fame we are turning into a population for whom the notion of personal privacy is no longer relevant? Are these technologies capable of being developed because we already have an altered and ever-altering view of the concept of 'privacy' or are we being forced to alter our concept of privacy to fit in with the evolving world around us? Looking to the future, will it be possible to expect technological progress to occur whilst maintaining the same levels of personal privacy that we have previously been accustomed to? From the often emotive media coverage of 'Big Brother' technologies (and indeed the very use of the term 'Big Brother') under which second generation biometrics invariably fall, it is clear that there are still plenty of people for whom the notion of personal privacy *does* matter. Furthermore, as pointed out earlier, the need to prevent sensitive data from being obtainable by people who could use it to discriminate against certain groups is vital. Nevertheless, it is clear that we live in a greatly altered world from when Pericles spoke of public and private spheres – in particular, the notion of increased vulnerability to security threats such as terrorists and rapidly spreading diseases is blurring the boundaries of what constitutes 'personal' and 'public' space. In the fight for justice and security of citizens, where do we draw the line? For instance, is the weapon loaded body of a terrorist still to be considered their individual private sphere or has it become public property?

Whilst we ponder the response to such questions, perhaps someone could undertake the task of penning an alternative Hippocratic Oath for use with biometric technologies.

References

Agrafioti, F., and D. Hatzinakos. 2008. ECG biometric analysis in cardiac irregularity conditions. *Signal, Image and Video Processing*. http://www.springerlink.com/content/w3104j4216806547. Accessed 8 June 2009.

Brownell, K., and R. Puhl. 2003. Stigma and discrimination in weight management and obesity. *The Permanente Journal* 7(3): 21–23.

Bruce, V., and A.M. Burton. 1993. Sex discrimination: How do we tell difference between male and female faces? *Perception* 22: 131–152.

Brunelli, R., and T. Poggio. 1992. HyperBF networks for gender classification. *Proceedings of DARPA Image Understanding Workshop*, San Diego, CA, 311–314.

Burton, A.M., V. Bruce, and N. Dench. 1993. What is the difference between men and women? Evidence from facial measurement. *Perception* 22: 153–176.

Coulston, J.C., and H. Dodge (eds.). 2000. *Ancient Rome: The archaeology of the eternal city*. New York: Oxford University Press.

Directive 95/46/EC of the European Parliament and of the Council of 24 October 1995 on the protection of individuals with regard to the processing of personal data and on the free movement of such data.

Dobzhansky, T. 1962. *Mankind evolving*. New Haven: Yale University Press.

Dobzhansky, T. 1973. *Genetic diversity and human equality. Facts and fallacies in the explosive genetics and education controversies*, The John Dewey Society Lecture. New York: Basic Books Inc.

Edelstein, L. 1943. *The Hippocratic Oath: Text, translation, and interpretation*. Baltimore: Johns Hopkins Press. Translation from the Greek; http://www.pbs.org/wgbh/nova/doctors/oath_classical.html. Accessed 7 Jan 2010.

Golomb, B.A., D.T. Lawrence, and T.J. Sejnowski. 1991. SEXNET: A neural network identifies sex from human faces. In *Advances in neural information processing systems*, vol. 3, ed. D.S. Touretzky and R. Lippmann, 572–577. San Mateo: Morgan Kaufmann Publishers.

Harland, C.J., T.D. Clark, and R.J. Prance. 2002. Electric potential probes – New directions in the remote sensing of the human body. *Measurement Science and Technology* 13: 163–169.

Herpin, N. 2005. Love, careers and heights in France, 2001. *Economics and Human Biology* 3(3): 420–449.

Heuer, R.J., R.T. Sataloff, J.R. Spiegel, L.G. Jackson, and L.M. Carroll. 1995. Voice abnormalities in short stature syndromes. *Ear, Nose, & Throat Journal* 74(9): 622–628.

Jankovic, J.J., and E. Tolosa (eds.). 2002. *Parkinson's disease and movement disorders*, 4th ed. Philadelphia: Lippincott Williams & Wilkins.

Judge, T.A., and D.M. Cable. 2004. The effect of physical height on workplace success and income: Preliminary test of a theoretical model. *Journal of Applied Psychology* 89(3): 428–441.

Khan, I. 2006. Multimodal biometrics – Is two better than one? *Auto ID & Security*. http://www.findbiometrics.com/Pages/multimodality.htm. Accessed 12 June 2009.

Lewontin, R.C., S. Rose, and L. Kamin. 1984. *Not in our genes*. New York: Pantheon Books.

Modi, S.K., and S.J. Elliott. 2006. *Impact of image quality on performance: Comparison of young and elderly fingerprints*, Purdue University, http://www2.tech.purdue.edu/it/resources/biometrics/publications/proceedings/ModiRASC2006.pdf. Accessed 12 June 2009.

Moghaddam, B., and M. Yang. 2000. Gender classification with support vector machines. IEEE International Conference on Automatic Face and Gesture Recognition, Grenoble, FR, 306–311.

Murray, T. 1987. Gifts of the body and the needs of strangers. *Hastings Center Report,* 30–38.

Ou, Y., X. Wu, H. Qian, and Y. Xu. 2005. A real time race classification system. *Proceedings of the 2005 IEEE International Conference on Information Acquisition*, June 27–July 3, 2005, Hong Kong and Macau, China.

Pericles, Plutarch lives. Trans. Bernadotte Perrin, 1916. Cambridge: Harvard University Press. London: William Heinemann Ltd.

Prance, R.J., S.T. Beardsmore-Rust, P. Watson, H.C. Harland, and H. Prance. 2008. Remote detection of human electrophysiological signals using electric potential sensors. *Applied Physics Letters* 93: 033906 (published online 25 July 2008) http://dx.doi.org/10.1063/1.2964185.

Puhll, R.M., and C.A. Heuerl. 2009. The stigma of obesity: A review and update. *Obesity*. doi:10.1038/oby.2008.636.

Ratha, N.K., J.H. Connell, and R.M. Bolle. 2001. Enhancing security and privacy in biometrics-based authentication systems. *IBM Systems Journal* 40(3): 614–634.

Revett, K. 2008. *Behavioral biometrics: A remote access approach*. Chichester: Wiley Interscience.

Stewart, D.W., and I.M. Martin. 1994. Intended and unintended consequences of warning messages: A review and synthesis of empirical research. *Journal of Public Policy and Marketing* 13(1): 1–19.

UK Passport Service. 2005. Biometrics enrolment trial report, May 2005, 8–9. http://www.passport.gov.uk/downloads/UKPSBiometrics_Enrolment_Trial_Report.pdf. Accessed 12 June 2009.

Wong, M., S.P. Choo, and E.H. Tan. 2008. Travel warning with capecitabine. *Annals of Oncology*. doi:10.1093/annonc/mdp278.

Wright, J.P., and P. Potter (eds.). 2000. *Psyche and soma: Physicians and metaphysicians on the mind-body problem from antiquity to enlightenment*. New York: Oxford University Press.

Zhang, D. (ed.). 2008. *Medical biometrics*. New York: Springer.

Part IV
New Biometrics in Context

Chapter 13
Security in the Danger Zone: Normative Issues of Next Generation Biometrics

Irma van der Ploeg

13.1 Introduction

For many years now, biometrics have been the subject of public debate, critical and ethical reflection, and regulatory efforts. The fact that certain developments in pattern recognition and sensor technologies enabled the use of human bodies for identification and authentication purposes in a wide variety of contexts triggered enthusiasm as well as worries from the beginning. Now, some two decades further on, we see not only a widespread and large scale use of this technology, but also a qualitative technical development that leads many to speak of the emergence of 'second generation biometrics'.

In this chapter, I discuss some key elements of this development of second generation biometrics that give rise to a new set of ethical and socio-political issues to be analysed and discussed. The two main elements of recent developments in biometrics I want to focus on here are first, the emergence of new biometric traits, often for multimodal use in combination with more 'classic' traits, in particular so-called soft biometrics and physiological biometrics, and second, the shift to embedded biometric systems, which includes an emphasis on distant sensing and 'passive' biometrics, and which forms a key element of the wider trends towards Ambient Intelligence (AmI) and ubiquitous computing (UbiComp); I will focus in particular on security related applications.

Together, these developments signify a new level of complexity not only of the technology, but also of the critical issues and policy challenges connected with it.

I will first discuss the emergence of '*soft biometrics*', i.e. the use of general traits such as gender, height, age, ethnicity or weight for automated classification or as supporting information in identification and authentication applications.

I. van der Ploeg (✉)
Infonomics and New Media Research Centre, Zuyd University,
Brusselseweg 150, 6217 HB Maastricht, The Netherlands
e-mail: irma.vanderploeg@zuyd.nl

E. Mordini and D. Tzovaras (eds.), *Second Generation Biometrics: The Ethical, Legal and Social Context*, The International Library of Ethics, Law and Technology 11,
DOI 10.1007/978-94-007-3892-8_13, © Springer Science+Business Media B.V. 2012

Such techniques, I will argue, reify social categorisations that are highly sensitive and essentially contestable, thereby closing them off from inspection, debate and contestation. Next I will focus on the shift towards *embedded systems* and Ambient Intelligence, in which biometrics are foreseen to play an important, facilitating role. Distant sensing and biometrics that require no conscious cooperation from subjects are at the centre of this development, giving rise to some obvious concerns about covert data capture, transparency and consent. Following that is a discussion of another set of new biometric traits that today are being researched and developed into authentication and assessment tools, namely physiological states and phenomena such as heart rate, body temperature, brain activity patterns, and pupil dilation. These '*under the skin biometrics*', so far applied mostly in various security contexts, bring a whole new range of body data within reach of data controllers, and especially when used in embedded and distant fashion, need to be assessed from social and ethical viewpoints urgently. Drawing the various types of developments discussed together, I next identify and articulate a highly normative assumption embedded in biometrics, that these various new technologies appear to be stretching to unprecedented extent: the *assumption of availability*. The very idea that one *can* direct biometric sensors and connected identification, inspection, and assessment tools at people's bodies, and register ever more, and ever more intimate aspects of even the general public, presupposes the body to be available in ways that perhaps ought not to be simply taken for granted.

13.2 Sensitive Categories, 'Partial Identities' and Soft Biometrics

Whereas biometrics is generally conceived of as using certain features of 'the' human body for IT-mediated authentication and identification, we are all aware that 'the' human body does not exist as such. People come in a variety of shapes, colours, genders, and ages, sharing, of course, most of their physiology and anatomy with most of humanity, but never all. Beyond the absolute unicity of each human body on which the very possibility of biometrics as an *identification* tool is premised, there are more general differences between people. These are commonly used to categorise people as belonging to specific social groups defined in terms of age, gender, ethnicity, (dis-)ability, and so on; moreover, the historical, ethical, legal and political significance of these differences is far reaching (Foucault 1975; Lacqueur 1990; Duden 1991; Schiebinger 1993).

In relation to biometric technologies, there are at least two different ways in which these differences matter:

On the one hand there is the matter of *exclusion* of certain categories of 'different' people from system use, because the systems can only cope with difference to a limited extent. In biometric discourse the set of problems connected with this issue is referred to with a number of concepts, such as, for example, 'usability', 'accessibility', 'failure to enroll', 'exception handling' and 'template aging'. In the discourses of social theory and politics, these matters invoke considerations in terms of distinctions made between *normal* and *abnormal* or *rule* and *exception.* Such notions, then, are connected with concepts like 'normalisation' and social inclusion and exclusion (Foucault 1977, 1979; Star 1991). 'Normalisation' here refers to the production and enactment of norms, through which the very distinction between what counts as normal and what as exception or deviancy is performed within and through technological practices. When a biometric system fails to cope with variations in human features falling outside a certain range, it thereby categorises and excludes, with more or less serious consequences to the people concerned.

Here, however, I will focus on another way in which the issue of bodily differences emerges in relation to biometric technologies. Recently, we have seen the emergence of biometric technologies that *use* the general differences mentioned, trying e.g., to *automatically classify* people in gender, age, and ethnic categories. So-called 'soft biometrics', i.e. a set of experimental biometric applications aiming at recognition of general body characteristics such as body weight, gender, age, or ethnicity, uses what, within technical discourse, are referred to as 'partial identities' in systems that are mostly in the research stage, with only a few applications in actual operation.[1] Here, questions about the *black-boxing* of contingent, perhaps unscientific and contestable, or even unethical *constructions* of those categories, by building them into the systems, may arise (Latour 1987; Bowker and Star 1999).

But first I want to bring back to mind how, in the 1990s initial worries about biometrics' potential for sensitive categorisations, even unconstitutional types of discrimination, were adamantly dispelled by biometrics' experts and advocates. The debate about this issue, as it took place then, underlines the significance of current developments.

> In 1998, J.L. Wayman, then director of the National Biometric Test Centre at San Jose State University, testified in a US congressional hearing: "We must note that with almost all biometric devices, there is virtual no personal information contained therein. From my fingerprint, you cannot tell my gender; you cannot tell my height; my age, or my weight. There is far less personal information exposed by giving you my fingerprint than by showing you my driver's license." (Castle 1998)

Despite such reassurances, the at that time still relatively unknown biometric technologies had some people worried enough to explicitly include the provision that "collection of a biometric identifier must not conflict with race, gender, or other anti-discrimination laws" in, for example, the proposals for the Californian Consumer Biometric Privacy Act'.(Biometrics in Human Services 1998)

> Although this was then dismissed by many biometrics advocates as being based on unfounded and ill-informed beliefs about biometrics, historian Simon Cole writes the following on the presumed 'emptiness' of fingerprints: Galton's [the founder of dactyloscopy] "regret," his failure to find the key to the code of heredity in fingerprint patterns, has been confused with

---

[1]For example, a Japanese software company has developed technologies named FieldAnalyst and Eye Flavour, which are automated market research systems that determine the gender and age of passers by, using facial pictures taken by cameras; they have been installed in several shopping malls in Japan. http://www.nec.co.nz/news/news.html

the notion that fingerprint patterns actually contain no information pertinent to health, ancestry or behaviour. But other researchers found rough correlations between fingerprint pattern type and ethnicity, heredity and even some health factors. These correlations, especially the ethnic ones, have proven robust and still hold up today. As with any correlation, they are not determinative; one cannot predict ethnicity from fingerprint pattern, but fingerprint pattern types do appear with different frequencies among different "ethnic groups" (as defined by researchers). True, not much has been done with these correlations. But the point is that the situation with fingerprint patterns and genes is fundamentally the same — correlations. It is not that fingerprint pattern correlations do not exist; rather, it is no longer scientifically acceptable to investigate them — unlike genetic correlations with so called ethnicity. In short, the perceived "emptiness" and harmlessness of fingerprint patterns is a social achievement, not a natural fact. (Cole 2006)

Half a decade after the congressional hearing, it became increasingly obvious that Wayman had not been entirely right: in an article on so-called 'soft biometrics', we read:

Many existing biometric systems collect ancillary information like gender, age, height, and eye color from the users during enrolment. However, only the primary biometric identifier (fingerprint, face, hand-geometry, etc.) is used for recognition and the ancillary information is rarely utilized. We propose the utilization of "soft" biometric traits like gender, height, weight, age, and ethnicity to complement the identity information provided by the primary biometric identifiers. (Jain et al. 2004) (*italics added, IvdP)*

In my view, today's developments in 'soft' biometrics, i.e. a set of experimental biometric applications aiming at using 'partial identities', and recognising general body characteristics such as body weight, gender, age, or ethnicity, signifies a new step in the informatization of the body that needs critical attention urgently (Van der Ploeg 1999, 2002, 2008). After all, the aimed for ability to distinguish concerns highly sensitive categories, many of which are overburdened with histories of discrimination of the worst kind.

In this context, however, they are called 'soft', because unlike biometric *identification* technologies, they focus on traits that do not single out one individual from all others, but on ones that are shared by large numbers of people.[2] From an identification perspective, however, that can be useful as supporting information or 'secondary mechanism', which, when used in conjunction with identifying biometric traits, can substantially improve the success rates of identification technologies. Also, they can be used to establish membership of a category (e.g. establishing adulthood) without actually identifying, for which reason a certain privacy-protective potential is sometimes attributed to these technologies (Li et al. 2009). Beyond this, however, the quoted article gives a few lines of research, which clearly point to more far-reaching potential applications:

Methods to incorporate time-varying soft biometric information such as age and weight into the soft biometric framework will be studied. The effectiveness of utilizing the soft

[2]Hence, we define *soft biometric traits* as *characteristics that provide some information about the individual, but lack the distinctiveness and permanence to sufficiently differentiate any two individuals.* The soft biometric traits can either be continuous (e.g., height and weight) or discrete (e.g., gender, eye color, ethnicity, etc.) (Jain et al. 2004, p. 732).

biometric information for "indexing" and "filtering" of large biometric databases must be studied. Finally, more accurate mechanisms must be developed for automatic extraction of soft biometric traits. (Jain and Lu 2004)

One could imagine useful applications of systems that can categorise, e.g., faces, according to gender or ethnicity, or 'filter' a database that way, for example when a reliable witness statement in a crime investigation would enable exclusion of particular categories from a police database search, thus saving valuable time. Another such example would be classifying subjects in broad age categories, in order to determine legal competence to apply for certain services, or buy certain products, while preserving anonymity. As mentioned, the potential to provide a broad categorical classification rather than full identification is often invoked as a privacy enhancing quality soft biometrics may provide (Li et al. 2009). On the other hand, I am convinced that it is not even necessary to spell out here the many situations imaginable in which filtering people out on the basis of their gender, age, or ethnic/racial background constitutes illegal discrimination. Therefore, the development of systems to automate this process must be considered inherently risky (Lyon 2003).

Also, and contrary to what their apparent self-evident reference in ordinary language and everyday life may lead one to believe, the reification of these categories and distinctions, as the history and philosophy of science have made abundantly clear, is essentially contestable and unstable (Lacqueur 1990; Schiebinger 1993). For example, the distinction between the male and female gender on a genetic level does not always match the one made on an endocrinological, anatomical, psychological, or socio-cultural level; and even when birth registered gender is taken as a reference point, a problem exists where even this is amenable to change during an individual's lifetime. In an exacerbated form, similar problems exist with ethnicity and race classifications, all of which have been proven to lack any indisputable basis in 'nature' (Hacking 1986; Harding 1993; Reardon 2004; Fausto-Sterling 2008).

A telling example of the way such important insights will be ignored in soft biometrics is the research by Jain and Lu (2004) into the development of a system for ethnic classification. In this study facial images from two separate databases, pre-defined as the 'Asian database' and the 'non-Asian database', were divided into a training set and a test-set. From that original, previously categorised training set, the system was to learn, and be able to continue classification of subsequently fed test images. Performance of the system was then evaluated (percentage of 'correct' classifications) by the researchers again. On the issue of which definition of ethnicity to use, and what criteria to apply, the authors state: "In this paper, we do not make a distinction between the terms 'ethnicity' and 'race', which are used to refer to people who share *common facial features that perceptually distinguish them* from members of other ethnic groups" (ibidem p. 1, *italics added IvdP, please note the circularity of the definition*), and: "Because the robust facial landmarks localization is still an open problem due to the *complex facial appearance in the real-world environment*, we do not utilize the anthropometrical measurements based classification scheme. Instead, we explore the *appearance-based* scheme, which has demonstrated its power in facial identity recognition" (ibidem p. 2, *italics added, IvdP*).

In other words, in order to render the great human diversity in 'the real world environment' technologically manageable in 'laboratory conditions', recourse is taken to a method based on *appearances*, on top of which, for the purposes of this study, "the task is formulated as a two-category classification problem, to classify the subject as an Asian or non-Asian"(*ibidem*, p. 2). Following the insights developed in Actor-Network theory based studies into the way laboratory developed technologies are made to 'work' outside the laboratory, such as, for instance (Latour 1983), there is good reason to worry about what such systems will require to happen with ethnic and racial categorizations in 'the real world' when people decide this needs to be made to work. What emerged from these studies is, one may remember, how 'real world' practices are *adapted*, rearranged and transformed, to copy the conditions under which technologies worked in 'the laboratory', in order to repeat functionality and successful performance elsewhere. This might mean that, for all practical purposes, and in as of yet unknown ways and situations, we may come to adapt our thinking about human diversity to fit the inbuilt categorizations and definitions of the systems (Hacking 1986).

Whereas the literature on the history of the categories of sex and gender, ethnicity and race has provided us with ample proof of their deeply problematic nature, this literature had the advantage of having scientists' texts, language, observable practices and visual representations as its relatively accessible objects of study. Crucially however, the development towards 'soft biometrics', implies that whatever definitions of the categories in question are going to be used, they will become inscribed in software and algorithms far more difficult to access and assess, and therefore, to contest. This fact renders this technology fraught with risks of various kinds, risks that will undoubtedly increase when these systems are applied in an embedded and 'unobtrusive' fashion.

13.3 Embedded Systems and Identification/Authentication

One of the most influential developments in ICTs today generally believed to be the shift away of computing power from PCs and desktop-configurations to the physical environment. Embedded software, ubiquitous computing, ambient technology, smart objects, and the emergence of 'the Internet of Things' are all terms denoting a particular aspect or view of our near future.

Due to developments in a.o. radio frequency identification (RFID), miniaturisation, wireless, sensor and networking technologies, people will be moving through and interacting with their physical environment in new ways. Objects themselves will interact and communicate, en send information about themselves, their users, or their environments to electronic networks and databases.

Thus we are concerned here with a paradoxical, dual development, namely the simultaneous 'shifting out' of computing power into the physical and built environment, and 'shifting in' towards the human body. Whereas on the one hand information and communication networks are extending, and becoming more pervasive in our daily activities and movements through space, on the other hand, through various technologies and applications among which crucially biometrics in its various guises, we become ever more intimately connected as embodied persons to these networks. As the European Group on Ethics argued in their Opinion on ICT implants in the human body,[3] becoming 'networked persons' is fraught with opportunities and threats. Threats, for example, to human dignity, freedom and autonomy, and inviolability of the human body. Indeed, what might be at stake could be as profound as a transformation of human embodied identity in the personal as well as the anthropological sense.

On the positive side, huge gains in convenience, efficiency, and safety are predicted to result from this; on a more bleak view, this could mean the ultimate track- and traceability and loss of privacy. In particular, the information on peoples' behaviours generated by these systems, will, in all probability, prove an invaluable and highly tempting resource for law enforcement and crime prevention and security policy, if not the primary aim of such systems. This would imply a (post-hoc) blanket recasting of all end-users, consumers, and citizens as suspects, which may render them vulnerable in unforeseen ways.

Many, if not most, of the (largely still only envisioned) AmI systems are 'user centric' and comprise personalised functionality. This means that people's interactions with such systems require them to be identified within the system, in order to enable it to retrieve the relevant personal profile and settings.

Moreover, for it to work effectively, AmI requires intuitive and convenient interfaces, meaning that identification and authentication are usually (half-) automated and as unobtrusive as possible, with little conscious effort from the user needed. This requirement makes identification technologies like biometrics and radiofrequency identification likely options in this context, but also puts such systems on a direct collision course with the data protection principle that personal data collection must always involve an aware and informed data subject (EU Data Protection Directive, art.7 par.1)

In addition to identification/authentication to access or activate the system, continued interaction or mere contact with the system generally results in registration of personal identifiers, data on movements, behaviour, location, etcetera, communicated to, and stored in databases. These personal data may be required for the improvement of the systems performance (by updating and perfecting the personal profile) but could, obviously, be used for other purposes.

With computing power 'retreating' into the background, i.e. the built environment, classic input devices such as keyboards, and the user interacting with the system by typing and clicking become obsolete, so that other input devices and interfaces between people and machines become necessary. In line with this aspect of the developments concerning embedded technology and ambient intelligence, is a number of remarkable calls to develop biometric systems that do not need the

[3]European Group on Ethics in Science and New Technologies (2005) Opinion on Ethical Aspects of ICT Implants in the Human Body, March 16.

active cooperation of people anymore. Biometric sensing from a distance appears high on the R&D agenda today, enabling the design of systems that can be applied without people even being aware that they are being identified, registered, or assessed.

For example, (Li et al. 2009) describe how the National Science and Technology Council Subcommittee in Biometrics identified in 2006 the research on sensor technology for biometrics at a distance as 'a primary research challenge' (National Science and Technology Council Subcommittee on Biometrics 2007), while 1 year later the BioSecure Network of Excellence published its Biometrics Research Agenda (The BioSecure Network of Excellence 2007), calling research in distributed sensor networks and the 'transparent use of biometrics, requiring no actions from the end-user in supervised or unsupervised ways' one of the most 'urgent' research topics.

Thus we see an increasing emphasis on the development of various wireless and remote sensors and highly sensitive cameras usable as input for biometric recognition, such as, for example electric-field sensors for physiological-signal acquisition, and iris scanners that can capture iris images from moving subjects.

In the language of biometrics, systems that do not require active and conscious cooperation are described as 'unobtrusive', 'user convenient', and 'transparent'. The latter term especially is rather intriguing, since it refers to a characteristic of systems that in other discourses, for example that of political accountability, would be qualified in absolutely opposite terms. Biometric applications for securing public spaces develop increasingly towards (potentially) covert biometric data capture, such as, for example, in applications like 'smart' video surveillance, or systems like 'Iris-on-the-move'.[4] Commonly, the vocabulary of 'convenience' is used here, but sometimes this takes the form of deliberately hiding the processes and technologies, such as in the example of the '*Smart Corridor*', developed by a Paris based company, and now in an experimental phase in a real life setting. This system consists of a 3–4 m long corridor to be installed at airports, where passengers previously identified at check-in are biometrically authenticated from a distance through iris and face recognition technology, while simultaneously being inspected by other sensors searching for hidden objects and traces of material that can be used to make explosives. In addition, an intelligent video surveillance system performs functions such as counting people, detecting abandoned objects and "suspicious behaviours such as customers performing seemingly unnatural actions such as a sudden U-turn" (Boussadia 2009, p. 8).

Besides saving expensive time and space needed for such security inspections, check points and queues, the big advantage for passengers is framed, once again, in terms of convenience. Not needing to wait in line surely will be appreciated by travellers. However, the various descriptions of the system given by its designers and producers, suggest how this '*convenience*' becomes almost indistinguishable from *active hiding*, which arguably carries a rather different set of connotations

[4]http://www.sarnoff.com/products/iris-on-the-move

(and motivations). In addition to being '*completely transparent* to the passenger' the system is described on the company's website as follows: "On the passenger side, the corridor will be neutral to avoid attracting passengers' attention. The control devices will be installed 'behind the scenes', on the other side of the corridor partitions. Passengers will feel just a slight brush of air activated by the system which collects molecules from materials that could be used to manufacture explosives", [5] and in an article by the project manager, we find: "For passengers, the technology behind the corridor will be *completely invisible*."(Boussadia 2009).

One may ask why completely hiding such an extensive inspection process should be considered as adding 'convenience'; would it really be that inconvenient to know or even notice what is going on behind those scenes? Perhaps symbolically, we can see here how technologists' 'transparency' is enabled by the very *in*transparency of the walls behind which the devices and processes are hidden, which could be indicative of the way this convenience offered to citizens may come at a cost.

13.4 Further Steps into Embedded Systems for Security: Physiology and Multimodality

Seen in this light the development of yet another type of biometrics, one that could be characterised as '*under the skin biometrics*' becomes all the more intriguing. Today there are concerted, often public funded research efforts to develop new biometric concepts involving (in some cases distant) registration of brain and heart activity patterns through electroencephalograms and electrocardiograms.[6] From these, personal profiles can be derived that subsequently can be used for 'unobtrusive' authentication (Riera et al. 2008a, b). Some workspaces, such as particular vehicles, laboratories, or control rooms, require high level security. To this end, new forms of access and operation control, authentication, and monitoring of persons present are being researched and developed in advanced new forms of multimodal biometrics. For example, authentication systems for high security control rooms for critical infrastructures, and for truck cabins have been developed using these techniques, based on the idea that security and safety can be enhanced by continuously authenticating who is present in these spaces or operating equipment and vehicles. Miniaturized, wireless electrode based sensors integrated in, for example, seats, hats (electroencephalogram) or shirts (electrocardiogram), allow a new form of continuous monitoring and/or authentication, that does not in any way

[5] http://www.thalesgroup.fr

[6] See for example the EC funded FP7 projects HUMABIO (http://www.humabio-eu.org/news.html) and ACTIBIO (http://www.actibio.eu:8080/actibio) Although within these research projects the terms ECG and EEG are used, these do not equal the measurements performed in medical context referred to by the same terms, which involve the use of officially gauged and certified equipment based on international standards.

interfere with work routines, and which is said to be more difficult to spoof as well as performing de-facto aliveness checks. Another new route in authentication is taken with the 'sensing seat', which measures the pressure distribution from the body of the seated person, after which a profile and a biometric signature are created. Though not unique enough for single modal use, this signature can increase the overall authentication accuracy when combined with other biometrics, in a way comparable to the soft biometric traits discussed above.

When considering possible issues emerging from such new systems, one needs to take into account that the context described so far concerns a working environment where compliance and consent to organisational security and safety policies can be demanded from personnel as a condition of employment. Especially when security and safety of the wider public are at stake in the organisation's activities, such as, for example, in nuclear plants, compliance and observation of stringent security measures unacceptable elsewhere should be enforceable.

However, the systems under discussion here, even though they are 'convenient' and unobtrusive, in that they do not disturb the person in their activities or require conscious effort, are, in another sense, quite *in*trusive. Having one's brain, heart, gestures and movements registered continuously gives a whole new meaning to the notion of 'worker surveillance'.

As always, a lot depends on the exact configuration of the system: how, where and what exactly gets to be registered, who will have access to the data, how and how long will they be stored, and to what extent is the subject informed and consenting freely. Be that as it may, such a system, and the various elements it consists of, take the possibilities of control over the worker and their body into a new plain. Not merely their privacy, but their very integrity as an embodied person is compromised in a way that requires protections to be implemented in the system set-up and its default operation. It is important to recognise that distinguishing between 'normal' and 'intended' use versus 'misuse' and 'abuse', and reserving ethical concern for the latter, is arbitrary and possibly naïve: these distinctions are not given with the technology itself, but are the result of discursive and practical negotiations, and therefore not power neutral. Of special concern should be the extent to which sensitive and intimate information about the worker, e.g. about their medical condition, could become available that in principle is unrelated and irrelevant to job performance, and therefore beyond an employer's need or right to know. In general, such intimate physical information becoming available at any level of management and system control, could possibly render these workers vulnerable to inadmissible forms of manipulation and pressure.

When considering the application of these and similar technologies for *public* spaces, the issues multiply. Next to the smart corridor discussed above, that involved the classic biometric identifiers face and iris, 'new security concepts' for public spaces are being researched in several places, that involve an extended set of sensors, measurements, pattern analyses and profiling techniques, focussing on pulse, body temperature, pupil dilation, gait and movement patterns, and voice pitch, in order to assess people passing through public spaces like airports (Burns and Teufel 2008). This may be not what most people understand as biometrics as such, which involves identification or authentication of a person by their unique bodily features. But even though it does not involve biometric identification, it does involve assessment and classification of subjects on the basis of some physiological or bodily characteristic. The aim here is to 'filter' out those deviating from a set of norms believed to distinguish the harmless and innocent from those that are not, and pulling them out for further inspection and interviewing, while easing security procedures for the rest of the public.[7] Comparable to the ideas underlying the polygraph, certain physical indicators of arousal, stress, and so on, are believed to correlate with so-called 'hostile intent'. Thus, the use of remote cardiovascular and respiratory sensors, remote eye trackers (camera plus software), thermal cameras, high resolution video, audio sensors, and other sensor types such as for pheromones, is hoped to enable registration of heart rates, body/skin temperatures, facial expressions, eye movements, pupil diameter, breathing rates, body movements, facial features and expressions, and voice pitch.

Again, the rationale given for such a system includes saving expensive time and space for the various organizations and agencies, both public and private, in charge of operating and securing the airport. In relation to this type of system, however, several specific issues comes to the fore.

First the *covert and distant* nature of data capture is obviously bringing the persons at whom the system is directed in a vulnerable position that contradicts many assumptions embedded in current discourse on privacy, data protection and user empowerment. For one thing, in contrast to most 'classic' biometric identification or authentication modalities, these persons can no longer in any meaningful way be construed as 'users' or 'end-users' at all. Even though *are* referred to sometimes as '*non-cooperative users*',[8] they rather seem to have become the *objects* of a new kind of physical search that defies many deeply ingrained values concerning bodily integrity, freedom from arbitrary inspection, and consent requirements.

Next, the fact that it is *body data* that are being captured increases the potential for ethical and legal problems, since body data should be considered highly sensitive, personal, and intimate; in certain ways more so than other identifiers like passports, names, social security numbers etcetera. In addition, the likelihood of collateral information on the person's state of health (certain features and values may be indicators of disease or illness) being collected this way should be considered as a serious privacy risk.

But perhaps more importantly, because by design completely intransparent, but at the same time highly consequential, the specific *normative nature* of these systems

---

[7]See for example the projects: (United States' Department of Home land Security) Hostile Intent; Future Attribute Screening Technology – FAST, and (EU-FP7) Suspicious and Abnormal Behavior Monitoring – SAMURAI.

[8]For example, in "The National Biometrics Challenge", we find this definition: "*Non-cooperative user: An individual who is not aware that his/her biometric sample is being collected. Example: a passenger passing through a security line at an airport is unware that a camera is capturing his/her face image.*" National Science and Technology Council Subcommittee on Biometrics (2007). The National Biometrics Challenge. Washington, DC: 1–19. p. 12.

ought to be questioned. In particular, the ideas concerning *normality* built into such systems carry great potential for legal and ethical problems, for instance regarding discrimination. A society in which information on physical state of arousal, facial expression, pitch of voice becomes routinely collected and used as input for surveilling public spaces is indeed stepping on slippery ground. Once a general awareness of the existence of such systems exists among the public, the occurrence of *anticipatory conformity* (Rouvroy 2009), will be a highly likely consequence: people will check themselves, up to their facial expressions, to avoid being harassed, which would constitute serious damage to our ideals of a free and open society.

Next to the assessment of particular measurements as deviant as such, the automated *interpretation* of such data in terms of elevated risk and behavioural intentions ('hostile thoughts') is fraught with problems. The interpretative leap from, e.g. accelerated heart rates or elevated body temperature, to particular states of minds, emotions, and even intentions, let alone hostile ones, is large, prone to error, and hardly scientific (National Academy of Sciences 2008). Moreover, individual baselines for such values might differ significantly, rather than being the same for every human being. Quite possibly, such differences will turn out to correlate with certain sensitive differences, such as gender, ethnicity, or age, giving rise to new occurrences of (indirect) categorical discrimination of historically already disadvantaged groups. Alternatively, but no less problematic, new categories and profiles may be developed, so that, depending on the tolerance range set for the system, certain individuals and categories of people may experience being flagged over and over again.

One may well ask whether there can ever be a legal and ethical justification for filtering out people for further inspection and interviewing based on such intimate bodily parameters. Moreover, the intent to detect mental and emotional states, based on their presumed correlation with these physiological signs, signifies the introduction of a form of privacy invasion of a new dimension altogether, one that will inevitably lead to people having to justify their emotional and mental states. At the same time, the chances that such technologies will really contribute to security goals, i.e., will actually be able to filter out those with immediate criminal intent, seem at best doubtful.[9]

13.5 The Assumption of Availability

This brings me to the recognition of a particular normative assumption built into biometric technologies in their various forms and applications. Elsewhere I have discussed the assumptions of similarity and stability of the human body on which biometrics are premised (Van der Ploeg 2010) But the techniques of distant sensing and physiological authentication show another highly normative assumption: the

[9]Ibidem.

very idea that one *can* direct such sensors, and do such measurements on people, be it in public spaces or in work environments, is indicative of the extreme extent to which people's bodies have come to be assumed to be *available* for biometric processing.

The moral significance of this point is not easy to clearly articulate, because it has less to do with the body as a physical object, but rather with the body as the 'site' of culturally and emotionally encoded embodied experience, and the way it is enrolled in biometric technological configurations (Haraway 1991; Bordo 1993; Hayles 1999; Merleau-Ponty (1945) 1962).

To illustrate the point for 'classic' biometrics, we can consider the study done by Den Hartog and colleagues (den Hartog et al. 2005). This study, exceptional as it is in focusing on the performance and usability of biometrics for very small children, shows the crucial role of this normative aspect in the performance of biometrics. One of the major conclusions in this study involving some 160 toddlers concerned not merely the often mentioned un-machine-readability of their fingerprints, but also the difficulty of getting the little ones to 'cooperate' sufficiently. Acquiring good quality fingerprints, or even facial photographs for that matter, requires a level of understanding and cooperation, a willingness to present the hand (not a clammy, clenched little fist) for processing, to sit still, to not cry and wriggle about, and, for today's passports' facial photographs, not laugh or smile – exigencies that turn out to be near impossible to demand from the very young. "You really have to like children" was one of the side comments of one of the researchers in the study, invoking in my mind quite amusing scenes.

What these difficulties point to, however, can be interpreted as illustration of the way in which biometric systems generally presuppose the *availability* of the body, or some part of it, for handling in certain ways, scanning and assessing it for particular purposes, as a prerequisite for the system to work. In the terms of Actor Network Theory, bodies are *enrolled* in technical networks, given a role in a quite particular *script*, and assumed to take a very specific position in the configuration that constitutes the technological system. In some cases this implies a visibility, in others touchability, and generally one can see here a demand for it to take certain postures, perform certain gestures, and allow or enable it to be processed in a number of ways. The requirements for ICAO compliant facial photographs for machine readable travel documents (MRTDs), for example, show a detailed list of prescriptions concerning posture, expression, what not to wear, and how to pull hair back to present ears (Home Office Identity and Passport Service). In the case of the toddlers it is easy to imagine how the requirements concerning the disciplined and docile availability of hands and faces is an unrealistic expectation that can become the cause of genuine distress, if, for example, an enrollment procedure in an unfamiliar environment, performed by a uniformed officer has to be repeated several times to acquire 'usable data'.

But one can also imagine the anxiety, distress, or feelings of humiliation and anger that compliance with the demands of the systems may instill, in, for instance, people deeply bound to certain cultural or religious norms of modesty and privacy, who are asked to remove veils in a public space, stick out their hands to be touched by a stranger. In biometrics' discourse 'acceptability' is the rather neutral term referring to this aspect, which in some ways is comparable to the problems the export of western medicine was confronted with in other cultures, because of its ways of examining bodies, or the distress caused in some indigenous peoples by photography. Both these historical phenomena caused surprise, because people simply had not been aware of the culture dependent normative assumptions relating to the body and embodiment hidden in their practice. Similarly, there is an extent to which biometrics can be said to impose a set of particular cultural norms regarding embodiment and compliance, specifically concerning the availability and usability of bodies on people who, for various reasons, may not share these norms. But even if they *are* shared, and no actual opposition will be encountered, that will not diminish in any way the normative nature of this taken for granted availability of people's bodies for overt or covert data capture, and everything following that.

The development of distant sensing and even 'under the skin biometrics' can, from one perspective, be perceived as convenient and unobtrusive, but from another it represents at least also a significant increase in the extent to which bodies are assumed to be available. Although such biometrics at a distance will not involve the body to be available in exactly the same ways as described above – to have it perform and behave in precisely circumscribed ways in order to render it machine-readable – it does not alter the fundamental assumption that bodies *are* available and accessible to this kind of treatment, inspection, and use. On the contrary, the trend towards multimodal and distant sensing indicates a decreasing sense of the necessity of actually asking people for their cooperation, and constitutes a good step towards involuntary, behind-their-backs scrutiny and virtual bodily searches. This threatens various ethical and legal principles, such as that of privacy, bodily integrity, self determination and freedom; beyond the level of individual rights, it might affect the quality of democratic societies, and the power relations that constitute them.

13.6 Conclusion: A Transparent and Unobtrusive Future?

Summarizing the arguments made, we have seen how the emergence of new biometric traits in conjunction with the shift towards embedded systems signify a new level of complexity, not only of the technology, but also of the critical issues and policy challenges connected with it. This new generation biometrics yields a picture of technical capabilities potentially deeply affecting embodied experience that clearly requires sustained critical research and assessment for years to come.

In particular, it was argued, these developments indicate the rapidly increasing extent to which human bodies are assumed to be available for automated recognition, inspection and assessment procedures. This 'assumption of availability' of human bodies for enrollment in technical configurations was coined to refer to the quite particular set of norms concerning the way bodies are by design simply assumed to be available and to comply with biometrics requirements. In discussing new developments in physiological 'beneath the skin' and distant sensing biometrics, the intensification and extension of this presumed availability was described, as it can be observed in its various dimensions: not only have we seen an extension of scale and ubiquity of systems, but also in number (multimodality) and range of different biometric traits used. Worth noting is the shift towards multimodal systems in this context, generally intended to increase reliability of the systems, but producing the concomitant effect of a prima facie intensification of monitoring and surveillance, and extended profiling potential.

With regard to the increasing range of biometric traits used in automated recognition and detection, the development of soft biometrics and electrophysiological biometrics were discussed in more detail. The first, soft biometrics – quite amenable to distant operation – was shown to entail the black boxing and automation of essentially contestable and socio-politically highly sensitive classifications and categorisations of gender, ethnicity, race, age, and so on. The second, the emergence of systems of 'under the skin biometrics', based on registration of physiological states and phenomena, was questioned with regard to its potential for creating disputable levels of worker surveillance and manipulation, whereas justifiability of its use in public spaces was suggested to be fundamentally doubtful.

This extension of the range of usable biometric traits comes, to some extent, from the incentive to improve the still far from perfect performance of 'classic' single mode biometric technologies: by using more than one, and 'supportive' additional information, the primary identification or authentication function can be improved. However, the pursuit of this goal has led to taking up types of biometric traits that, far from solving a mere technical problem, actually bring biometric endeavours in uncharted social, ethical and legal danger zones.

Taken together, the developments concerning the 'next generation biometrics' discussed here signify the deeply political and ethical nature of the incremental informatization of the body we witness today (Van der Ploeg 2002). It is therefore crucial that the interdisciplinary study and analysis of the intricacies of these technologies keeps pace with this development. Difficult though this may be, given their highly complex, formalised, abstract, and ever more deeply embedded, software encoded character, we must continuously assess precisely which norms, which definitions, and which aspects of our embodied identities are being woven into the very fabric of our ever smarter environment.

Acknowledgement Funding of the research for this paper was partly provided by the European Research Council and the European Commission, both under the European Community's Seventh Framework Programme (FP7/2007–2013), DigIDeas Project/ERC Grant Agreement 201853, and HIDE project/EC Grant Agreement 217762

References

Biometrics in Human Services. 1998. 2(2).

Bordo, S. 1993. *Unbearable weight. Feminism, western culture, and the body*. Berkeley: University of California Press.

Boussadia, K. 2009. The evolution of airport screening technology. *Biometric Technology Today* 17(2): 7–8.

Bowker, G.C., and S.L. Star. 1999. *Sorting things out. Classification and its consequences*. Cambridge, MA/London: MIT Press.

Burns, R.P., and H. Teufel. 2008. *Privacy impact assessment for the future attribute screening technology project*. Washington, DC: Department of Homeland Security.

Castle, M.N. 1998. *Hearing on biometrics and the future of money*. Washington, DC: Committee on Banking and Financial Services.

Cole, S.A. 2006. The myth of fingerprints. Gene watch http://www.gene-watch.org/genewatch/articles/19-6Cole.html. Accessed 25 June 2009.

den Hartog, J.E., S.L. Moro-Ellenberger, et al. 2005. *How do you measure a child? A study into the use of biometrics on children*. Delft: TNO.

Duden, B. 1991. *The woman beneath the skin. A doctor's patients in eighteenth century Germany*. Cambridge, MA/London: Harvard University Press.

Fausto-Sterling, A. 2008. The bare bones of race. *Social Studies of Science* 38(5): 657–694.

Foucault, M. 1975. *The birth of the clinic: An archeology of medical perception*. New York: Vintage/Random House.

Foucault, M. 1977. *The history of sexuality vol. 1: The will to knowledge*. London: Penguin.

Foucault, M. 1979. *Discipline and punish. The birth of the prison*. New York: Vintage/Random House.

Hacking, I. 1986. Making up people. In *Reconstructing individualism*, ed. T.C. Heller, M. Sosna, and D. Wellbury, 222–236. Stanford: Stanford University Press.

Haraway, D.J. 1991. *Simians, cyborgs, and women: The reinvention of nature*. London: Free Association Books.

Harding, S. (ed.). 1993. *The 'racial' economy of science*. Bloomington: Indiana University Press.

Hayles, K.N. 1999. *How we became posthuman. Virtual bodies in cybernetics, literature, and informatics*. Chicago: Chicago University Press.

Home Office Identity and Passport Service Passport Photographs. London http://www.direct.gov.uk/en/TravelAndTransport/Passports/Applicationinformation/DG_174152. Accessed 28 June 2009.

Jain, A.K., and X. Lu. 2004. Ethnicity identification from face images. In *SPIE International Symposium on Defense and Security: Biometric Technology for Human Identification*. Orlando.

Jain, A.K., Sarat C. Dass, et al. 2004. Soft biometric traits for personal recognition systems. In *International Conference on Biometric Authentication*. Hong Kong.

Lacqueur, T. 1990. *Making sex. Body and gender from the Greeks to Freud*. Cambridge, MA/London: Harvard University Press.

Latour, B. 1983. Give me a laboratory and I will raise the world. In *Science observed*, ed. K.D. Knorr-Cetina and M. Mulkay, 141–170. London: Sage.

Latour, B. 1987. *Science in action: How to follow scientists and engineers through society*. Cambridge, MA: Harvard University Press.

Li, S.Z., B. Schouten, et al. 2009. Biometrics at a distance: Issues, challenges, and prospects. In *Handbook of remote biometrics for surveillance and security*, ed. M. Tistarelli, S.Z. Li, and R. Chellappa, 3–21. London: Springer.

Lyon, D. (ed.). 2003. *Surveillance as social sorting: Privacy, risk, and digital discrimination*. London/New York: Routledge.

Merleau-Ponty, M. (1945) 1962. *Phenomenology of perception*. London: Routledge and Kegan Paul.

National Academy of Sciences. 2008. *Protecting individual privacy in the struggle against terrorists. A framework for program assessment*. Washington, DC: The National Academies Press.

National Science and Technology Council Subcommittee on Biometrics. 2007. The national biometrics challenge, 1–19. Washington, DC.

Reardon, J. 2004. Decoding race and human difference in a genomic age. *Differences* 15: 38–65.
Riera, A., A. Soria-Frisch, et al. 2008a. *Multimodal physiological biometrics authentication. Biometrics: Theory, methods, and applications*. Piscataway: Wiley/IEEE.
Riera, A., A. Soria-Frisch, et al. 2008b. Unobtrusive biometric system based on electroencephalogram analysis. *EURASIP Journal on Advances in Signal Processing* 2008: 1–8.
Rouvroy, A. 2009. *Governmentality in an age of autonomic computing. Technology, virtuality, utopia*. Computer Privacy and Data Protection, Colloquium on Autonomic Computing, Human Identity and Legal Subjectivity. Brussels.
Schiebinger, L. 1993. *Nature's body. Gender in the making of modern science*. Boston: Beacon.
Star, S.L. 1991. Power, technology and the phenomenology of conventions: On being allergic to onions. In *A sociology of monsters: Essays on power, technology, and domination*, ed. J. Law, 26–56. Oxford: Basil Blackwell.
The BioSecure Network of Excellence. 2007. The biosecure research agenda. http://biosecure.it-sudparis.eu/AB/index.php?option=com_content&view=article&id=67&Itemid=36.
Van der Ploeg, I. 1999. Written on the body: biometrics and identity. *Computers and Society* 29(1): 37–44.
Van der Ploeg, I. 2002. Biometrics and the body as information: Normative issues in the socio-technical coding of the body. In *Surveillance as social sorting: Privacy, risk, and automated discrimination*, ed. D. Lyon, 57–73. New York: Routledge.
Van der Ploeg, I. 2008. Machine-readable bodies: Biometrics, informatisation and surveillance. In *Identity, security and democracy*, NATO science series, ed. E. Mordini and M.S. Green, 85–94. Amsterdam/Lancaster: Ios Press.
Van der Ploeg, I. 2010. Normative assumptions in biometrics: On bodily differences and automated classifications. In *The impact of europe on eGovernment,* ed. S. van der Hof, and M. Groothuis. The Hague: TMC Asser Press, IT&Law series.

Chapter 14
The Dark Side of the Moon: Accountability, Ethics and New Biometrics

Juliet Lodge

Abbreviations

DG	Directorate General
ECG	Electrocardiography
EEG	Electroencephalogram
EU	European Union
FBI	Federal Bureau of Investigation
FTSE	Financial Times and the London Stock Exchange
ICT	Information communication technology
ID	Identity document
LIBE	European Committee on Civil Liberties, Justice and Home Affairs
PET	Privacy enhancing technologies
RFID	Radio frequency identification
UK	United Kingdom
US	United States

14.1 Introduction

2009 marked increasing public coverage of the potential of new technologies, from expanding capabilities offered by the net and mobile phones to robotics. The proponents, vendors and public and private sector agencies advocating their use to enhance the administration of their businesses make various claims as to why they

J. Lodge (✉)
Jean Monnet European Centre of Excellence, Institute of Communication Studies, University of Leeds, Leeds LS2 9JT, UK
e-mail: j.e.lodge@leeds.ac.uk

E. Mordini and D. Tzovaras (eds.), *Second Generation Biometrics: The Ethical, Legal and Social Context*, The International Library of Ethics, Law and Technology 11,
DOI 10.1007/978-94-007-3892-8_14,

are necessary, 'life-enhancing', security and privacy enhancing and generally desirable. Efficiency in the administration of government has been sought through collaboration for over 15 years (European Commission 1992). It is commonly measured in relation to cutting the administrative burden, software costs, vendor independence and the deployment of human resources.[1] For the public, efficiency is generally promoted in terms of gains in convenience for citizens accessing services or making online transactions. The ethical implications of rolling-out such applications are rarely debated for either able-bodied citizens or for those less able to reach informed decisions and exercise individual consent.

Public claims made for these technologies are invariably couched in terms of their efficiency gains, therapeutic benefits, and their impact on expediting transactions of all kinds. Inter-operability is the holy grail. The emphasis is on efficiency and speed gains for mutual benefit. These range from the exchange of medical information about an individual to cross-border commerce, banking, travel card purchases, and online shopping (European Parliament 2008a, b, c, d). It includes automated recognition of a person with a subcutaneous RFID chip (tracking dementia patients). It covers convenience and paperless purchasing for allowing automated recognition of the embedded chip in a person ordering a drink in a bar and charging their account accordingly. The implanted chip, remote identification and authentication of individual 'identity' to allow access or confirm entitlement to private and public services and territorial and digital spaces has multitudinous possibilities for use. Adding biometrics (most commonly fingerprints (Wein and Baveja 2005), facial images and sometimes iris patterns) is claimed to significantly boost confidence in the authenticity of the person using them to prove they are who they claim to be for that particular purpose.

The bionic ear and eye, robotic soldiers, understanding of boosting resistance in combat, pilot free surveillance drones, implants to restore mobility and gene therapy all promised strides towards semi-utopia if not nirvana. The dystopian dark side of the moon was largely invisible, or dismissed. If you are who you say you are (Lodge 2007), what do you have to fear or hide from allowing a biometric smart card to disclose and confirm that?

This chapter examines the challenges for those required to demonstrate public accountability for the use they make of identity verification techniques for public policy purposes, ranging from border controls to accessing public services. It concludes that accountability mechanisms are not yet sufficient to the task of retaining public trust in either the technologies or in those deploying them in the name of security. To better understand this, the chapter begins with a brief overview of where biometrics is located in the discourse on security and liberty, since liberty is impossible without security and biometrics are used as a tool to reduce insecurity.

---

[1] See for example http://www.eu-fasteten.eu

14.2 Security v Liberty and the Role of Biometrics

The examination of biometrics is often located within frameworks of the technologisation of the political and surveillance and security on the one hand, and identity on the other hand. In the first instance, theoretical propositions relate to the notion of the pan-opticon and the all-seeing state, to policing and tracking at macro and micro levels. In the second, identity is located in the discourse on the philosophical traditions and norms of identity, linked to psychological conceptions and diachronic and synchronic reconfigurations of the self, and the self in relation to the state. Political scientists emphasise power and order, global civil society, and the relationship between the state and the individual, the exchange of fielty for protection within defined 'state' borders. The link between the two is illustrated in the discourse over exceptions to the rule of transparency and openness in government: exceptionalism validates secrecy and opacity, often for operational reasons, in respect of those items that can be labelled defence and security sensitive. Yet, knowledge is the key to liberty. The new technologies for gaining and commodifying information, as the basis of 'knowledge', potentially challenge not just our understanding of security and liberty, but also of responsibility, and credible accountability.

Surveillance and security theoretical propositions have led to a focus on political violence and terrorism, and the extent, nature and consequences of surveillance for security. Socio-legal studies focus on their implications for social control, data protection and privacy, and for intellectual property rights. Politico-legal studies hone in on the implications for constitutional and institutional practices of democracy; the tension between norm diffusion and confusion and the impact on society. Political economists explore competition policy, public procurement, and the mutual dependencies of ICT industries on governments in their quest to garner shares of the global market for their new 'security' goods. Political theorists and philosophers explore the relationship between freedom, liberty and data protection. Vagueness over what liberty, security, transparency and accountability would resemble in practical terms where egovernance is practised mean that the packaging of liberty and security in over-simplistic claims about securitising liberty by applying technology to tracking identity go unchallenged at worst and at best are submerged by unwarranted inferences about assumed homogeneous cultural values.

In the context of security and surveillance, identity is determined by reference to nationality, ethnic grouping, behaviour (particularly that deemed subversive, criminal, suspect or 'radical') and associated profiling. This differs from identity that is self-conceived and derived from kinship, family, community, culture and nationality, although it may overlap to a degree with the latter. More specifically, in the context of internal and external security, biometric identity is defined in a template resulting from the processing of raw data through an algorithm, set to specific standards, to allow a mathematical profile or 'solution' to map a biometric identifier, such as a fingerprint. This in turn is embedded in a document that can be automatically read by a machine. The test in this case may be whether the biometric presented (a fingerprint) matches that stored in the biometric identifying document.

The use of biometric identifiers for automated access challenges conceptions of identity and de-territorialises both borders and security. The structural relationship between the citizen and the state is tested and mediated by public-private partnerships, by inadequate technical security architectures which undermine data protection law, and which together challenge the balance between liberty and security. From a scientific viewpoint, the question ceases to be one of 'identity' as framed by traditional social theories and what a biometric identifier tells us about identity and processes of authenticating identity. The question is how the application of inter-operable technologies to the realisation of a common security goal shared by EU governments has a transformative impact on both our understanding of the democratic bargain, on accountability and privacy and on e-government public-private partnerships, especially in a supranational, cross-border setting.

14.3 Defining and Using Biometrics

At the level of policymaking, biometrics is a term that is employed loosely but instrumentally by state authorities concerned with securitising borders against illegal crossing of those borders into defined territorial spaces. Biometric characteristics are also seen as tools to establish the unique identity of an individual. Possession of and the ability to verify that uniqueness then becomes the key by which the owner accesses a range of private services (such as commercial transactions), public services (such as online registration of civil matters, like births, car tax, and e-administration of socio-economic welfare entitlement information and application forms), and mixed private-public services (such as health insurance, medical records and increasingly passports and other civil state documents administered on the state's behalf by outsourced agents). The territorial location of the point at which a biometric is presented has particular relevance in relation to the use of biometrics for the crossing of state borders. However, a biometric identity may also be used for transactions in non-territorial, digitised space where presentation, verification and authentication and exchange may occur in one or several different jurisdictions.

For our purposes, a biometric is a measure of a physical attribute of an individual, such as his iris, his face, his fingerprint. As such, the biometric identifier is a tool, a short-cut to establishing and verifying a person's claim that they are who they claim to be. The question is then whether or not a given fingerprint, for example, matches one previously given by the person offering it. Checking and verifying the match can be done by a human, as it has been traditionally done at passport border checking posts, or now, increasingly in prospect, automatically by scanners, ambient intelligence, or by automated methods when presented with machine-readable documents in which biometric data are embedded.

Whereas the EU use of the term biometrics is precise – a measurement, the US definition is ambiguous and connotes profiling, intelligence gathering, and new data creation (US Department of Homeland Security 2007). This makes the deployment and exchange of biometric information and documents for security purposes far more controversial since the assumptions underlying the differential definition raise different aspirations and expectations as to their legitimate use and re-use. The former is associated primarily with verifying 'identity', the latter with profiling (and the assumptions inferred from specific profiles and different distinctions made between information and intelligence) and hence with justifying a range of sometimes vague policy goals framed as essential for maintaining security.

EU biometric discourse centres on migration, the detection and prevention of the abuse of the right to asylum and visa fraud. That in the US is mired in the 'war on terror' reflected in measures to expand pan-Opticon technologies to track individuals (Echelon and the Spy in the Sky scenarios) and to store information in, and allow the FBI inter alia remote access to, large databases or using new through-the-wall surveillance techniques. This reflects very different conceptions of the purpose, purpose limitation, proportionality and functions of using and storing biometric data. In the EU, this translates into verification of identity and concerns with ownership, privacy and data protection. In the US, this relates to data mining, profiling and associated means of using 'intelligence data' to detect people suspected of criminal behaviour. In both, the nature of the ICTs used for both purposes lends itself to function creep and to data mining and data coupling techniques which potentially compromise individual liberty. Belated concerns with privacy impact assessment, privacy compliance and peer reviewed audit trails do not disguise a growing accountability deficit. This is accompanied by a sense that the essence of territorially defined exercises of responsible, democratic government is being leeched of authority in e-space where traditional guardians of territorially defined sovereign space and associated public goods externalities not only share and argue over authority (as in supranational settings like the EU) but fail to recognise the challenges posed by ICTs and jump on biometric identifiers as a panacea for processes that are far more complex and require far more radical responses. Can authorities used to using state borders to override other divisions within them exercise authority and require accountability from potentially semi- or invisible non-state actors?

Part of the problem is that the traditional guardians of accountability – parliaments – have been absent from scrutinising e-government for too long. In the EU, e-government was conceptualized as a response to the eEurope 2005 and Lisbon agendas, bureaucratised, and compartmentalized and susceptible to government initiatives to promote innovation for economic, and later, societal gain (Deutscher Bundestag 2005). This followed logically from EU initiatives on knowledge management, interoperability and pan-European services, identity management, secure e-government and e-democracy. Well-intentioned technocratic stimuli were organisationally dispersed within the EU Commission, bureaucratically packaged as anondyne technologies, and separate from citizen experience of either fraud, identity theft, and surveillance. Indeed, in states where citizens see e-government as surveillance, the purpose of parliament seems distorted and eclipsed by a discourse centred on the proportionality principles regarding the enrolment, storage, handling, transfer and access to personal data. The automaticity of machine-led transactions removed the human face, and the experiential basis for trust and accountability. Consequently, the human face of political oversight necessary to foster and sustain citizen trust went missing. Moreover, the concepts of 'transparency' and 'trust' are being re-configured in ways that beg *ethical* questions about public consent and the ability to give consent to the manipulation of personal data for imprecise, unknown and unknowable public policy purposes by invisible, sometimes automated, systems, including those drawing on second generation predictive biometrics based on data-mined to create profiles for risk assessment.

The transformational capacity of ICTs must be re-appraised if trust in credibly accountable government is to be restored. This is about more than transparency codes. In 2007, the EU Commission's DG Information Society stressed that e-government services needed to be strongly underpinned by citizen trust in government transparency. E-government was equated firstly with accessible online visibility, greater efficiency and effective e-administration and online transaction of government service access and delivery (European Commission 2007). It was very narrowly linked with 'e-participation' – vaguely seen as trusted e-voting; and thirdly with e-communication, often in defined spaces allowing for the expression of a view or question rather than reflection and critical discourse. The roll-out of e-government and associated expansion in initiatives to prepare citizens to access and participate in the 'information society' was fuelled by the rhetoric of market liberalism that privileged a business model over representative democracy (Freiheit and Zangl 2007). Inevitably, transparency and accountability were re-configured not just as multi-level, dispersed or privatized responsibilities but as quick-fix micro-management peer reviews.

14.4 Biometrics and Governance

It is important to note that the ethical use of biometrics is no longer a contestable proposition. How and why they are used and with what purpose and impact raises serious questions, however. Moreover, whereas the European approach to biometrics was to see a digi-biometric as a measurement of a characteristic biological feature of a person (their iris, fingerprints, hand or voice print), the United States had a broader interpretation that encompassed behavioural traits. As will become apparent, the second approach seems to be becoming the norm. Second generation biometrics rely not simply on the accuracy of a single biometric, such as a finger print, but on multi-modal techniques that use a combination of biometric traits and/or other authenticity tests.

Second generation biometrics might indeed enhance a border post's confidence in the veracity of a person presenting his biometrics at an entry point in support of his claim to be who he says he is. However, whereas the technology of second generation, multi-modal biometrics has progressed, the same cannot be said (yet) of accountability mechanisms. Legislation at all levels is still playing catch-up to scientific and technological innovation.

Legislation covers different eventualities and divides between that framed by data protection and privacy laws derived from intellectual property rights; and that designed to cover how data is managed. This aims at curtailing systematic abuses of personal data by monitoring and preventing unauthorised use, re-sale or data mining. More recently, in the light of numerous public scandals over the loss of data, much of it sensitive and identifying individuals in ways which breach their personal privacy and details used to prove they are who they say they are, attention has focused on the internal mismanagement of the handling of data and the ease with which huge files can be copied onto memory sticks, corrupted by malware and data sniffing Trojans, dispatched on disks and sent or lost in transit, kept on laptops subsequently lost or stolen or captured in cyber-space. Data management handling failures are widespread. Criticism and penalties from data protection offices[2] sit alongside steps to establish good practice guides and more rigorous in-house management and either person or system-based audit trails. None is sufficient and codes of practice remain imperfect and ambiguous as the first British standard for personal data handling launched in June 2009 shows.[3]

The audit trail approach is tantamount to a commodification and privatisation of accountability. In some cases the onus is put firmly on the individual whose data is being manipulated.[4] However, monitoring and retention of telephony and internet traffic question the nature and locus of accountability and responsibility in the private space of the home. Outside, industry, either by itself or acting in private-public partnerships with government administration, oversees itself. National ombudsmen and information commissioners set standards for public administration. All too often, governments evade them if not completely then partially. Good intentions are undone by the insynchronicity between administrations, parliaments, officials and private agencies. The result is the paradoxical pursuit of (and investment in) getting ever-more

[2]The number of data breaches reported to the British Information Commissioner's Office (ICO) in 2008 soared to 277 since HMRC lost 25 million child benefit records in 2007. They include 80 reported breaches by the private sector, 75 within the NHS and other health bodies, 28 reported by central government, 26 by local authorities and 47 by the rest of the public sector. The ICO investigated 30 of the most serious cases. Information Commissioner's Office, London, 29 October 2008. Sixth months later, more trusts were under investigation for non-compliance, and data loss theft. Scottish Government 2009.

[3]Guide to data protection compliance launched in June 2009 British Standard BS 10012, *Data protection – Specification for a personal information management system*. http://www.bcs.org/server.php?show=nav.10666

[4]http://www.dsk.gv.at/site/6286/default.aspx This also provides a footnote: **Please note:** To access your data according to sect. 26 of the Datenschutzgesetz 2000, you must, to prevent abuse, prove your identity in an appropriate manner to the agency that is responsible according to data protection law, the Data Controller [Auftraggeber]. This means you should append a photocopy of an official identity document that carries your signature (e.g. a passport).The request for information must be made in writing on paper or per fax. Requests by telephone or e-mail may not be answered. You can learn more about the right to information on data processed by the Austrian police authorities on the website of the Federal Ministry of Interior (Bundesministerium für Inneres) (German only).The data protection commission only has the power to examine the rightfulness of the information given or the rightfulness of a refusal to give information according to sect. 26.

citizens online to expedite the transaction of government business without a commensurate investment in ICTs for administrations and parliaments themselves. This is outside the scope of this chapter but is the context within which the policy for using biometrics to authenticate citizens for multifarious purposes proceeds. It is the framework within which differential ICT systems with different levels of secure architectures work and seek to collaborate in exchanging information, including biometric information, about individuals. Weak or inadequate security architectures compromise both the security and reliability of data and biometric data on which we rely and on the basis of which further decisions are made, and highlight the gaps in policy and ICT investment which undermine the credibility of the policies and claims made for using biometrics in all manner of transactions. It is therefore important to recall this when considering the advantages and disadvantages of biometrics when used for the purposes of automated information exchange.

14.5 The Infallibility Trap

The alleged uniqueness of a biometric trait leads to the misleading inference that it is a means of proving without doubt that a person presenting that trait is who he claims he is. From this, it is a short step to claiming, as governments routinely do, that biometrics enhance security. However, the reliability of a biometric to confirm an identity claim can be compromised by technical and non-technical factors. The former include the reliability and compatibility of scanners and the setting of the algorithms for verification, corruption of telephony systems along which data is sent, transmission interference, data storage, corruption and degradation. The latter include environmental factors such as dust and heat, and behavioural factors like stress, agitation and sweating.

Uni-modal biometrics, such as fingerprint or iris scan, are open to forgery or coerced re-use. They can be combated to some extent. For example, the opportunity to use latent fingerprints left on a scanner can be minimised by sweep techniques that clean it before subsequent use, or which require liveness checks with the subject providing a side or roll-shot of a stipulated finger (which may be a different one to that offered), or pulse oximetry or deeper fat checks of the finger to minimise the likelihood of the finger presented being live only by virtue of a cardiac pump action beneath it. Iris digital images can be use to create fake identities when placed over a subject's eyes. Digital image liveness checks using papillary hippos check involuntary changes in the size of the pupil and iris. Disguised voices can also create 'genuine' biometrics based on false identities that can be checked by other voice liveness checks. All are time-consuming and so expensive and intrusive.

Consequently, steps have been taken to boost the accuracy and reliability of biometrics and applications using biometric verification and authentication. At the technical level, this relates to the underlying system requirements and security architectures, the weightings given to settings for matching biometrics presented to those stored (and how they are adjusted in different places), and ensuring that a system does not necessarily accept a fingerprint that is re-submitted ten times (excluding the person on each sequential submission) as genuine when it may have been spoofed or used with a fake ID.

Given the technical cost of developing new systems for matching, and the costs of steganography (whereby data is hidden inside the data carrier), multi-modal second generation biometrics have growing appeal. Combining fingerprint with iris recognition, for example, makes forgery less simple. Nor is it perhaps not surprising that government authorities (and others, notably in the medicare and insurance sectors) have an interest in extracting supporting information about a person from a file that contains more information than his fingerprint. The operational imperatives for justifying this in policing, crime prevention and health care are compelling. There is a growing industry in private sector protection of such information, including ensuring compliance with regulations and data management based on real time data replication in real time management scenarios.[5] This out-sourcing of responsibility shifts accountability too.

Stepping outside the first wave of biometric endorsement, the second generation biometric technologies, and especially the way in which they might be deployed, reveal a trend towards the creation of systems of continuous monitoring and authentication of individuals by ambient intelligence means of which the individual may be scarcely, if at all, aware. What is the potential impact of supposedly unobtrusive biodynamic monitoring on society? Has the dark side of the moon already eclipsed the promised enlightenment of first generation biometrics to boost citizen security and so provide the environment for sustainable prosperity and endeavour? Have the new technologies led to ever-watchful, anxious rather than secure citizens?

14.6 Forgetting Big Brother

The association of Big Brother with surveillance cameras has diverted attention away from more pervasive ICT intrusions recently as public authorities and businesses have dismissed legitimate concerns and pressed ahead with the introduction of ICTs to expedite the administration of their business. The negative criminal associations of first generation biometrics exemplified by the fingerprint, facial recognition and surveillance for tracking (Soros 2007) were discounted as passports and travel documents, school registers (Working Party 2009, Gillies 2008), libraries, etc. adopted biometric technologies using them to expedite convenience for consumers and for managers and administrators of bureaucratic processes. The gain was claimed to be one of efficiency (speedier transactions leading to shorter processing time and hence cost cuts) and added value by improving the 'experience' of the transaction by the consumer (freeing time for him to do other things, and certainty in the infallibility of the transaction). Impelled by an unthinking logic of

[5] Such as Double-take. www.doubletake.com

speed equating with a better 'life', ICTs that accelerated bureaucratic processes for private and public purposes were welcomed and adopted by governments, and sometimes without sufficient critical understanding of their implications for privacy by public authorities, including primary schools. For example, in one 300,000 sized town in Northern England, 15 of 97 primaries jumped on the biometric ID bandwagon.[6] Fears about fraudulent use of biometrics, malware, spoofing, corrupted RFIDs and phishing were voiced but never sufficiently addressed. Instead, developers of biometric identification technologies (a lucrative market) downplayed the difficulties associated with the exclusion of an individual whose identity had been forged. They promoted multimodel biometric identification schemes: the exclusion forever on the basis of a forged uni- or bimodal biometric (face and fingerprint, for example) would thereby become a relic of the past.

Second generation biometrics progressed from the crudity of unimodal iris and fingerprint recognition to bimodals, typically combining the two (as in the new generation travel documents). Developers investigated the robustness of heat imaging, palm, scalp and vein prints, neural patterns (Poulos et al. 2001), gait and voice recognition. The individuality of the latter, it was claimed by mobile companies, offered the prospect of greater security than the more easily spoofed fingerprint, even though voice recordings could fool a mobile into accepting the 'authenticity' of a caller. Few noted the relative insecurity of data transmission via analogue lines which account for over half of lines in even some of the most industrially advanced states.

On the one hand, this means that the advocacy of the adoption of second generation multi-modal biometrics is in part a response to persisting infrastructural weaknesses rather than simply a response to the commercial driver behind developers' inventions and marketers' goal to gain ever increasing market share. On the other hand, it raises serious questions about the underlying assumptions about the relationship between science and technology and good government.

The proposal to capture and combine physiological features of an individual – their heart beat and blood flow using EEG and ECG and brain scan technologies – raises serious questions that go way beyond issues about the reliability of the technology. These include: issues of the associated securitisation grabs of medical technologies; the justness, legitimacy and discriminatory power of these technological means of identifying and authenticating individuals; the implications for individuals and for society of combining these 'unobtrusive' technologies with ambient intelligence tracking and surveillance; and the implications for the sustainability of systems based on a territorial rule of law when digi-borders escape the accountability checks and balances enforceable by systems of justice purporting to uphold principles of liberty, equality, fairness and justice. The linkage of biometric with other surveillance-type data makes it possible, in theory, to identify the whereabouts of everyone with a specific physical characteristic, such as blue eyes. On the dark side of the moon, the blue-eyed could be targeted for all manner of benevolent or malevolent action.

---

[6]Interview data, March 2009.

Proponents of these multimodal biometrics advocate their use to minimise blended threats, digital hijacking and cut risks in critical situations, by for example monitoring changes in body sway, seat-based anthropometrics, physiological and behavioural profiles subject to change under stress, and especially in the event of substance abuse. Such use is readily justifiable with respect to surgeons, pilots and others having access to critical infrastructures, for example (Damousis et al. 2008). However, it is improbable to assume that their adoption for one purpose would not be extended to other purposes, especially as the technology becomes cheaper and relatively obsolete.

This begs serious questions about the impact of ICTs for identity on society. The therapeutic medical purposes of brain scans as a diagnostic tool in the identification and remediation of disease and injury are being captured for an entirely different purpose – albeit, a diagnostic one of a sort. As a biometric, its purpose was not as simple as confirming the identity claimed by a person submitting a biometric in support of his claim to be who he claimed to be with associated rights to enter a given territorial, bureaucratic or commercial space. Instead, it was re-configured as a forensic tool to bring as close a possible match between a 'behaviour state' that could be deduced from brain activity and seen as symptomatic of intentionality. The securitisation of medicine proceeds apace, this time not simply through the gains of robotic soldiers and determination of the site of brain injury, but to reflect psychological states associated with criminal or 'unacceptable/risky' intent.

On the dark side of the moon, moreover, it is not human evaluation and intuition that determines and assesses the riskiness of the behaviour, but 'intent' is digitally inferred from algorithms where open-ended ambiguity is excluded. Instead, it is fixed according to a set of pre-determined digitally constructed 'solutions' and 'answers'. Ask the wrong question, get the wrong answer but one that is digitally 'correct'. The 'answer' expressed as an abstract specification of an algorithm becomes 'more pure or real than any instance of its implementation (the thing explained) (Vaden 2004). This scenario approximates decisionmaking under crisis, where speed and accelerating decisiontaking are the norm. It is instructive to consider the lessons that can be learned for this.

14.7 Digitised Decisionmaking, Time and Decisionmaking

Extolling automated information exchange indicates that function creep persists, legal regulation notwithstanding. Micro-managerialist audit trails reflect a privatisation of public accountability. Worse still, converging support for inter-operability and the associated reliance on identical and/or compatible ICTs aggravates this especially when individuals are unable to inspect or are unaware of who is accessing, handling and using their data for whatever purpose, and group-think besets the procurers of ICTs for public policy purposes. The mantra of inter-operability conflicts with technologies to combat non-consensual data linkage.

Who's in control? If it is supposed that the impact of ICTs in any field is not neutral, questions must be asked about the source of the motor for their accelerating roll-put and adoption by private and public sector authorities. The days of ICT adoption being simply a cool status symbol, a manifestation of modernity and power have passed. Their potential to be used for particularistic interests has been recognised. Within the nebula of the discourse about the benign impact of ICTs and their claimed benefits to boosting 'citizen' participation, lurks the dust of mixed purpose use, the abrogation of the precautionary principle, the lie of disembodied information, the reality of unobservable data mining, the erosion of the principle of consent as the levels of application criss-cross leisure, pleasure, domestic convenience and bureaucratic efficiency fields. Inter-operability is the new utopian goal. As yet, the need to think about appropriate means of governing inter-operable spaces remains territorially grounded instead of being mapped onto the user-communities and appropriate requirements for ensuring benign intentionality.

The (in)security associated with inter-operability has been addressed primarily with reference to the need to construct robust security architectures to minimise the danger of intrusion, phishing, data mining, data re-configuration, exploitation by third agents here and 'out there', data re-selling, error, fraud and misuse. Accordingly, codes of practice and strong audit practices for data handlers and data bases have their advocates. Data protection laws are re-visited and improved, most recently by the proposals of the European Parliament's LIBE Committee, and the responses of the EU Ombudsman to the Commission's proposals on access to documents (European Parliament 2008a, b, c, d). The individuals are provided with rights of redress: to correct incorrect data entries about themselves. The universality of this right is undermined in practice by the legal costs associated with doing so and by the fact that invoking it relies on an individual's capacity, mental and otherwise, to do so. Such legal provisions are necessary but not sufficient conditions to protect individual's data. Intellectual and ownership rules, moreover, mean that data that is reconfigured and recast becomes the property of those who recast it, rather than the person who provided at least part of the data. The commodification and commercialisation of data reduces the ability of the individual to control it in general and in particular where that data is out-sourced or sub-contracted and re-sub-contracted to territories where laxer data protection provisions exist. Discrepancies arise over any assumed status of privacy arising in privileged legal or medical contexts (Harting 2009); and over ambiguous interpretations of the notion of consent.

Inter-operability advocates, moreover, have focused on 'privacy enhancing technologies' (PETs). PETs aim to ensure that incursions on individual privacy is minimised technically. Yet, casual web surfers' information is regularly automatically forwarded from sites which host advertisements without explicit consent or warning. Not until April 2009 did the Commission launch infringement proceedings against the UK following complaints over Phorm's behavioural advertising that mines web users' surfing to target them. The Commission had since July 2008 been in touch with the UK over the UK's implementation of the ePrivacy and personal data protection rules that require EU states to ensure, inter alia, the confidentiality of communications by prohibiting interception and surveillance without the user's consent. Structural problems in the way in which the UK implemented EU rules highlight a general problem (European Commission 2009).

Ambiguity persists PETs do not place ownership of the data in the hands of the individual, or insist upon the right of the individual to be consulted, and to give consent to, individual data being transferred to or shared with other agencies on the decision of an existing agency to whom that individual has provided data for a specific purpose. Data linkage is common, corrodes privacy, principles of purpose limitation, data minimisation and data subject awareness of the use to which his data is being put. This breaches principles regarding data protection and consent that are at the heart of EU provisions on access to documents whereby the Commission has to obtain the agreement of the third party to enter a sensitive document in the register.[7]

Moreover, the out-sourcing of data processing itself involves different types of obligation on the data processor. The third party may not have any rights or duties other than processing the data (the Auftragsverarbeitung in Germany) or may act solely upon instructions of the sender. This contrasts with the third party assuming an independent role vis-à-vis the data (the Funktionsübertragung) under which it is obliged to adhere to specific legal rules, such as the requirement for explicit consent of the consumer regarding the use of his data. For the individual, the problem lies in him blindly agreeing in advance to data sharing or being simply informed. However, few appreciate the inherent dangers associated with this and accept the consequent intrusions of targeted spam, profiling and tracking. While this has been sold to consumers as a way of preventing unsolicited spam, it rests on data mining and profiling techniques which are infinitely expandable, as the recent Swift, Phorm and mobile phone commerce attest. Opting into (or forgetting to opt-out of) advertising does not constitute informed consent.

Automated information exchange, and cross-departmental, cross-agency and cross-border inter-operability compromise privacy. Imprecision, different clustering of data, of entering data, making it partly or fully visible to someone seeking access create false certainties in the digital realm that are rarely matchable in the real world. There is a real risk that digital 'certainties' will not be scrutinised sufficiently and that absolute truths will be inferred from digitally 'correct' documents. In the digi-space, there is no room for ambiguity, flexibility, critical thought and reflection. The imperative of speed is paramount.

However, the logic of this discourse is de-contextualised. Digital solutions only exist if the ICTs behind them work and can be implemented. Their reliability depends on conditions that allow them to function and be implemented. Critical infrastructures and ICTs can fail completely, or partially (as in the case of malicious attacks, such as those that shut down Estonia's egovernment, police, banking and online media in 2008) compromising or degrading the code, making information sharing less controllable. The relative reliability of 'answers' is context dependent, and on how data is inputted, handled, stored and ultimately interpreted by humans

[7]See paragraph 30 of the staff guide to public access to Commission documents, Article 29 Data Protection Working Party, Opinion 3/2009 on the Draft Commission Decision on standard clauses for the transfer of personal data to processors established in third countries, under Directive 95/46/EC, adopted 5 March 2009, 00566/09/EN, WP161.

for good or ill. The growth in business in identity and access management has not been matched by appropriate visible mechanisms of public accountability.

Whole processes may be sub-divided and parts out-sourced in fuzzy private public partnerships, or pure private arrangements where the digital is incorporated into different bundles of, and appropriated as, a commodity.

An example takes the ethical implications of using one biometric (e.g. iris recognition) distinctions further. The implications of using this to discriminate between 'us' (those able to enrol into the fast-lane automated iris recognition for swifter processing at border controls, as in airports – despite the relatively high error rates) for immigration purposes is less problematic than more generalised differentiation among groups of people who consider themselves all as 'us'. These will inevitably create divisions, and categorise individuals again into 'them' and 'us', and lend themselves to closed and potentially unaccountable abuses. The public has a disingenuous sense of individual anonymity in cyber-space and generally a poor understanding of the impossibilities of ensuring effective accountability.

Even when medical diagnostics like radiography are outsourced beyond the state to compensate for staff shortages, patients are unaware of where and by whom their data are being viewed. At heart is not a zero-sum game of full disclosure to third parties or none, but the need to reconsider the type, kind, extent, duration and retention of data that should be minimised and purpose limited. Why? Both because privacy and data protection are legal undertakings and because as yet neither the public nor the private sectors have adequately dealt with the issue of the vicarious liability of authorised or non-authorised slippery sub-sub-contracted third parties for amassing, transmitting or using such data. The impact of society has yet to be understood. The questions that remain concern (i) the impact on social capital: trust and reciprocity; (ii) the motor behind ICT adoption and (iii) the locus of accountable, trusted authority in digi space.

14.8 Power and (Un)Accountability

The new ICTs, from ambient intelligence to second generation biometrics in smart cards for inter-operable purposes and environments are open to being seen as instruments of diplomacy in general as well as new tools of (in)security and fear. Their advocates affirm their contribution to security, equating the introduction of biometrics with an argument promising certainty in respect of unknown or unpredictable threats that generate fear. Associated with this is an individualisation of responsibility for personal- and possibly communal – security. Many governments rolling out digital biometric identity documents require their citizens to have one, pay for one and renew them periodically on pain of penalties. While this is questioned in terms of ability to pay (the poor, disadvantaged, socially excluded, illegal migrants, trafficked people and elderly), much less attention has been paid to the impact on the vulnerable (non-discrimination rules may be essential to ensure respect for personal integrity given that data protection laws refer primarily to personal data not the person or groups of persons associated with the concept of marginalisation) (Karanja 2008).

Issues of consent too are critical. Children (unable to give consent in their own right, although the authorities say the ID uses must be simply explained to them) or the infirm and the handicapped arguably find their limited autonomy even further eroded by the armies of invisible people having automated access to their personal data. Such vulnerable people are obviously not able to take personal responsibility for their own security or privacy let alone the security and biometric enrolment responsibilities required of all citizens required by law to hold biometric ID cards. The infirm can be especially vulnerable moreover to the side-effects of treatment that damage their finger-prints making them 'appear' as security alerts when crossing borders (Wong et al. 2009). There is clearly a need to disentangle the legitimate benefits of biometrics from an (in)security discourse echoing the rationale of military security requirements and open to dystopic interpretation. Moreover, where either dual-use scientific research or obsolete discoveries of military and medical research technologies are being rolled out now for civilian use, it is imperative that appropriate accountability mechanisms are in place first. Therein lies utopia.

Uncritical acceptance of the privatisation of accountability and of the administration of government undermines and weakens democratic governance. The Data Protection and privacy commissioners alone are neither the appropriate nor adequate locus of authority capable of routinely scrutinising and demonstrating that political control of government is feasible and effective. Politico-legal benchmarking may be a step towards identifying and remedying some current deficiencies, but the e-commodification of personal information in fuzzy e-space is insufficiently susceptible to ethical practices, visible, authoritative public regulation (Kapstein 1997) and accountability. Private redress is notoriously socially divisive, unequal and difficult to obtain even by those with the greatest capacities to do so. Worse still, governments' ad hoc approach leads to incompatible and contradictory positions being taken.

14.9 Beyond Public Accountability and Audit: The Privatisation of (Un)Accountability?

The logic of intelligence services permeates the quest for inter-operability that rests on the existence of data bases, data linkage, automated data exchange and access without the knowledge or consent of the data subject. The Phorm and Swift cases illustrate how fast the traditional vehicles of public accountability and management audit trails have escaped effective public scrutiny. As a result, working practices have evolved which prioritise speed over critical reflection of how time impacts on decisionmaking. This is risky since the known dangers are extensive and go unheeded. The need for and adoption of standard operating procedures for automated information exchange define what and how work is done. ICTs structure relations. Space for critical reflection, common sense, questioning, reordering priorities and assumptions on which information (including the choice of which uni or multimodal biometric) is collected, stored or shared shrink.

The speed of technology has an impact akin to that known in crisis decisionmaking. What this means is that pre-determined 'solutions' are likely to be sought. These in turn are likely to have been pre-determined and structured by what the ICTs require, allow, and admit into the data fields. In these scenarios there is no room for hesitation, laterality, or the introduction of alternatives that contravene the pre-chosen preferences. If event x matches specific criteria logged into the programme, then outcome y is likely regardless of countervailing evidence that may be highly relevant but scoring too low a probability to be selected by the programme.

14.10 Impact of Group Think

The more a programme predicts or recommends 'solution' x that is the preferred option of the chief decisionmaker or executive, the less likely it is that inferiors will challenge it if their own interests could be compromised by challenges. These interests extend to their own career prospects, the resources available to them in resource distribution and re-distribution, their understanding of which colleagues/associated or adjacent departments might expect them to support that position in order to achieve their own departmental goals, and possibly as a quid pro quo in mutual trading and bureaucratic bargain brokering. In short, the well-known risks of Groupthink are likely to be exacerbated in systems that increasingly rely on automated information exchange.

At present ICT enabled decisionmaking is undertaken by computers that are not (yet) as able as humans to discriminate in the information they hold to deal with unknown risks and manage countervailing evidence to accepted 'wisdom' or dogma as predicted by their programming.

But their 'decisions' are ultimately uncontestable. Computers cannot be held (yet) accountable to the data subjects whose 'information' they process for unclear, invisible and insecure purposes. Data protection acts are essential but insufficient safeguards for the individual and for society against potential abuses by the 'owners' and 'users' of data mined from and re-configured for and by various sources. Democratically elected institutions as yet are at a severe disadvantage over private organisations at home and in third states who reconfigure, sell, use, and alter information about individuals. The former cannot exploit databases holding information in quite the same way, so speedily and evasively as some private agencies have been able to.

The consequence is that a form of security surveillance becomes domesticated, privatised and individualised with the purveyors and users of the ICTs acting either uncritically or in ways that erode regulatory norms and codes. The normalisation of biometric identification for trivial purposes (library cards in junior schools) softens up individuals to accepting surveillance using second generation biometrics (such as the smart voice recognition of mobile phones) for undefined purposes by unknown agents. The temptation is for the media to 'blame biometrics' and to ignore the gains they can bring to verifying identification. The issue is therefore not whether biometrics are good or bad per se but the way in which they become embedded in all kinds of documents created and used for completely different purposes.

The digi-society's expansion is not matched by the identification of trusted loci of authority in digi-space. Instead, digi-space becomes ever more contested and contestable. Its regulation depends on constructs of the knowledge society and ownership of digi-tools and outputs. This flies in the face of scientific advance where progress is a shared good not something to be compartmentalised and used to instrumentalise power.

Worse still, the locus of public and private accountability are becoming diffuse and mixed. Public authorities, by out-sourcing and sharing functions with private companies in public-private arrangements, opt for full or semi-privatisation of accountability. Such accountability is then rendered into codes of practice and management procedures divorced from the logic of shared ideals over resource allocation, visions of fairness, justice and universality, responsibility and responsiveness. Automated e-administration does not require mediation by ideology or aggregated majority views to make citizens comply with bureaucratic desiderata (such as completing certain types of forms to get smart cards that enable access to services). All that is needed is that bureaucrats, those responsible for buying, running and renewing the ICTs agree that it is a 'good' thing that enables them to do their job or abdicate it to computers and automated decisionmaking. Information management and control can be prioritised as motors of policy.

But behind that is the dark side of the moon. In the domestication of security and the securitisation of the prosaic mundane activities of daily life, can the driver behind this be identified? Or is the impact on society more likely to be an ineluctable erosion of the ideals of trust and reciprocity, ethical responsibility and accountability at the heart of social capital?

14.11 Impact on Society

The use of ICTs for the administration of governmental, commercial and personal daily activities is now the norm. The transformative impact on society is shown by social networking, by safer internet initiatives, and by the acceleration of all kinds of transactions. States embraced broadband to provide information online and in edemocracy chimera (Deutscher Bundestag 2005). The impact on the structures of government is less well-documented. Crude measures might include the readiness of legislators, inter alia, to use ICTs for managing and exchanging information as part of their legislative scrutiny roles. Parliaments seem to lag behind in the same way that the judiciary did in the adoption and use of ICTs for processing securely legal documents and information across departments and territorial borders. At the same time, they imperfectly regulate it and associated ID management codes.

A more telling indicator of the potential transformative power of the adoption of ICTs for identity management purposes is provided by answers to the question of the ultimate locus of authority and responsibility for identity documents. For example, in Germany, following the Government's initiative of 7 October 2008, the issue of such documents will be an exclusive federal responsibility from November 2010 (Rossnagel and Hornung 2009). The potential for inter-operability between egovernment, e-commerce, e-health and the citizen register has yet to be fully exploited. However, a citizen's 'official identity' is potentially infinitely manipulable. Whereas, the German government appears to have greater safeguards to strengthen internet security and responsibility and to prevent discrimination against citizens refusing to have e-IDs compared to the British, the trend is similar and all manner of transactions for those without them, for whatever reason, will become more difficult and protracted. Differentiation and discrimination hit the most vulnerable and those least able to take compensatory steps or seek redress. Privacy and human integrity and dignity are being compromised no matter how 'unethical' that may seem.

The potential impact on society of such developments is disturbing not because e-identities per se are becoming widespread but because of the wider context within which they are being selected and rolled out. The border between the private and public domain is being steadily eroded. Legislation brought in with good intentions – such a transparency and freedom of information regulations – had a different purpose than that to which they are being used. The context was the imbalance between the executive and elected representatives to access and therefore effectively scrutinise draft legislation. This was behind the transparency initiative of the EU in 1994, and a prelude to the European Parliament acquiring rights of co-decision with the Council of Ministers. In short, this was construed as part of procedural reform designed to facilitate structured inter-institutional change, power-sharing and mutual openness in the name of improved – meaning better informed – decision making on behalf of citizens (European Parliament 2008a, b, c, d; Lodge 2007).[8]

In the age of inter-operability and information sharing in public and private and mixed arrangements, the benign intent can be subverted by those seeking both openly and legitimately to have access to information for honourable purposes as well as by those seeking to misuse it. Examples of how easy it is for unknown individuals to ignore purpose limitation principles to expand data mining tools, such as those provided routinely to councils, by companies and credit checking bodies like Experian and Equifax, grow daily. At the height of the recession, Experian out-performed the FTSE all share index. Its profits rose 8% and its share price rose 80% from 2008's low[9] with overseas operating margins rising to 23.3% and expected to climb as Experian cuts data many ways to sell onto to different clients.

---

[8]World e-Parliament Report 2008 Vice-president of the European Parliament Mechthild Rothe told the UN in 2007 that "e-parliament" strategies need to guarantee "a high level of IT security," to respect the privacy of the personal data when engaging with citizens. The European Parliament's ICT use – RRS feeds, podcasts, online streams of plenary sessions in 23 languages, including some committee meetings is not typical of parliaments. Over 85% do not 'communicate' with citizens in this way http://www.ictparliament.org/index.php?option=com_content&task=view&id=245

[9]The Times, 21 May 2009, p. 61. The share price low was £2.50 and in May 2009 was £4.93. Capita's low as £5.50 against a £7.73 high.

It is no longer plausible to argue that the checks on who accesses them is sufficient a guarantee that the systems will not be abused. They are. It is too late to insist that the biometric e-ID is a sufficient safeguard against unauthorised people accessing the system, especially when access credentials are loosely defined (as they are in England) to allow excessive access even to children's files by all manner of public service personnel, from schools, to health and social workers and associated agencies.[10] The claim of privacy and security enhancement is disingenuous. Safeguards like encryption against forgery and copying of biometrics do not yet provide sufficient security when all the associated information is held either centrally (as in England's Schools Information Management System) or is diffused, ambient, or interrogatable by insecure ambient means. The problem lies not with the biometric per se but rather with their contextualisation and their use as policy tools.

14.12 Using Biometrics Differently?

Is there an alternative? Can the benefits that inter-operability supposedly offers be used differently? Instead of requiring biometrics to be used to verify that a person is who he claims he is, the question should put the onus on the person asking for information to prove that he has a *legitimate* right and authority to ask someone to prove *something* about themselves. For example, for civil purposes a supermarket check-out might have to prove to a customer wanting to buy alcohol that s/he is entitled – by virtue of his job which he would have to confirm using his ID authentication – to request that information, and that information only from the customer. The customer would show only that element of his eID showing whether he was over 18 or not. Some supermarkets already have their own voluntary store cards for similar and targeted marketing purposes. This psychic ID would ensure respect for purpose and data minimisation principles (Birch 2009), boost the integrity of e-privacy and put the data subject back in control of his data, its disclosure or partial disclosure. This is a different principle to using eIDs than the ad hoc introduction more commonly used with crude effect. For instance, biometric eIDs for new air-side employees being gradually introduced in 'pilot schemes' lead to highly visible discrimination among staff.

Without critical reflection about how and to whom everyone's personal data is opened, there is a grave risk that privacy will become a myth. The cynical approach of public and commercial enterprises to persuading young people, from school age upwards, to divulge excessive personal information storable in RFID smart cards, inures the next generation to uncritically assuming that these ICTs are neutral in their impact on the individual, on society, on the structure of democracy, on the need

[10] See guidance on the application of the Data Protection Act 1998 to the use of biometrics in schools on http://schools.becta.org.uk

for, and sustainability of, public mechanisms of political accountability and therefore on their future (in)security. The question is not whether or not biometric identification per se is advantageous and desirable or not. The answer depends both on the context of its deployment and the association of that biometric with other data from which all manner of unknown inferences may be made. Few would argue with using robust multi-modal biometric identifiers to facilitate trusted policing and justice agencies as access tokens to enhance their operational effectiveness. However, the manner in which areas of activity become subject to 'security' cannot be separated from the ethics of using such ICTs given the speed with which domestic policy areas have become securitised and, in the public eye, 'criminalised'. Nor can the out-sourcing of policing to voluntary or 'community neighbourhood officers', monitoring petty crime and anti-social behaviour be ignored when they, too, are likely to seek and acquire partial access to data about people in their community and neighbourhood. Is such comprehensive, universal access ethical?

Moreover, the impact on the socially disadvantaged, the elderly, infirm, handicapped is to place them at a greater disadvantage of neither being able to refuse consent to information sharing about them among extensive public private services nor to being able to check its veracity as their circumstances change. As the (in) securitisation of the domestic arena of everyday activities progresses, those in authority can arbitrarily redefine the borders as to who and is not entitled to enjoy specific benefits and access to services. Do adequate ethical codes exist and provide sufficient checks to detect and curtail an abuse of such power? Is the technology up to the job? The next generation smart card authentication, identity-aware networks, and hosted identity and ambient intelligence and out-of-hand authentication might precipitate the extension of existing arrangements. In the rush to sell cloud computing, profitability is prioritised over privacy and security. Is this ethical and moral?

The digitisation of society and of the human challenges the values that allow the liberal development of science and technology. Many human actions may be capturable, and people may well become attached to haptic advances, robots and robotic dogs, but as yet those things that differentiate the human from the algorithmic response to decide and balance good and evil, the border between (in)security and (un)accountability require human interfaces and human, real time reflection, reciprocity and moral, ethically informed judgement not just technological 'solutions'. It is ethically irresponsible of governments to create ICT systems – whether using biometrics or not – that exchange personal data in a fundamentally flawed and insecure manner, and which thrive on weakly accountable public-private arrangements driven and overseen by non-publicly accountable agencies and corporations.

The political claims made for these systems to the public are disingenuous and incredible. Speed and convenience of access do not compensate for underlying infrastructure insecurities often concealed from the public. There is low awareness among citizens as to flaws in the installation of the technologies (e.g. the risks of too close parallel telephone cables) and the manner of collecting, collating, storing, reselling, mining and re-configuring and managing the processing of data. If technologically it is possible to create secure data bases, and use multi-model biometric applications effectively, the real issue is one of access policies and contextual integrity.

Who accesses data in them, with what and why? This begs fundamental questions about the nature of society that relies on administrative ICT-led transactions and interaction and appears to have decreasing trust in the ability and commitment of government to work for the common good. Biometrics facilitates not just social sorting, that we may find desirable (in keeping 'them' out, as in border controls) and allowing 'us' to create and maintain our 'in groups', (by inference the good guys). IBM over 60 years ago facilitated social sorting using punch cards. This is nothing new. Nor are biometrics. What is new is the space within which, and speed with which, decisions about various types of socially sorted data can be made, information transacted and good or ill done to individuals differentiated by often undisclosed criteria with or without their consent. That raises a moral question and poses an ethical dilemma about the nature of governance in digi-space and for humankind. The new biometrics accelerate the refinement of social sorting and impact on the ensuing greatly dissected societies more immediately than before. It is imperative to revisit the relationship between the individual, community and society, and good, open, responsible and accountable governance (Lodge 2009a, b).

Too often governments wriggle out of responsibility by claiming they are 'working towards' it without dealing with core ethical issues. Plausible claims about data privacy and robust audits to prevent (or at least track) data mining by the British National Health Service are advanced alongside implausible claims by the eHealth agenda of cross-border information exchange (including data on prescriptions and insurance). The Australian government's 2009 proposals justifying information sharing 2009 in terms of combating fraud were opposed by doctors concerned about the impact on patient confidentiality and privacy. The British government also proposed allowing access and selling patients' medical records (Gillies 2008) to a wide variety of government and private bodies in the name of enhancing security. Simultaneously, it advocated a shift to the use of open source software (OSS) in February 2009 claiming that continuous upgrading of software to combat security breaches, coupled with robust vetting and management of data handlers at all levels was essential. The tension between technical and managerial priorities ostensibly to boost system security seem only too easily to be sacrificed to unethically exploiting citizen data for the state's financial gain, even when the claim is dressed up in the rhetoric of security. This ambiguous and loose use of the term 'security' seems, if anything, to engender insecurity.

Baked-in privacy, contextual integrity, and security architectures must prioritise security against malware intrusion and the principle of putting the individual in control of accessing and permitting access to all or part of his data. Could the decline in trust be stalled if governments chose to prioritise an ethical principle over economic and bureaucratic expedience, and if they took responsibility for insisting on such systems? Not tackling this challenge now is risky. ICT vulnerabilities are believed to endanger individuals' integrity and their essence symbolized by their biometrics. If the credibility of and locus of authoritative sources of accountable information holders, providers and scrutineers are hazy, the credibility of claims made about the trust-worthiness of public and private companies, parliaments and the ICT providers and auditors will inevitably crash. If this is compounded by a pecuniary approach to exploiting personal data for government or invisible private gain, public trust in public accountability will vanish. For constitutionalists, the assumed bargain between the state and citizen, aggravated by legal uncertainty, ICT insecurities and public feelings that privacy and the assumed idea that an individual has control over his physical self will be broken no matter how ubiquitous 'surveillance' in its many guises, and no matter how strongly claims are made regarding the gains of biometrics for identification to legitimize access to data.

Is there a future for the exercise of voice by parliamentarians? If parliaments reassess their traditional communication function, may be. The European Parliament's Civil Liberties Committee has been the most articulate and vociferous critic of EU national governments in the Council of Ministers and of the Commission. Attempted sleights of hand by them have not gone unheeded, possibly because the European Parliament's ingrained need to be vigilant of the legal base of legislative proposals (European Parliament 2008a, b, c, d). National parliaments could learn much from being more robust and prepared to embarrass executives over their failures to safeguard citizens. To do that, however, they must inspect current practice and work with MEPs. A return to the tradition of in depth scrutiny of government is vital and imperative to encourage open and informed debate about what data people want shared and how effective access and appropriate safeguards can be ensured. Security is no longer simply a matter of safeguarding territorial integrity. ICTs' ubiquitous impact on citizens' daily lives and geo-cyber attacks on critical infrastructures are not amenable to the traditional defences offered by international treaties even if international principles of sovereignty regarding territorial integrity and political independence are breached. E-networks link up public and private organisations and services in ways that so far escape effective political oversight and control. Whether out-sourced or not, public-private egovernment relationships erode parliamentary authority lead to and semi- privatisation of accountability and almost imperceptibly by default of privacy.

The impact on society and on concepts of community, social good, democratic government are ill-understood. Instead, attention too readily switches to devising yet more penalties for breaches that by their nature are uncontrollable in digi-space or where data controllers and purveyors can relocate at will to territorial regions that do not share European values and ideas about privacy, individual human rights, equality and justice. However, in the short term, politicians should follow the European Parliament's example, even before the ratification and entry into force of the Lisbon treaty, and devise ways of identifying, inspecting, publicising, communicating and politically and financially penalising those whose lax ICT management practices and unthinking espousal of the mantra of 'interoperability' endanger individual and ultimately collective security. It should be made clear to the public that there is no point having secure data lockers (secure servers) when machines that access them are inherently insecure. The ICT purveyors and politicians must accept responsibility for insisting in security all along the chain. Without political insistence, the developers will proceed as they have in the past to maximize profit regardless. The sharing of data by public agencies or in mixed private-public arrangements risks leading by default to uncritical acceptance of the privatisation of accountability and sloppy privatization of responsibility for maintaining individual privacy. Parliaments need to re-assert their traditional communication and education functions vis-à-vis the public to inform and promote intelligent debate about government and the management of their digitised data for public policy purposes. Without constant vigilance and pressure from accountable parliaments, ombudsmen and data protection supervisors, a tragedy of the increasingly re-bordered unaccountable domestic and internationalised cyber-spaces of the ICT commons is inevitable. Privatised (in)security, (un)accountability and (un)privacy presage differentiation, discrimination and divided societies where invisible, rather than secret, arbitrary e-power endangers all. The ethical use of biometric identifiers has to be a step to rectifying such imbalance.

References

Birch, D. 2009. Psychic ID: A blueprint for a modern national identity scheme. *Identity Journal* 1(1): 189–201.

Damousis, I.G., D. Tzovaras, and E. Bekiaris. 2008. Unobtrusive multimodal biometric authentication: The HUMABIO project concept, EURASIP. *Journal on Advances in Signal Processing*, Article ID 265767, doi:10.1155/2008/265767s.

Deutscher Bundestag. 2005. 15.Wahlperiode, Bericht des Ausschusses für Bildung, Forschung und Technikfolgenabschätzung (17. Ausschuss) gemäß § 56a der Geschäftsordnung Technikfolgenabschätzung Internet und Demokratie – Abschlussbericht zum TA-Projekt "Analyse netzbasierter Kommunikation unter kulturellen Aspekten", Drucksache 15/6015, 17 Oktober 2005.

European Commission. 1992. *A new European instrument to combat fraud and trafficking: The customs information system (CIS) memo.* Brussels, 17 September 1992.

European Commission. 2007. *Culture and society: e-Government projects, putting citizens first "Example Projects"*, Brussels 2007. Available at: http://ec.europa.eu/information_society/tl/soccul/egov/projects/index_en.htm.

European Commission. 2009. Press release IP/09/570 telecoms: *Commission launches case against UK over privacy and personal data protection,* Brussels, 14 April 2009.

European Parliament. 2008a. *Resolution on the proposal for a decision of the European Parliament and of the Council on interoperability solutions for European public administrations* (ISA) COM(2008)0583-c-0337/2008-2009/0185(COD).

European Parliament. 2008b. *LIBE Committee: Transparency and public access to documents: Some aspects concerning e-transparency in the EU institutions and the Member States*, PE393.285, March 2008.

European Parliament. 2008c. *Resolution on the proposal for a decision of the European Parliament and of the Council on interoperability solutions for European public administrations (ISA).*

European Parliament. 2008d. LIBE Committee, Report with a proposal for a European Parliament recommendation to the Council on the problem of profiling, notably on the basis of ethnicity and race, in counterterrorism, law enforcement, immigration, customs and border control (2008/2020(INI)) 12 February 2008.

Freiheit, J., and F.A. Zangl. 2007. Model- based user – Interface management for public services. *The Electronic Journal of e-Government*, 5 (1): 53–62. Available online at www.ejeg.com.

Gillies, Al. 2008. The Legal and Ethical Changes in the NHS Landscape accompanying the Policy Shift from Paper-Based Health records to electronic health records. *Studies in Ethics, Law and Technology*, 1: 1–13. http://www.bepress.com/selt/vol2/iss1/art4.

Harting, N. 2009. Datenschutz und Anwaltsgeheimnis. *IT-Recht Kompakt* 138.
Kapstein, E.B. 1997. Regulating the Internet, President's Commission on Critical Infrastructure Protection. Available at: http://chnm.gmu.edu/cipdigitalarchive/object.php?id=170.
Karanja, S.K. 2008. Privacy and protection of marginalised social groups. *Studies in Ethics, Law and Technology* 2: 1–22. Available at http://www.bepress.com/selt/vol2/iss3/art510.2202/1941-6008.1063.
Lodge, J. (ed.). 2007. *Are you who you say you are? The EU and biometrics borders*. Nijmegen: Wolf Legal Publishers, pp 15–30.
Lodge, J. 2009a. E-Government, security and liberty in the EU: A role for National Parliaments? *Romanian Journal of European Affairs* 9: 5–19.
Lodge, J. 2009b. Access to information: Problems of accountability and transparency in a digital age, presentation to the Public Hearing of the LIBE committee.
Poulos, M., N. Rangoussi, N. Alexandris, and A. Evangelou. 2001. On the use of EEG features towards person identification via neytral networks. *Medical Informatics and the Internet in Medicine* 26(1): 35–48.
Rossnagel, A. and G. Hornung. 2009. 'Ein Ausweis für das Internet', DOV, 2009, 301.
Scottish Government. 2009. Draft guidance for education authorities, consultation analysis report, February 2009.
Soros, G. 2007. Open Society Institute & Soros Foundations Network. Retrieved from Open Society Justice Initiative: http://www.soros.org/initiatives/justice.
US Department of Homeland Security. 2007. *Privacy office annual report to Congress, July 2006–July 2007,* Washington, DC: US General Accounting Office, 2003. '*Information Security: Challenges in Using Biometrics',* GAO-03-1137T, Washington, DC; available at http://www.gao.gov/new.items/d031137t.pdf.
Vaden, T. 2004. Digital nominalism. *Ethics and Information Technology* 6: 223–231.
Wein, L.M., and M. Baveja. 2005. Using fingerprint image quality to improve the identification performance of the U.S Visitor and Immigrant Status Indicator Technology. *National Academy of Sciences of the United States of America* 102: 7772–7775.
Wong, M., S.-P. Choo, and H. Tan. 2009. Travel warning with capecitabine. *Annals of Oncology* 20(7): mdp278v1–mdp278.
Working Party. 2009. Data Protection Article 29, Opinion 2/2009 *on the protection of children's personal data (General Guidelines and the special case of schools)*, Adapted 11 February 2009, 398/09/EN, WP 160.

Chapter 15
Conclusions

Emilio Mordini

The position we have taken in editing this book is that for better or worse, biometric technologies are here to stay. The best answer to those who fear an Orwellian future is not merely to provide optimistic reassurances (which rarely reassure anyone), but rather to engage with the technology and seek to ensure that biometric identification systems are developed in positive ways.

15.1 Biometrics Go Well Beyond Security

As a scientific discipline biometrics date back to nineteenth century, a century that was passionate about the idea of measuring human attributes for whatever purposes. Biologists gave empirical foundation of Darwin's theory of natural selection by looking for biometric patterns in natural populations. Medical doctors and psychologists provided a sound scientific foundation to modern medicine and experimental psychology by looking for change of biometrics and psychometrics over time and in different health conditions. Measurable physical and behavioral attributes were also explored to develop instruments to ground racial classifications as well as physiognomy, which attempted to estimate character and personality traits. The usage of biometrics for individual identification and law enforcement was expression of the *zeitgeist*, characterized by medicalisation of social deviance and crime. Bertillon, the French police officer who first introduced an identification system based on physical measurements, was associated by his contemporaries to Luis Pasteur, the famous French microbiologist, and Bertillon's identification system was initially categorised as a tool for identifying mentally disturbed people (Bertillon 1984).

E. Mordini (✉)
Centre for Science Society and Citizenship, Piazza Capo di Ferro, 23,
Rome 00147, Italy
e-mail: emilio.mordini@cssc.eu

E. Mordini and D. Tzovaras (eds.), *Second Generation Biometrics: The Ethical, Legal and Social Context*, The International Library of Ethics, Law and Technology 11,
DOI 10.1007/978-94-007-3892-8_15,

This has driven various scholars – whose approaches explicitly or implicitly refer to Michel Foucault – to include biometrics among those disciplines which aim to manage and control the population (Piazza 2000). Biometrics are often conceived by critical scholars as integral to strategies – like health records, sanitation studies, mortality statistics, genealogical records, demographics, and so forth – which concern the practice of power on people's body. This analysis has been reinforced by the post 9/11 biometric wave, when politicians have marketed biometrics as the panacea of national security and for the prevention of terrorist's threat. This has been mirrored by industrial policies focusing on security applications. Access control, at any level, both physical and logical, is undoubtedly an important element of security schemes. Also there is no doubt that identity screening and verification can increase security. Yet biometrics have not been the dramatic breakthrough in security they were claimed to be. Stronger identification does not necessarily mean more security. Strictly speaking, identity biometrics simply provide a high level of assurance that a person is (or is not) a person whose biometric details are already on file. The identity (sameness) of biometrics, does not, in and of itself, tell us anything about the character, intent or entitlements of the individual other than that she is or is not truthful about who she is.

Provided that there is a "surveillance apparatus" biometrics do not belong entirely to that apparatus, or, at least, could hardly be considered a building block of that apparatus. Biometrics cannot respect its promises as far as security is concerned, but they can do a lot of other things. Biometric technologies can make easier and more reliable all human activities in which is important to recognize, identify or authenticate, an individual. To be sure, they also include repression of crime (but they hardly include prevention of crime, which is hardly an activity which concern single individual). Yet most activities made easier by biometric applications have little to do with law enforcement. They are, for instance, trade and financial transactions, contract stipulations, exercise of civil, social, and political rights, and most online activities. Biometrics also promise to become essential to smart environments, and in doing so to contribute to create a new category of objects, say, objects which are able to recognise their human "master". We call animism the tendency to regard inanimate objects as living and conscious. Animism is as old as humans, it can be traced back in almost all ancient religions and mythologies. It is universal and deeply rooted not only in the history of human evolution but also in contemporary world. Rather than the "primitive", "childish" superstition of attributing life to the lifeless, animism could be understood as a response to universal semiotic anxieties about where and how we could draw boundaries between persons and things. These very boundaries are now threatened by new and emerging biometrics, which can contribute to generate a collective feeling to live in an animistic world that Marc Pesce (2000), one of the early pioneers in Virtual Reality, calls 'techno-animism'. This is likely to be the main cultural challenge that new and emerging biometrics are going to pose us. These technologies – together with progress in embedded technologies, nanotechnologies, and context aware, adaptative, software – are destined to re-create our environment. They are going to give voce – to say – to objects and to project our civilisation in a new animistic culture. "I'm increasingly convinced that, as networks of smart objects permeate our environment, people's attitudes toward technology will become more animist. In other words, we'll start to anthropomorphize our stuff [...] When this happens, we'll stop expecting our tools to be mechanical and predictable and will begin to expect more complex, intuitive capabilities from all of them, even the dumb ones. This sounds far out and spacey, but I think it's right around the corner. This kind of intelligence is already starting to leak into mainstream products, and I bet that designers will have to think about it seriously within the next five years" (Kuniavsky 2003).

15.2 Privacy Is Larger Than Data Protection

Biometrics of second generation are collecting by default a huge amount of data, and are designed in a way which makes it difficult to respect the fair informational principles of data minimization, purpose specification, and proportionality. Would it possible to limit data collection? Yes, to a certain extent this is possible and should be done. Yet on other hand, the essence of new biometric technologies is, to say, to over collect data, which are meaningful and tend to create a rich web of meanings around the data subject. Without such an "over-collection of data" the same notion of second generation biometrics would fade away. In the book's introduction we have briefly discussed the paradigm shift provoked by the "semiotic implications" of new biometrics. The best way to deal with them is to distinguish between privacy and personal data.

What is privacy? Etymologically the word derives from the Latin *privatus*, past participle of *privo*, "I deprive", "I cut away". Privacy thus refers to the state of something that is separated, secluded from others. It refers to the state of being set apart, belonging to oneself, in contrast to the state of being public or common. Privacy is primarily a moral concept. It involves claims about the moral status of the individual self, about its dignity and relation to others. Philosophers and legal theorists, for example, tend to talk about privacy in terms of ideas like 'inviolate personality'. This moral core, it is argued, is the origin of social values such as autonomy, integrity, independence. Such values form a foundation for contemporary notions of human rights, citizenship and civic obligation in European public affairs.

In the 1950s, Hannah Arendt (1958) was one of the first scholars to observe the importance of privacy. She argued that privacy guarantees psychological depth, providing individuals the space to separate from others; it establishes boundaries between the internal and external worlds, which are of paramount importance in order to fix identity; it preserves the sacred and mysterious spaces of life. Some phenomena are different when lived in a condition of seclusion, so to preserve this difference individuals who need to exist organized in groups also need to protect the existence of separated spaces in their lives. Arendt's defense of the importance of the private sphere warns about dangers arising from the evaporation of the private and reminds of how of twentieth century totalitarian social orders sought to rob people of their privacy in order to better control them. Indeed the reaction against an imposed nakedness in concentration camps and other physical degradations related to attempts to annihilate privacy can be taken as one root of the new European self-understanding following World War II. The foundation of the EU notion of the citizen, and thereby of the legitimacy of the European political institutions, also rests with a new appreciation of privacy conceived as control over personal information,[1] respect for human dignity (Bloustein 1964), respect for intimacy,[2] promotion of personhood (Pound 1915; Fried 1970; Reiman 1976).

The concept of "personal data", which originated in the 1980s, is still related to the notion of privacy but it should not be confused with it. The idea of personal data comes from the increasing capacity of new electronic devises to turn continuous, non measurable, properties into discrete, measurable, quantities. The theoretical legitimacy of such an operation is efficaciously discussed by Ghilardi and Keller in their chapter. The development of the ability to elicit automatically quantities from almost anything, individuals included, has been a key historical event. It represents a shift from personal knowledge understood as self-knowledge (attained by introspection) to personal knowledge understood as knowledge about the self (attained by technical instruments). Knowledge about oneself becomes detachable from the person and marketable. Privacy gives increasingly way to personal data protection, which focuses only on ownership of knowledge about oneself, disregarding any other relevant aspect. Finally data protection generates a technical conception of privacy, framed chiefly in terms of risk management and technical ability to protect or to penetrate the private sphere. Privacy is not the same as anonymity, while framing everything in terms of data protection ends up turning all issues in an issue concerning various degree of anonymity. This leads to a series of considerations about data protection and information disclosure.

Data protection is twofold a tricky game. First it is almost impossible to truly limit the amount of data generated by new info-communication technologies. For many reasons, which range from technical reasons to economic and market motives, we are producing more and more data in a race which appears, for now, endless. "Digital data is now everywhere—in every sector, in every economy, in every organization and user of digital technology (…) The ability to store, aggregate, and combine data and then use the results to perform deep analyses has become ever more accessible as trends such as Moore's Law in computing, its equivalent in digital storage, and cloud computing continue to lower costs and other technology barriers (…) The ability to generate, communicate, share, and access data has been revolutionized by the increasing number of people, devices, and sensors that are now connected by digital networks. In 2010, more than four billion people, or 60% of the world's population, were using mobile phones, and

[1]Perhaps the best examples is the definition of privacy given by William Parent, who defines it as "the condition of not having undocumented personal information known or possessed by others". See Parent (1983) and Westin (1967).

[2]Julie Inness defines privacy as "the state of possessing control over a realm of intimate decisions, which includes decisions about intimate access, intimate information, and intimate actions". See Inness (1992).

about 12% of those people had smartphones, whose penetration is growing at more than 20% a year. More than 30 million networked sensor nodes are now present in the transportation, automotive, industrial, utilities, and retail sectors. The number of these sensors is increasing at a rate of more than 30% a year" (Manyika et al. 2011).

Moreover electronic data is by default public. No data that can be truly protected in the electronic, online, world. Everything has already been disclosed if it is digital because of the dimension of the online world and its nature. In the digital world too many actors share "confidential" information so that information is unavoidably destined to become public because of an amplification phenomenon. Electronic information is disseminated and easily copied and duplicated, and, in such a way, it becomes accessible and somehow indelible. This makes pure data protection an illusory endeavor. What is really important, and feasible, is control over collection and usage of data. "People are willing to share all sorts of information as long as they are in control" (Schneier 2006). This could be done by empowering data subjects. A good example on the way in which this strategy can be fruitfully applied is provided by multiple biometrics. Attempts to minimize data collected by multiple biometric systems are certainly legitimate but one should be aware that either they will end up by simply preventing the system working, or they are destined to fail. Yet one can imagine multiple biometrics *à la carte*, say, systems in which the data subject can autonomously decide to opt out from any specific biometric subsystem, by switching off related sensors. This could in principle allow each subject to tailor the system according her own preferences, by selecting what biometric features will be captured and merged and what will be not. Of course there are several technical problems to be solved before one could implement such a solution, yet this is an opportunity created by biometrics of new generation that should not be overlooked.

On the positive side, new biometrics can dramatically improve privacy, in terms of respect for people mental and physical integrity and intimacy. The unobtrusive nature of these technologies can allow people to appreciate the benefit of staying in tailored environments, which recognise the person, without obliging the subject to undergo to humiliating, intrusive, procedures. Of course such a dramatic divarication between privacy and data protection is likely to be another puzzling and difficult issue to be faced in the next future.

15.3 Ethical Issues Are Not Necessarily Barriers, They Could Be Also Drivers

Usually one thinks of privacy and ethical considerations as a barrier to innovation, as though respecting ethical values and democratic principles were in contradiction with improving technology performance (Drucker 1985). This is misleading. There is undoubtedly a certain tension between technology innovation, and respect for rights and values. Some elements of this tension can be even irreducible, yet ethics and privacy can be still conjugated with innovation.

Innovation is driven by several factors, one of them could become the need to promote a better performance within a framework dictated by ethical and privacy considerations. Next generation biometrics can become an example of innovative technology driven by privacy and ethical considerations. This is facilitated by the very nature of second generation biometrics, say, its symbolic dimension. In other words, taking into consideration ethical and cultural variables could generate a competitive advantage for those industries which are able to address properly these aspects. One can imagine a world in which biometrics are chosen by private costumers and governmental agencies not only because of their technical dependability, but also because of their ethical and cultural trustworthiness. Ethical and cultural awareness could become integral part of the system and generate new benefits. This is a totally new field of research, but I guess that it is the logic consequence of the current approaches based on ethics and privacy by design, which are dominating the current debate on privacy and new technologies. In such a sense, I would like to conclude this book by advocating a paradigm shift. We have started by discussing "identification technologies", it is now necessary to introduce a new concept, "advanced human recognition". I think that we are on the verge of a revolution which is leading us from "identification technologies" to "advanced human recognition". It is difficult to predict the cultural, ethical, social, legal, consequences of this revolution, which is still in its initial phase. Yet if we are concerned about ethical and societal implications of next generation biometrics, we would do better to form a clearer picture of this revolution and start thinking of the necessary new policy framework. If this book contributes to such an endeavour, we are satisfied.

References

Arendt, H. 1958. *The human condition*. Chicago: Chicago University Press.

Bertillon, S. 1984. *Vie d'Alphonse Bertillon, inventeur de l'anthropométrie*. Paris: Gallimard.

Bloustein, E.J. 1964. *Privacy as an aspect of human dignity: An answer to dean Prosser*. New York: New York University, School of Law.

Drucker, P.F. 1985. *Innovation and entrepreneurship*. New York: Collins.

Fried, C. 1968. Privacy (a moral analysis). *Yale Law Journal* 77: 475–493.

Inness, J. 1992. *Privacy, intimacy and isolation*. New York: Oxford University Press.

Kuniavsky, M. 2003. User expectations in a world of smart devices, *Adaptative Path, Oct. 17*. http://www.adaptivepath.com/publications/essays/archives/000272.php.

Manyika, J., M. Chui, B. Brown, et al. 2011. *Big data: The next frontier for innovation, competition, and productivity*. New York: McKinsey Global Institute.

Parent, W. 1983. Recent work on the concept of privacy. *American Philosophical Quarterly* 20(4): 341–355.

Pesce, M. 2000. *The playful world: How technology is transforming our imagination*. New York: Ballantine Books.

Piazza, P. 2000. La fabrique"bertillonienne"de l'identité, *Labyrinthe*, 6 | 2000, Thèmes (n° 6), http://labyrinthe.revues.org/index400.html.

Pound, R. 1915. Interests in personality. *Harvard Law Review* 28: 343.

Reiman, J. 1976. Privacy, intimacy and personhood. *Philosophy and Public Affairs* 6(1): 26–44.

Schneier, B. 2006. Lessons from the Facebook Riots, *Wired*, September 21.

Westin, A. 1967. *Privacy and freedom*. New York: Atheneum.

Greek DPA ACTIBIO

On 12-08-2010 The Hellenic Data Protection Authority (HDPA) authorized the installation of two pilot applications of the ACTIBIO project (Unobstructive Authentication Using Activity Related and Soft Biometrics).

Given this is, to our knowledge, the first official statements of a European DPA on second generation biometrics, we think that it may be of some interest to the reader.

– The Editors

E. Mordini and D. Tzovaras (eds.), *Second Generation Biometrics: The Ethical, Legal and Social Context*, The International Library of Ethics, Law and Technology 11,
DOI 10.1007/978-94-007-3892-8,

Athens, 12-08-2010
Ref.No..: G/OUT/5019/12-08-2010

HELLENIC REPUBLIC
DATA PROTECTION AUTHORITY

Address: **KIFISIAS1-3**
115 23 Athens
Telephone: **210–6475601**
FAX: **210–6475628**

DECISION 57/2010

The Hellenic Data Protection Authority (HDPA), consisting of the President Mr. Ch. Geraris, the regular members Mr. L. Kotsalis, Mr. A. Papaneofytou, Mr. A. Prassos and Mr. A. Roupakiotis, and the alternative member Mrs G. Pantziou and G. Lazarakos substituting the regular members Mr. A. Pomportsi and Mr. A. Metaxa, who could not be present due to an impediment, although they were formally invited, convened on 8/7/2010, following an invitation by the President in order to examine the case that is reported herein below. The rapporteur Ms. A. Bourka, IT auditor, and Ms. G. Palaiologou, employee of the Administration and Economics Department, as a secretary, were also present without the right of vote.

The HDPA took under consideration the following:

The Informatics and Telematics Institute of the National Center for Research and Technological Development (EKETA/HIT) and the security company, G4S TELEMATIX A.E., partners in the funded by the European Commission research project, ACTIBIO, (Unobstructive Authentication Using Activity Related Soft Biometrics), submitted to the HDPA the notification with ref.no. GN/IN/785/23-06-2010, asking permission for installing and operating two pilot biometric access control systems as well as for the operation of a specific system capable of recognizing malicious events in critical infrastructures within their facilities and for a limited time.

According to the aforementioned notification, the project ACTIBIO, aims at the development of a novel multimodal biometric system that is based on the combination of biometric data extracted from physiological features of the user and characteristics that are related to his/her behaviour and/or responses to specific environmental stimuli ("behavioural" characteristics). Specifically, the user's unique and final "biometric signature" extraction is based on static and dynamic face characteristics (e.g. grimaces), the way a subject walks (gait recognition), activity-related characteristics while the subject is performing specific work-related tasks (for instance the way that the user holds the handset, or the way that he/she typing a text on the computer), as well as anthropometric and other characteristics that are related with his/her external appearance (e.g. height, weight, eye colour, hair colour, presence of moustache, glasses, etc.). The above method is examined experimentally as an alternative/complementary solution of the standard biometric practices (based on fingerprints, iris scanning or facial geometry) aiming at a more efficient and accurate biometric data extraction and use, particularly with regard to their use in access control systems in the context of critical infrastructures that met high security requirements.

Within the frame of ACTIBIO project, the installation of the two pilot applications will be in accordance with the notification in order to test the experimental method and investigate possible errors or problems that may arise during its operation. Pilot tests will be realised via the execution of specific scenarios of use (use cases) in predefined areas/infrastructures. Particularly, the first pilot application will be realised at the central station of the G4S TELEMATIX SA company, where alerts and signals from mobile and fixed security and control systems (e.g. banks, cash and valuables transportation, shipments of high value and/or increased hazard) are received with the aim of intervention of G4S for handling security incidents. The aforementioned area is considered critical and access on the control room is permitted only to specific authorized personnel. The second pilot trial will be performed in an equipped research vehicle of CERTH/HIT in order to simulate vehicles of high security (for instance vehicles that transfer money to a bank) that require strong authentication mechanisms for driver identity verification as well as for driver and cargo safety. In both cases, the two pilot trials are composed of two basic components:

(a) An experimental biometric system, which is composed of several separate authentication modules that extract physiological and behavioural characteristics (e.g. face recognition, activity-related recognition or gait analysis). The biometric system will be used for control access to pilots' installations areas (concerning the first pilot test, physical and logical access to computer systems of G4S TELEMATIX SA central station will be tested, while for the second pilot, the physical access to the equipped research vehicle of EKETA/HIT, will be tested). The trial operation shall be structured as follows: during the registration in the biometric system, participants will be asked to perform – with the guidance of project's researchers-, certain daily activities (for the first pilot: e.g. walking, phone conversation, handling the control panel of the central station, typing on a computer, communicate via microphone panel and for the second pilot: seat on the driver's seat, handling gear and handbrake). The proposed biometric system based on the aforementioned multimodal biometric method, will extract the corresponding biometric data (e.g. dynamic facial features, gait profile, etc.) in order to produce the "biometric signatures" of the participants. The raw biometric data (e.g. images, gait and weight signals, etc.) will be deleted after the extraction of the "biometric signatures". The "biometric signatures" will be initially encrypted and afterwards stored in smart cards as well as in a database (different for each pilot), along with the encoded personal data information of each participant. The latter data will be included in specific forms completed by the participants before the beginning of the pilot trial (provide personal information such as full name, age, job details, physiological characteristics, etc.). Following the registration phase and for the special needs of the pilot trial cases, participants will be asked to perform in these pilot sites control access use cases in order to assess the validity and effectiveness of the "biometric signatures" as novel authentication mechanisms. Specifically, according to the first pilot scenario, the indicated scenarios will be related to the authorized access of specific computers that operate within the central station of G4S TELEMATIX S.A.

(b) A specific (non-biometric) subcomponent capable of detecting malicious attacks, that based on image analysis (acquired by a camera) and physical characteristics (such as weight) will be able to detect situations in which an individual of a controlled room is on risky circumstances (e.g. if the user raises his/her hands, appropriate blinking of eyes) and will transmit the relevant alarm signal of danger (e.g. via a specific sound or alarm). For the pilots' needs, participants of the experiment will be asked to simulate, with the guidance of project's researchers, events that imply danger, in order to check the system proper functioning (namely if the required alarm signal has been transmitted). This specific component of the system does not perform raw image recording neither extracts any human's physiological or behavioural characteristic and finally allows the individual to activate and deactivate the specific component whenever he/she wishes.

The trial in the first pilot application will be performed with the participation of G4S TELEMATIX S.A company employees on a voluntary basis. The pilot system will not replace existing access control systems that are already in use at the premises of G4S TELEMATIX S.A, whilst the refusal of employees to participate in the experimental application will not be notified by any means in the company administration. The second pilot trial will take place with the participation of volunteers who are already registered, with their consent, to the relevant research CERTH/HIT database (most of these people have participated in the past in similar programs of the institute). In both pilot applications trials, volunteers will be informed before their participation in the experiments and upon having given their explicit and informed consent (formal copies of the information and consent forms have been submitted for both pilot trials with the above notification).

Pilot applications testing and analysis of results will be performed within the period of seven (7) days. After that period, all data of the pilot trial (cards, databases, forms with personal data information of participants) will be destroyed.

The HDPA, after having examined the foregoing information, listened to the rapporteurs and discussed the facts thoroughly,

DELIBERATED ACCORDING TO THE LAW

1. The article 2 of Law 2472/1997, stipulates that "personal data" is "any information relating to the data subject". "Data subject" is the "any natural person to whom such data refer and whose identity is known or may be found, i.e. his/her identity may be determined directly or indirectly, in particular by reference to an

identity card number or to one or more factors specific to his/her physical, psychological, mental, economic, cultural, political or social identity". As the HDPA has repeatedly stressed via previous decisions, biometric data constitute personal data since they may be considered both as content of information characterizing specific biological traits, physiological characteristics and/or personal features unique to an individual, as well as a way to connect this piece of information to a specific individual and thus may operate as a means of identification of this person.

2. In the case of ACTIBIO project, the two pilot applications (in G4S TELEMATIX central station and the equipped research vehicle of CERTH/HIT) consist of an experimental biometric system and a specific module of detecting abnormal events such as malicious attacks. Controller of the first pilot is G4S TELEMATIX SA, while controller of the second is CERTH/HIT. Concerning the experimental biometric system, and according to the above, it is performed processing of personal data of the participants in the trials and particularly biometric data are collected that correspond to the combination of physiological and "behavioral" characteristics. Additionally, within the framework of this experimental application, specific documents are kept with personal information of the participants (full name, business status, personal characteristics), which are filled by them before the execution of the pilot trials. Regarding the abnormal event detection module, its operation is based solely on the acquisition of images and signals from sensors (without raw data recording) that are automatically processed (without intervention of a person) and based on their analysis it is feasible to activate specific alarm signals in case of an abnormal event. Therefore, within the framework of this module, no processing of personal data is performed in the terms of Law 2472/1997. The biometric systems of the two pilot applications trials are independent.
3. According to the article 4, par. 1 of Law 2472/1997, criterion for the lawfulness of each processing is the principle of proportionality, according to which it must be examined each time whether the processing of specific personal data is essential for the purpose aimed at and that this purpose cannot be met by less onerous means. Specifically, personal data (a) must be collected fairly and lawfully, for specific, explicit and legitimate purposes and to be processed fairly and lawfully in relation to the purposes for which they are processed; (b) must be adequate, relevant and not excessive in relation to the purposes for which they are processed, (c) should be accurate and properly updated, and (d) shall be kept in a form which permits identification of data subjects for no longer than the period required for the purposes for which they were collected or processed.
4. Within the framework of the ACTIBIO project, the nature of the purpose of personal data processing of the two biometric systems is the exclusive scientific research of a novel multimodal biometric authentication technology that is based on the combination of human's physiological and "behavioural" characteristics. Taking into account that (a) the aforementioned research procedure which aims at the improvement of biometric systems effectiveness and accuracy, when they are utilized in real operation conditions and scenarios, (b) the fact that the effectiveness and accuracy of biometric systems are crucial factors for their efficient

operation, and which in special cases of high security environments and infrastructures can be considered as necessary mechanisms for the enhancement of security of individuals and commercial goods, and (c) the fact that the pilot data are preserved only for seven (7) days, and after this period of time are destroyed, it is decided by majority voting, that the aforementioned principle of proportionality is met.

5. It should be noted that in real operation conditions (e.g. if the purpose was not for research only but it concerned for instance the control access to infrastructures), and in the terms of article 4 of Law 2472/1997, the lawfulness of the above processing should have been re-examined, taking into account the following criteria: (i) the criticality and the advanced security measures for the specific infrastructures, (ii) intrusiveness into humans' privacy that the proposed biometric modalities could pose in relation to other biometric technologies (e.g. fingerprint, iris, etc.) and (iii) usage of user-friendly biometric architecture models that protect humans' privacy (e.g. biometric signatures stored to smart cards, encryption of biometric data). In accordance to these criteria, the Data Protection Authority (HDPA) has already examined applications of biometric systems in real operation conditions, and particularly access control systems for monitoring the presence of workers in working environments and is some cases HDPA has imposed the cessation of the respective biometric systems as exceeding the purpose of personal data processing (see decisions 245/9/2000, 52/2003, 50/2007, 74/2009), whilst in other cases has permitted their operation under specific conditions (e.g. decision 9/2003, 52/2008, 56/2009). It must be pointed out that the aforementioned criteria have not been investigated thoroughly as the foreseen pilot applications are based on experimental use cases, that could not be the same as in real conditions. Indicatively, it is pointed out that in real operation conditions it would be more effective to store the biometric data to smart cards (and not in a database), as smart cards provide to the subjects a more effective way of controlling their personal data. In the frame of the specific experimental conditions of the pilot trial the data processing in a central database can not be considered as not an effective approach, due to the fact that the processing is prerequisite to conduct the research. According to article 10 of Law 2472/1997, the Controller will take all the technical and organizational measures for the security of the data.
6. In the context of the two pilot application trials, biometric data processing is being performed via the written consent of subjects. According to article 2, of Law 2472/1997, data subject consent is considered to be "any freely given, explicit, and specific indication of will whereby the data subject expressly and fully cognizant signifies his/her informed agreement to personal data relating to him/her being processed". In the framework of ACTIBIO project consent is free since the data subjects voluntarily participate in the pilot trials. Particularly, in the first pilot application, pilot participants are employees of the G4S TELEMATIX S.A. that declare to the ACTIBIO project responsible their consent to participate in the pilot trials, without communicating their consent to participate or not to the trial by any means to the company management committee. The experimental

pilot application will operate on a trial basis and shall not replace the existing access control systems which are already in use at the premises of G4S TELEMATIX S.A. In the second pilot application, participants are subjects that have been registered with their written consent in the research database of CERTH/HIT and after the official invitation of the latter, have freely declared their consent to participate in the context of the research project. Furthermore, the consent for both application trials is explicit and express as it is given via filling in special forms (different for each pilot application). By these form, volunteers are also informed about the processing of their personal data, the identity of the Controller, the purpose of the processing, as well as the way in which data subjects may exercise their rights of access and objection according to the Law 2472/1997 (articles 12 and 13). It is noted that there are no other recipients of the personal data except for the authorized employees of the Controllers who belong to the research team of the ACTIBIO project. The Hellenic Data Protection Authority decided by majority that legally, according to the article 5 par 1. of the Law 2472/1997, personal data will be under process with the consent of the participants.

FOR THESE REASONS

The Hellenic Data Protection Authority decided that the installation of the aforementioned biometric system exclusively for scientific research purposes does not contravene the provisions of Law 2472/1997, insofar as the following terms are fulfilled:

1. The Controllers (G4S TELEMATIX S.A., EKETA/HIT) are obliged to notify the Authority about the destruction of biometric data as well as any other personal data retained within the framework of the pilot biometric system after the end of the 7 days period which is required for the accomplishment of the processing purpose. The notification should be take place within fifteen (15) days after the destruction of the data. The Controller should submit the date destruction protocol to HDPA.
2. Any expansion of operations or other technical change or change procedures relating to the application and operation of the biometric system, without the HDPA's prior notification and approval is forbidden.

The President	**The Secretary**
Christos Yeraris	**Georgia Palaiologou**

Index

E. Mordini and D. Tzovaras (eds.), *Second Generation Biometrics: The Ethical, Legal and Social Context*, The International Library of Ethics, Law and Technology 11,
DOI 10.1007/978-94-007-3892-8, © Springer Science+Business Media B.V. 2012

B